Universitext

Sylvestre Gallot
Dominique Hulin
Jacques Lafontaine

Riemannian Geometry

Third Edition

With 58 Figures

 Springer

Sylvestre Gallot

Université Grenoble 1
Institut Fourier, C.N.R.S., UMR 5582
Rue des Maths 10
38402 Saint-Martin d´Hères
France
e-mail: Sylvestre.Gallot@ujf-grenoble.fr

Dominique Hulin

Université Paris XI
Département de Mathématiques
Bâtiment 425
91405 Orsay CX
France
e-mail: dominique.hulin@math.u-psud.fr

Jacques Lafontaine

Université Montpellier II
Département de Mathématiques, C.N.R.S., UMR 5149
Place Eugène-Bataillon
34095 Montpellier CX 05
France
e-mail: jaclaf@math.univ-montp2.fr

Mathematics Subject Classification (1980): 53-XX, 58-XX

Library of Congress Control Number: 2004106886

ISBN 3-540-20493-8 Springer Berlin Heidelberg New York
ISBN 3-540-17923-2 1st edition Springer-Verlag Berlin Heidelberg New York
ISBN 3-540-52401-0 2nd edition Springer-Verlag Berlin Heidelberg New York

Springer is a part of Springer Science+Business Media
springeronline.com
© Springer-Verlag Berlin Heidelberg 1987, 1990, 2004
Printed in Germany

Cover design: Erich Kirchner, Heidelberg
Data conversion by LE-TeX Jelonek, Schmidt & Vöckler GbR, Leipzig using a Springer LaTeX macro package

Printed on acid-free paper   41/3142ud - 5 4 3 2 1 0

Preface to the third edition

Many years have passed since the first edition. However, the encouragements of various readers and friends have persuaded us to write this third edition.

During these years, Riemannian Geometry has undergone many dramatic developments. Here is not the place to relate them. The reader can consult for instance the recent book [Br5]. of our "mentor" Marcel Berger. However, Riemannian Geometry is not only a fascinating field in itself. It has proved to be a precious tool in other parts of mathematics. In this respect, we can quote the major breakthroughs in four-dimensional topology which occurred in the eighties and the nineties of the last century (see for instance [L2]). These have been followed, quite recently, by a possibly successful approach to the Poincaré conjecture. In another direction, Geometric Group Theory, a very active field nowadays (cf. [Gr6]), borrows many ideas from Riemannian or metric geometry.

But let us stop hogging the limelight. This is just a textbook. We hope that our point of view of working intrinsically with manifolds as early as possible, and testing every new notion on a series of recurrent examples (see the introduction to the first edition for a detailed description), can be useful both to beginners and to mathematicians from other fields, wanting to acquire some feeling for the subject.

The main addition to this new edition is a section devoted to pseudo-Riemannian geometry: studying what happens when positive-definiteness is dropped is much more than a free intellectual exercise. Most of the fascinating and counter-intuitive features of General Relativity come from Lorentz geometry, i.e. pseudo-Riemannian geometry of signature $(n, 1)$. Moreover, compact Lorentz manifolds, even though they may not be interesting in Relativity, have beautiful dynamical properties. This new section is written in the same spirit as the rest of the book, with numerous examples and exercises.

Here and there, some other new topics have been added: an introduction to isosystolic inequalities, a description of the geodesic flow from the Hamiltonian point of view, and some minor points such as the equality case in the Bishop-Gromov inequality.

Much of this material, even if not new, might not be so easy to find in book form.

It is a pleasure to thank Luc Rozoy for pointing out various mistakes, and Marc Herzlich for many valuable suggestions. This work was made possible by the help of the "Institut de Mathématiques et Modélisation de Montpellier[1]" and the "Centre National de la Recherche Scientifique."

---

[1] UMR 5149

Preface to the second edition

In this second edition, the main additions are a section devoted to surfaces with constant negative curvature, and an introduction to conformal geometry. Also, we present a -soft- proof of the Paul Levy-Gromov isoperimetric inequality, kindly communicated by G. Besson. Several peoples helped us to find bugs in the first edition. They are not responsible for the persisting ones! Among them, we particularly thank Pierre Arnoux and Stefano Marchiafava.

We are also indebted to Marc Troyanov for valuable comments and suggestions.

Introduction

This book is an outgrowth of graduate lectures given by two of us in Paris. We assume that the reader has already heard a little about differential manifolds. At some very precise points, we also use the basic vocabulary of representation theory, or some elementary notions about homotopy. Now and then, some remarks and comments use more elaborate theories. Such passages are inserted between *.

In most textbooks about Riemannian geometry, the starting point is the local theory of embedded surfaces. Here we begin directly with the so-called "abstract" manifolds. To illustrate our point of view, a series of examples is developed each time a new definition or theorem occurs. Thus, the reader will meet a detailed recurrent study of spheres, tori, real and complex projective spaces, and compact Lie groups equipped with bi-invariant metrics. Notice that all these examples, although very common, are not so easy to realize (except the first) as Riemannian submanifolds of Euclidean spaces.

The first chapter is a quick introduction to differential manifolds. Proofs are often supplied with precise references. However, numerous examples and exercises will help the reader to get familiar with the subject.

Chapters II and III deal with basic Riemannian geometry of manifolds, as described in the content table. We finish with global results (Cartan-Hadamard, Myers' and Milnor's theorems) concerning relations between curvature and topology. By the way, we did not resist the temptation to give an overview of recent research results. We hope the reader will want to look at the original papers.

Analysis on manifolds has become a wide topic, and chapter IV is only an introduction. We focused on the Weitzenbock formula, and on some aspects of spectral theory. Our Ariadne's thread was what our "Mentor" Marcel Berger calls "la domination universelle de la courbure de Ricci", discovered by Gromov in the seventies. Chapter V deals with more classical topics in Riemannian submanifolds.

The reader will find numerous exercises. They should be considered as a part of the text. This is why we gave the solutions of most of them.

SOME HISTORICAL AND HEURISTIC REFERENCES

Riemannian Geometry is indeed very lively today, but most basic notions go back to the nineteenth century.

At the beginning of each chapter, we endeavored to explain what is going on. As a complement to these short introductions, we strongly recommand the following references.

- E. Cartan: La Géométrie des espaces de Riemann [Ca]

- M. Spivak: Differential geometry (t.2) [Sp], which contains in particular a commented translation of Riemann's dissertation.

- P. Dombrowski: 150 years after Gauss' "disquitiones generales circa super-ficies curvas" [Dom]

- M. Berger: La géométrie métrique des variétés riemanniennes [Br3]

- C. W. Misner, K. S. Thorne, J. A. Wheeler: Gravitation ([M-S-T]). (I (J.L.) found their point of view of physicists very illuminating!)

ACKNOWLEDGEMENTS

During the preparation of this book, the help of our host institutions, the Institut Fourier of Grenoble (S.G.), the Centre de Mathématiques de l'Ecole Polytechnique (D.H.), and the Unité Associée au C.N.R.S. no. 212 (J.L.), was decisive. At the Ecole Polytechnique, Thomas Ehrhardt taught us the wonderful TeX system. Particular thanks are due to him.

Contents

1

Differential Manifolds

"The general notion of manifold is quite difficult to define precisely. A surface gives the idea of a two-dimensional manifold. If we take for instance a sphere, or a torus, we can decompose this surface into a finite number of parts such that each of them can be bijectively mapped into a simply-connected region of the Euclidean plane." This is the beginning of the third chapter of "Leçons sur la Géométrie des espaces de Riemann" by Elie Cartan (1928), that we strongly recommend to those who can read French. In fact, Cartan explains very neatly that these parts are what we call "open sets". He explains also that if the domains of definition of two such maps (which are now called *charts*) overlap, one of them is gotten from the other by composition with a smooth map of the Euclidean space. This is just the formal definition of a differential (or smooth) manifold that we give in 1.A, and illustrate by the examples of the sphere, the torus, and also the projective spaces.

Charts allow to generalize to manifolds the classical notions of differential calculus, such as vector fields and differential forms. But in fact the rule of the game is to dispense oneself from using charts as soon as possible, and to work "intrinsically". The main purpose of this chapter is to show how to do that. However, we do not claim to give a complete formal exposure about manifolds. For that, the reader can look at [Wa], [Le], the first chapter of [Sp], and [La3].

1.A From submanifolds to abstract manifolds

1.A.1 Submanifolds of Euclidean spaces

1.1 Definition. A subset $M \subset \mathbf{R}^{n+k}$ is an *n-dimensional submanifold of class C^p of $\mathbf{R}^{n+k}$* if, for any $x \in M$, there exists a neighborhood U of x in $\mathbf{R}^{n+k}$ and a C^p submersion $f : U \to \mathbf{R}^k$ such that $U \cap M = f^{-1}(0)$ (we recall that f is a submersion if its differential map is surjective at each point).

1.2 Examples. a) The sphere S^n defined by

$$S^n = \{x = (x_0, ..., x_n) \in \mathbf{R}^{n+1} / f(x) = x_0^2 + ... + x_n^2 - 1 = 0\}$$

is an n-dimensional submanifold of class C^∞ of $\mathbf{R}^{n+1}$. The map $f : \mathbf{R}^{n+1} \to \mathbf{R}$ defined above is indeed a submersion around any point of S^n, its differential map at x being $df_x = 2\sum_{k=0}^n x_k dx_k$.

b) Similarly, the hyperboloid H_c^n defined by

$$H_c^n = \{x = (x_0, ..., x_n) \in \mathbf{R}^{n+1} / g(x) = x_0^2 - ... - x_n^2 = c\}$$

is for $c \neq 0$ an n-dimensional submanifold of $\mathbf{R}^{n+1}$ since the differential map of g is never zero on H_c^n. But the differential map of g at the origin is zero: actually, H_0^n is not a submanifold (exercise)

c) Be careful: if $M = \{x \in \mathbf{R}^{n+k} / f(x) = 0\}$, where $f : \mathbf{R}^{n+k} \to \mathbf{R}^k$ is C^p but is not a submersion at every point of M, M may although be a submanifold of $\mathbf{R}^{n+k}$. Consider for example

$$M = \{x \in \mathbf{R}^2 : f_1(x) = x_1^3 - x_2^3 = 0\} = \{x \in \mathbf{R}^2 / f_2(x) = x_1 - x_2 = 0\}.$$

d) The n-torus T^n defined by

$$T^n = \begin{cases} \{(z_1, \cdots, z_n) \in \mathbf{C}^n / \mid z_1 \mid^2 = \cdots = \mid z_n \mid^2 = 1\} \\ \{(x_1, \cdots, x_{2n}) \in \mathbf{R}^{2n} / (x_1^2 + x_2^2 = \cdots, x_{2n-1}^2 + x_{2n}^2 = 1\} \end{cases}$$

is a submanifold of $\mathbf{R}^{2n} \simeq \mathbf{C}^n$.

e) The special orthogonal group

$$SO(n) = \{A \in M_n(\mathbf{R}) / \det(A) = 1 \quad \text{and} \quad {}^t AA = Id\}$$

is an $\frac{n(n-1)}{2}$ dimensional submanifold of $M_n(\mathbf{R}) \simeq \mathbf{R}^{n^2}$. Indeed

$$GL_n^+(\mathbf{R}) = \{A \in M_n(\mathbf{R}) / \det(A) > 0\}$$

is an open subset of $M_n(\mathbf{R})$. Define

$$f : Gl_n^+(\mathbf{R}) \to \text{Sym}(n) \quad \text{by} \quad f(A) = {}^t AA - Id,$$

where $\text{Sym}(n)$ is the set of symmetric (n,n) matrices. Then $SO(n) = f^{-1}(0)$ and f is a submersion at any point of $SO(n)$ since for $A \in SO(n)$,

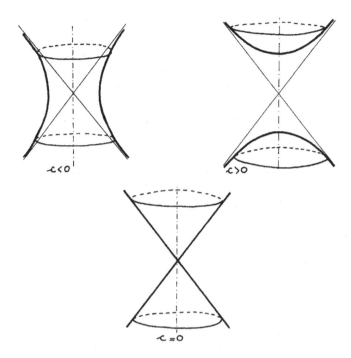

Fig. 1.1. H_c^n, according to the sign of c

$df_A(H) = {}^t\!AH + {}^t\!HA$: in particular for a symmetric matrix S we have $\frac{AS}{2} \in T_A(Gl(n, \mathbf{R}))$ and $df_A(\frac{AS}{2}) = S$.

1.3 Proposition. *The following are equivalent:*

i) *M is a C^p submanifold of dimension n of $\mathbf{R}^{n+k}$.*

ii) *For any x in M, there exist open neighborhoods U and V of x and 0 in $\mathbf{R}^{n+k}$ respectively, and a C^p diffeomorphism $f : U \to V$ such that $f(U \cap M) = V \cap (\mathbf{R}^n \times \{0\})$.*

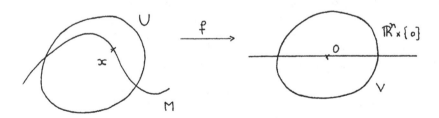

Fig. 1.2. Property ii) of 1.3

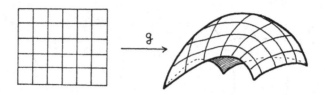

Fig. 1.3. property iii) of 1.3

iii) *For any x in M, there exist a neighborhood U of x in $\mathbf{R}^{n+k}$, a neighborhood Ω of 0 in $\mathbf{R}^n$, and a C^p map $g : \Omega \to \mathbf{R}^{n+k}$ such that (Ω, g) is a local parametrization of $M \cap U$ around x (that is g is an homeomorphism from Ω onto $M \cap U$ and $g'(0)$ is injective).*
Proof. Use the implicit function theorem (see for example [Sp]). ∎

1.4 **Examples.** a) The torus T^2 can be viewed as the image of the map $g : \mathbf{R}^2 \to \mathbf{R}^4$ defined by

$$g(\theta, \phi) = (\cos \theta, \sin \theta, \cos \phi, \sin \phi).$$

This map is an immersion (its differential map is of rank 2 everywhere) and a local homeomorphism: indeed, g induces a map $\hat{g}$ defined on $\mathbf{R}^2/(2\pi \mathbf{Z})^2$, which is bijective and continuous from $\mathbf{R}^2/(2\pi \mathbf{Z})^2$ (compact) onto $g(\mathbf{R}^2) = T^2$: hence $\hat{g}$ is an homeomorphism, and since the canonical projection p is a local homeomorphism, so is g.
b) The torus T^2 as the image of the map $h : \mathbf{R}^2 \to \mathbf{R}^3$ defined by

$$h(\theta, \phi) = \big((2 + \cos \theta) \cos \phi, (2 + \cos \theta) \sin \phi, \sin \theta\big).$$

Proof as in a).
c) It is necessary to assume that g is an homeomorphism in 1.3 iii). For example, the image of the map $g :] - \frac{\pi}{2}, \frac{\pi}{2}[\to \mathbf{R}^2$ defined by

$$g(t) = \sin t \cos t \, (\cos t, \sin t)$$

is not a submanifold of $\mathbf{R}^2$, although $g'(t)$ is never zero: indeed if U is a small open ball around zero in $\mathbf{R}^2$, then $U \cap M \setminus (0,0)$ has always four connected components.
In iii), g' has to be injective: for example, the image of the map $t \mapsto (t^2, t^3)$ from $\mathbf{R}$ to $\mathbf{R}^2$ is not a submanifold of $\mathbf{R}^2$.
d) The image of the map $g : \mathbf{R}^2 \to \mathbf{R}^3$ defined by

$$g(u, v) = (\sin u \cos v, \sin u \sin v, \cos u)$$

is a submanifold of $\mathbf{R}^3$ although the differential map of g is not of rank two for $u \equiv 0[\pi]$: it is the sphere S^2. To prove it is a submanifold around the poles, we must use another parametrization, for example

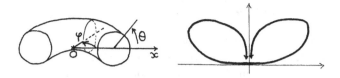

Fig. 1.4. Examples 1.4 b) and 1.4 c)

$$(x, y) \rightarrow (x, y, \sqrt{1 - x^2 - y^2}).$$

e) The image of the map $g : \mathbf{R} \rightarrow \mathbf{R}^4$ defined by

$$g(t) = \big(\cos t, \sin t, \cos(t\sqrt{2}), \sin(t\sqrt{2})\big)$$

is not a submanifold of $\mathbf{R}^4$, see 1.10.

1.5 Proposition. *Let $M \subset \mathbf{R}^{n+k}$ be an n-dimensional submanifold of class C^p. Let U_1 and U_2 be two neighborhoods of $m \in M$. If (Ω_1, g_1) and (Ω_2, g_2) are two local parametrizations of $U_1 \cap M$ and $U_2 \cap M$ respectively, then $g_2^{-1} \circ g_1$ is a C^p diffeomorphism from $\Omega_1 \cap g_1^{-1}(U_2)$ onto $\Omega_2 \cap g_2^{-1}(U_1)$.*
Proof. Exercise for the reader. ∎

Hence if $(g_i : \Omega_i \rightarrow U_i)$ is a family of local parametrizations for a submanifold M of $\mathbf{R}^{n+k}$, the domains of which cover M, we can define a differentiable structure on M: we just endow each $U_i \cap M$ with the *differentiable structure* inherited from Ω_i, and carried by the parametrization (or chart) g_i. From the previous proposition, we know that the differentiable structures so defined on $U_i \cap M$ and on $U_j \cap M$ do coincide. This idea leads to the notion of abstract manifold.

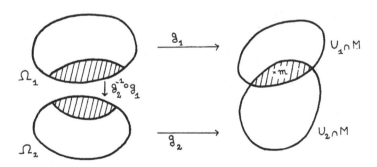

Fig. 1.5. Changing of parametrization

1.A.2 Abstract manifolds

Let M be a Hausdorff topological space such that any point of M admits a neighborhood homeomorphic to an open set in $\mathbf{R}^n$: roughly, such a space is get by gluing together small open subsets of $\mathbf{R}^n$ (imagine the structures get by soldering or gluing together metal plates).

We decide to carry the differentiable structure from the open subsets of $\mathbf{R}^n$ onto these neighborhoods in M. But the differentiable structures so defined on these neighborhoods are not a priori consistent (as it was the case for a submanifold of $\mathbf{R}^n$), that is do not automatically coincide on the intersections of two of them: in the case of submanifolds of $\mathbf{R}^n$, the ambient differentiable structure of $\mathbf{R}^n$ forced consistency.

1.6 Definition. A C^p *atlas* on a Hausdorff topological space M is given by:
i) an open covering U_i, $i \in I$, of M;
ii) a family of homeomorphisms $\phi_i : U_i \to \Omega_i$ (where the Ω_i are open subsets of $\mathbf{R}^n$), such that for any i and j in I, the homeomorphism $\phi_j \circ \phi_i^{-1}$ *(transition function)* is in fact a C^p diffeomorphism from $\phi_i(U_i \cap U_j)$ onto $\phi_j(U_i \cap U_j)$.

Remark. The hypothesis that M is Hausdorff is necessary, and is not a consequence of the other assumptions. The $(U_i, \phi_i)_{i \in I}$ are *charts* for M. It is clear that M is *locally compact*. If M is not compact, a further assumption is very useful. Namely, every connected component of M will be assumed to be *countable at infinity (i.e. countable union of compact sets)*. This property allows some constructions ("partitions of unity", cf. **H**) which turn out to be very important. In fact, it is satisfied for all the manifolds of in usual mathematical life.

1.7 Definition. Two C^p atlases for M, say $(U_i, \phi_i)_{i \in I}$ and $(V_j, \psi_j)_{j \in J}$, are C^p *equivalent* if their union is still a C^p atlas, that is if the $\phi_i \circ \psi_j^{-1}$ are C^p diffeomorphisms from $\psi_j(U_i \cap V_j)$ onto $\phi_i(U_i \cap V_j)$.

1.8 Definition. A *differentiable structure of class C^p* on M is an equivalence class of consistent C^p atlases. As usual the equivalence class will be known by one representative element. That is why a *differentiable manifold* will be a Hausdorff topological space, together with an atlas.

Remark. When the topological space M is connected, the integer n in the definition does not depend on the chart and is the *dimension* of the manifold M. The proof is easy for C^p manifolds with $p \geq 1$, but is in the C^0 case a consequence of the difficult property of "domain invariance", which says the following (cf. [Spa]):

If $U \subset \mathbf{R}^n$ is open and $f : U \to \mathbf{R}^n$ is one-one and continuous, then $f(U) \subset \mathbf{R}^n$ is open.

Exercise. Check the same statement for smooth maps.

IN THE SEQUEL, IF NOT SPECIFIED EXPLICITLY, THE MANIFOLDS WILL BE CONNECTED, SMOOTH (C^∞), AND COUNTABLE AT INFINITY.

Remark. It was not necessary in the previous construction to begin with a topological space: it is indeed possible to define simultaneously on M a topological and a differentiable structure by using "bijective charts" (no structure) satisfying suitable conditions. But in this case, the conditions to be satisfied, and the proofs are more subtle ([B-G], 2.2). For example, see the construction of the tangent manifold (1.30).

1.9 Definition. Let M^d be a smooth manifold. A subset $N \subset M$ is a *submanifold* of M if for any $n \in N$, there exists a chart (U, ϕ) of M around n such that $\phi(U \cap N)$ is a submanifold of the open set $\phi(U) \cap \mathbf{R}^d$.

1.10 Examples. a) An n-dimensional submanifold of $\mathbf{R}^{n+k}$ is canonically equipped with an abstract manifold structure of the same dimension (use 1.3 iii)).

b) The square in $\mathbf{R}^2$ is of course not a C^∞ submanifold of $\mathbf{R}^2$ (even not of class C^1!). It can although be equipped with an abstract smooth manifold structure, which can be embedded in $\mathbf{R}^2$, but this embedding is of course not given by the inclusion: just parametrize the square by the circle as $\phi(x) = \frac{x}{\|x\|}$, and carry the differentiable structure of S^1 onto the square. The map ϕ is then a differentiable embedding of the square in $\mathbf{R}^2$. Note that the topologies induced by the inclusion in $\mathbf{R}^2$, or by the manifold structure coincide since ϕ is an homeomorphism.

c) Back with the example 1.4 e). The image M of the curve c in $\mathbf{R}^4 = \mathbf{C}^2$ defined by $c(t) = (e^{it}, e^{ait})$ is a submanifold of $\mathbf{R}^4$ if and only if the constant a is rational. Indeed, if $a = \frac{p}{q}$, then M is compact and equal to $c([0,q])$: apply 1.3 iii).

Assume then that a is not rational, and notice that M is included in the torus T^2 defined in 1.2 d). Hence if M was a submanifold of $\mathbf{R}^4$ it would be of dimension 1 or 2. But M cannot be 2-dimensional: it would contain a non void-open subset of the torus and we would get a contradiction by noticing that $T^2 \setminus M$ is dense in T^2. On the other hand if U is a neighborhood in $\mathbf{R}^4$ of $x \in M$ and if a is irrational, then M is dense in the torus and 1.3 ii) is not satisfied: in fact, $U \cap M$ has an infinite number of connected components. Still $c'(t)$ is never zero, but c is not a local homeomorphism from $\mathbf{R}$ on its image. However M can always be equipped with an abstract manifold structure, with the unique chart $(\mathbf{R}, c)$: M is then merely the real axis with its canonical differentiable structure. In the case a is rational, this structure coincides with the structure of submanifold of $\mathbf{R}^4$. In the case a is not rational, the topology of this manifold is not the topology induced by $\mathbf{R}^4$.

d) We want to equip the *torus* T^n defined at 1.2 d) with an atlas consistent with its structure of submanifold of $\mathbf{R}^{2n}$. Define $f : \mathbf{R}^n \to \mathbf{C}^n$ by

$$f(x_1, ..., x_n) = (e^{ix_1}, ..., e^{ix_n}) :$$

T^n is the image of f. Let $\bar{x} = f(x)$ be a point of the torus, with $x = (x_1, ..., x_n)$, and

$$\Omega = \,]x_1 - \pi; x_1 + \pi[\times \cdots \times \,]x_n - \pi; x_n + \pi[\,;$$

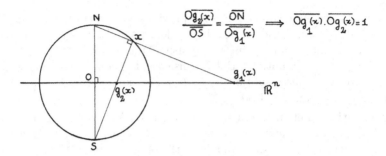

Fig. 1.6. Stereographic projection

the restriction of f to Ω is an homeomorphism on its image $U = f(\Omega)$: let g be the inverse map. If (U_1, g_1) and (U_2, g_2) are two such charts which domains intersect, the restriction of $(g_2 \circ g_1^{-1})$ to each connected component of $g_1(U_1 \cap U_2)$ is a translation and hence is smooth.

The smooth manifold structures on T^n defined by the atlas of all these charts, or by the ambient structure of $\mathbf{R}^{2n}$ (submanifold structure) are consistent since the maps g^{-1} are local parametrizations (1.3 iii)).

e) We now want to equip the *sphere* S^n with an atlas consistent with its structure of submanifold of $\mathbf{R}^{n+1}$. Two charts will be enough: they are called the *stereographic projections*. Denote by N and S the north and south poles of S^n (that is N=(1,0,..,0) and S=(-1,0,...,0)) and define

$$g_1 : U_1 = S^n \setminus \{N\} \to \mathbf{R}^n \quad \text{and} \quad g_2 : U_2 = S^n \setminus \{S\} \to \mathbf{R}^n$$

$$\text{for} \quad x = (x_0, ..., x_n), \quad \text{by}$$

$$g_1(x) = \frac{(x_1, ..., x_n)}{1 - x_0} \quad \text{and} \quad g_2(x) = \frac{(x_1, ..., x_n)}{1 + x_0}.$$

The maps g_i (i=1,2) are homeomorphisms from U_i onto $\mathbf{R}^n$ and $g_2 \circ g_1^{-1}$ is the inversion from $\mathbf{R}^n \setminus \{0\}$ onto itself defined by $y \to \frac{y}{\|y\|^2}$ and hence is a diffeomorphism.

The maps g_i^{-1} are local parametrizations for S^n: the differentiable structure induced on S^n by these two charts is consistent with its structure of submanifold of $\mathbf{R}^{n+1}$ (note also that the fact that $g_2 \circ g_1^{-1}$ is a diffeomorphism would be a direct consequence of this consistency (1.5)).

f) Recall that the *hyperboloid*

$$H^n = \{x \in \mathbf{R}^{n+1} / x_0^2 - \cdots - x_n^2 = 1, x_0 > 0\}$$

is a submanifold of $\mathbf{R}^{n+1}$; its structure of abstract manifold can be described by the unique chart $f : H^n \to \mathbf{R}^n$ defined by

$$f(x_0, \cdots, x_n) = (x_1, \cdots, x_n).$$

This structure is consistent with its submanifold structure by 1.3 iii).

g) The *real projective space* $P^n\mathbf{R}$ is, as a topological space, the quotient of $\mathbf{R}^{n+1}\backslash\{0\}$ by the equivalence relation which identifies proportional $n+1$-uples. The restriction of this relation to S^n identifies the points x and $-x$ (antipodal points): the real projective space is then homeomorphic to the quotient of S^n by the action of the group $\{Id, -Id\}$. Denote by p the quotient map. It is easy to check that $P^n\mathbf{R}$ is Hausdorff and hence compact.

We now want to equip $P^n\mathbf{R}$ with an atlas (U_i, ϕ_i) $(i = 0, ..., n)$, and hence with an abstract manifold structure consistent with its topology. Let

$$V_i = \{x \mid (x_0, ..., x_n) \in \mathbf{R}^{n+1}/x_i \neq 0\} \quad (0 \leq i \leq n)$$

and define maps $\Phi_i : V_i \to \mathbf{R}^n$ by

$$\Phi_i(x) = \left(\frac{x_0}{x_i}, \cdots, \frac{\widehat{x_i}}{x_i}, \cdots, \frac{x_n}{x_i}\right)$$

where the sign ^ means that the term must be omitted. The map Φ_i is continuous and

$$\Phi_i(x) = \Phi_i(y) \quad \text{if and only if} \quad p(x) = p(y).$$

Therefore Φ_i yields a bijective and continuous map

$$\phi_i : U_i = p(V_i) \to \mathbf{R}^n,$$

with

$$\phi_i\big(p(x)\big) = \left(\frac{x_0}{x_i}, \cdots, \frac{\hat{x}_i}{x_i}, \cdots, \frac{x_n}{x_i}\right).$$

The map ϕ_i is also open since if O is open in U_i, $\phi_i(O) = \Phi_i(p^{-1}(O))$ is open. Hence ϕ_i is an homeomorphism from U_i onto $\mathbf{R}^n$. The inverse map is given by

$$\phi_i^{-1}(y_0, \cdots, y_{n-1}) = p(y_0, \cdots, y_{i-1}, 1, y_i, \cdots, y_{n-1}).$$

Also the transition functions $\phi_j \circ \phi_i^{-1}$ are diffeomorphisms from $\phi_i(U_i \cap U_j)$ onto $\phi_j(U_i \cap U_j)$: we have indeed for $y_j \neq 0$:

$$(\phi_j \circ \phi_i^{-1})(y_0, \cdots, y_{n-1}) = \left(\frac{y_0}{y_j}, \cdots, \frac{y_{i-1}}{y_j}, \frac{1}{y_j}, \cdots, \frac{\hat{y}_j}{y_j}, \cdots, \frac{y_{n-1}}{y_j}\right).$$

Hence we get a structure of smooth manifold on $P^n\mathbf{R}$.

For $P^2\mathbf{R}$, we can have an intuitive vision of these charts: they consist in representing a landscape in $\mathbf{R}^3$ on a 2-dimensional plane (for example on a painting or a plate). All the points that belong to a same straight line from the eye of the painter or the photographer will be represented by a unique point (at least if the painter uses the perspective rules!).

h) The *complex projective space* $P^n\mathbf{C}$ is obtained by replacing $\mathbf{R}$ by $\mathbf{C}$ in the preceeding construction.

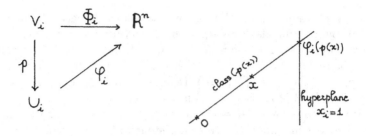

Fig. 1.7. A chart of the real projective space

Like in the real projective case, we prove that $P^n\mathbf{C}$ is Hausdorff and denote by p the quotient map (note that $P^n\mathbf{C}$ is compact since it is equal to $p(S^{2n+1})$). The constructions and the proofs are identical to the preceding ones (g), except that we now work with complex numbers in place of real numbers. The complex projective space is then equipped with an abstract smooth manifold structure of (real) dimension $2n$. We will meet in the following (cf. 2.29) more natural and convenient definitions for this manifold.

1.11 Exercise. Define the Möbius band M as the topological quotient of $[0,1] \times \mathbf{R}$ by the equivalence relation R which identifies the pairs $(0,t)$ and $(1, -t)$. Show that M can be equipped with a structure of a smooth manifold consistent with its topology.

1.12 Definition. A smooth manifold M is said to be *orientable* if there exists an atlas for M such that the jacobians of all the transition functions are positive (given an orientation on $\mathbf{R}^n$, the transition functions preserve this orientation: then we can orient "continuously" all the "tangent spaces" to M (see 1.30)). Such an atlas is an *orientation atlas*.

1.13 Exercises. a) Show that the Möbius band defined in 1.11 is not orientable.

b) Show that $P^2\mathbf{R}$ can be obtained by gluing together a disk and a Möbius band along their boundary, and deduce from a) that $P^2\mathbf{R}$ is not orientable.

c) Is a submanifold of an orientable manifold automatically orientable?

d) Show that the product of two orientable manifolds is orientable.

e) Let f be a submersion from $\mathbf{R}^n$ to $\mathbf{R}$; show that for any $y \in f(\mathbf{R}^n)$, the submanifold $f^{-1}(y)$ is orientable.

1.A.3 Smooth maps

To define them, it is necessary to use charts and go back to open sets in $\mathbf{R}^n$.

1.14 Definition. Let X and Y be C^p manifolds. A continuous map $f : X \to Y$ is C^k, for $k \leq p$ if for any $x \in X$ there exist charts (U, ϕ) and (V, ψ) for X and Y around x and $f(x)$ respectively, with $f(U) \subset V$, and such that the map

$$\psi \circ f \circ \phi^{-1} : \phi(U) \to \psi(V)$$

(that is f read through the charts) is C^k (in the usual sense, since $\phi(U)$ and $\psi(V)$ are open sets in numerical spaces).

$$
\begin{array}{ccc}
U & \xrightarrow{\;\;f\;\;} & V \\
\phi\downarrow & & \downarrow\psi \\
\phi(U) & \xrightarrow[\psi \circ f \circ \phi^{-1}]{} & \psi(V)
\end{array}
$$

This definition does not depend on the charts we choosed around x and $f(x)$ since the transition functions are C^p diffeomorphisms. This is just diagram-chasing.

1.15 Definition. A smooth map $g : \mathbf{R}^n \to \mathbf{R}^p$ is a *submersion* (resp. an *immersion*, resp. *a local diffeomorphism*) at $x \in \mathbf{R}^n$ if its differential map $D_x g$ at x is surjective (resp. injective, resp. bijective) from $\mathbf{R}^n$ to $\mathbf{R}^p$.

1.16 Definition. Let M and N be smooth manifolds. A smooth map $f : M \to N$ is an *immersion* at $m \in M$ if for a chart (U, ϕ) for M around m, and a chart (V, ψ) for N around $f(m)$, the map $\psi \circ f \circ \phi^{-1}$ is itself an immersion (usual sense!). This definition does not depend on the chart.

1.17 Proposition (canonical form for immersions and submersions).
Let $p \le n$ be two integers:
i) *Let $f : \mathbf{R}^n \to \mathbf{R}^p$ with $f(0) = 0$ be a* submersion *at 0. There exists a local diffeomorphism ϕ around 0 in $\mathbf{R}^n$, with $\phi(0) = 0$ and such that*

$$f \circ \phi(x_1, ..., x_n) = (x_1, ..., x_p).$$

ii) *Let $f : \mathbf{R}^p \to \mathbf{R}^n$ with $f(0) = 0$ be an* immersion *at 0. There exists a local diffeomorphism ψ around 0 in $\mathbf{R}^n$, with $\psi(0) = 0$ and such that*

$$\psi \circ f(x_1, ..., x_p) = (x_1, ..., x_p, ..., 0, ..., 0).$$

Proof. See [Bo]. ■

1.18 Definition. A map $f : M \to N$ is a *submersion* (resp. an *immersion*, resp. is *a local diffeomorphism*) if it has this property at any point of M.
The map f is a *diffeomorphism* if it is bijective, and if f and f^{-1} are smooth. The map f is an *embedding* if it is an immersion and if it is an homeomorphism on its image.

1.19 Examples. a) If M is a submanifold of $\mathbf{R}^{n+k}$ and if $F : \mathbf{R}^{n+k} \to \mathbf{R}$ is smooth, the restriction f of F to M is also smooth. Consider indeed a local parametrization of M around m: it yields a chart for M around m, and the map f read through this chart is $F \circ g$ which is smooth. Hence f is smooth.

b) We now use charts and build a diffeomorphism g between S^2 and $P^1\mathbf{C}$. The complex projective line is equipped with an atlas containing two charts:

$$(U_1, \phi_1) \quad \text{with} \quad U_1 = P^1\mathbf{C} \setminus \{p(0,1)\} \quad \text{and} \quad \phi_1(p(\xi)) = \tfrac{\xi_2}{\xi_1}$$
$$(U_2, \phi_2) \quad \text{with} \quad U_2 = P^1\mathbf{C} \setminus \{p(1,0)\} \quad \text{and} \quad \phi_2(p(\xi)) = \tfrac{\xi_1}{\xi_2}$$

The sphere S^2 is also equipped with an atlas containing two charts:

$$(V_1, \psi_1) \quad \text{with} \quad V_1 = S^2 \setminus \{(0,0,1)\} \quad \text{and} \quad \psi_1(x,y,z) = \tfrac{(x,y)}{1-z}$$
$$(V_2, \psi_2) \quad \text{with} \quad V_2 = S^2 \setminus \{(0,0,-1)\} \quad \text{and} \quad \psi_2(x,y,z) = \tfrac{(x,y)}{1+z}$$

The transition functions are respectively

$$\phi_2 \circ \phi_1^{-1}(\xi) = \frac{1}{\xi} \quad \text{and} \quad \psi_2 \circ \psi_1^{-1}(u) = \frac{u}{\| u \|^2}.$$

It is natural to try and build a map g which sends V_i to U_i ($i = 1, 2$). We just set

$$g(x,y,z) = p(x+iy, z) \quad \text{if } z \neq 1$$
$$g(x,y,z) = p(z, x-iy) \quad \text{if } z \neq -1$$

It is now clear that g is well defined and smooth. Its inverse is obtained by taking the map

$$H : (u,v) \mapsto \left(\frac{2u\bar{v}}{|u|^2 + |v|^2}, \frac{|u|^2 - |v|^2}{|u|^2 + |v|^2} \right),$$

Indeed, H is a smooth map from $\mathbf{C}^2 \setminus \{0\}$ to S^2, which goes to the quotient and defines a smooth map $h : P^1\mathbf{C} \to S^2$. The restriction of H to S^3 is known as *the Hopf fibration*. See in 2.32 the Riemannian analog of this result.

1.20 Exercises. a) Prove that, if $f : N \to M$ is an embedding, then $f(N)$ is a submanifold of M.

b) Prove that a proper and injective immersion $f : N \to M$ is an embedding.

c) Show that there is no immersion of S^1 into $\mathbf{R}$.

d) Let $f : N \to M$ be an immersion. Assume that M and N have the same dimension and show that f is open. Deduce that there is no immersion of S^n into $\mathbf{R}^n$.

1.B The tangent bundle

1.B.1 Tangent space to a submanifold of $\mathbf{R}^{n+k}$

1.21 Definition. Let M be an n-dimensional submanifold of $\mathbf{R}^{n+k}$, $m \in M$ and $v \in \mathbf{R}^{n+k}$; v is said to be *tangent* to M at m if there exists a curve c drawn on the submanifold in $\mathbf{R}^{n+k}$ such that $c(0) = m$ and $c'(0) = v$.

Remark. The vector $c'(t)$ is well-defined, since the differentiation of $c(t)$ occurs in the numerical space.

1.22 Theorem. *The set of tangent vectors to M at m is an n-dimensional linear subspace of $\mathbf{R}^{n+k}$, that we denote by $T_m M$.*
Proof. We use 1.3 ii). Let f be a diffeomorphism from a neighborhood U of m on a neighborhood V of $0 \in \mathbf{R}^{n+k}$, with

$$f(M \cap U) = (\mathbf{R}^n \times \{0\}) \cap V.$$

The set of tangent vectors at 0 to $f(M \cap U)$ is $\mathbf{R}^n \times \{0\}$, and since the curves drawn on $M \cap U$ and $f(M \cap U)$ are exchanged by f, we have

$$T_m M = (D_m f)^{-1}(\mathbf{R}^n \times \{0\}). \blacksquare$$

1.23 Theorem. *Let M be an n-dimensional submanifold of $\mathbf{R}^{n+k}$, $m \in M$ and U be a neighborhood of m in $\mathbf{R}^{n+k}$.*
i) *If $M \cap U = f^{-1}(0)$, where f is a submersion of U on $\mathbf{R}^k$, then*

$$T_m M = \mathrm{Ker}(D_m f).$$

ii) *If f is a diffeomorphism from U onto a neighborhood V of $0 \in \mathbf{R}^{n+k}$ with*

$$f(U \cap M) = V \cap (\mathbf{R}^n \times \{0\}),$$

then
$$T_m M = (D_m f)^{-1}(\mathbf{R}^n \times \{0\}).$$

iii) *If (Ω, g) is a parametrization of $M \cap U$ such that $g(x)=m$, then*

$$T_m M = D_x g(\mathbf{R}^n).$$

Proof. ii) has already been proved in 1.22.
i) For a curve c drawn on M we have $f(c(t)) \equiv 0$ and, by differentiating, $D_m f.(c'(0)) = 0$. This yields an inclusion, and then equality since the two vector spaces have the same dimension.
iii) If γ is a curve drawn on Ω, $g \circ \gamma$ is a curve drawn on M and $D_x g.(\gamma'(0)) \in T_m M$. Conclude by equality of the dimensions. $\blacksquare$

1.24 Exercises. a) Compute the tangent space to S^n at x.
b) Compute the tangent space to the hyperboloid

$$H = \{(x, y, z, t) \in \mathbf{R}^4 / x^2 + y^2 + z^2 - t^2 = -1\}$$

at the point (x, y, z, t).
c) Compute the tangent space to $SO(n)$ at Id and $g \in SO(n)$.

1.B.2 The manifold of tangent vectors

The problem in the following is that if a curve c is drawn on an abstract manifold M we don't know how to "differentiate" c and hence cannot define directly the tangent vector at the point $c(t)$ by "$c'(t)$".

1.25 Definition. Let M be an abstract manifold and $m \in M$. A *tangent vector* to M at m is an equivalence class of curves $c : I \to M$, where I is an interval containing 0 and such that $c(0) = m$, for the equivalence relation $\sim$ defined by

$$c \sim \gamma \quad \text{if and only if in a chart } (U, \phi) \text{ around } m,$$

$$\text{we have} \quad (\phi \circ \gamma)'(0) = (\phi \circ c)'(0).$$

This notion does not depend on the chart since

$$(\phi \circ c)'(0) = (\psi \circ \phi^{-1})'(\phi(m)) \cdot (\phi \circ c)'(0).$$

The tangent space to M at m, let $T_m M$, is the set of all tangent vectors to M at m. It has an n-dimensional linear space structure (just carry by a chart (U, ϕ) the linear structure of the tangent space to $\mathbf{R}^n$ at $\phi(m)$: this structure does not depend on the chart since $(\psi \circ \phi^{-1})'(\phi(m))$ is a vector space isomorphism).

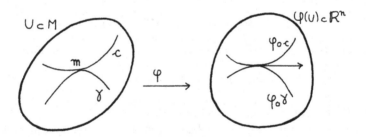

Fig. 1.8. A tangent vector

1.26 Exercise. Show that in the case M is a submanifold of $\mathbf{R}^n$, the previous definition is consistent with the definition of the tangent space to a submanifold.

We now give an equivalent definition of tangent vectors to an abstract manifold.

1.27 Definition. Let M be a manifold, $m \in M$, (U, ϕ) and (V, ψ) be two charts for M around m. Let u and v be two vectors in $\mathbf{R}^n$ (you have to consider them as tangent vectors to $\mathbf{R}^n$ at $\phi(m)$ and $\psi(m)$ respectively). We say that

$$(U, \phi, u) \sim (V, \psi, v) \qquad \text{if and only if} \quad (\psi \circ \phi^{-1})'_{\phi(m)} \cdot u = v.$$

A *tangent vector* to M at m is then a class of triples (U, ϕ, u) for the equivalence relation $\sim$.

In the first definition, a vector $\xi \in T_m M$ is represented by a curve c. In the second one, this vector is represented in the chart (U, ϕ) by a vector $u \in \mathbf{R}^n$: we get the equivalence of the two definitions by considering $u = (\phi \circ c)'(0)$ (check there is no problem when changing of chart).

If we choose a chart (U, ϕ) around m, we get with the help of the second definition a distinguished isomorphism between $\mathbf{R}^n$ and $T_m M$, let

$$\theta_{U, \phi, m} : u \to \mathrm{class}(U, \phi, u).$$

1.28 Exercise. Consider the sphere S^2 with the two stereographic charts (1.19 b). Let $u = (u_1, u_2)$ be a vector in $\mathbf{R}^2$. Let $\xi \in T_m S^2$ be the tangent vector represented by u in the chart (V_1, ψ_1): by which vector is ξ represented in the chart (V_2, ψ_2)? Go through the computation when $m = \left(\frac{\sqrt{3}}{2}, \frac{1}{2} \right)$, and compute the coordinates of ξ in $\mathbf{R}^3$.

1.29 Definition. We *denote* by TM the disjoint union of the tangent spaces to M at all the points of M: TM is the *tangent bundle* of M (see 1.33 for an explanation of this expression).

1.30 Theorem. *Let M be an n-dimensional C^p manifold ($p \geq 1$). Then its tangent bundle TM can canonically be equipped with a $2n$-dimensional and C^{p-1} abstract manifold structure.*

Proof. Denote by $\pi : TM \to M$ the map which sends a vector $\xi \in TM$ on the point m where it is based (that is $\xi \in T_m M$). Do not forget that TM does not carry yet a topology! For each chart (U, ϕ) of M, we define a "chart" (bijective, no regularity) $(\pi^{-1}(U), \Phi)$ for TM by

$$\Phi(\xi) = \left(\phi \circ \pi(\xi), \theta^{-1}_{U, \phi, \pi(\xi)}(\xi) \right).$$

We now equip TM with a topology such all the Φ are homeomorphisms, deciding that a fundamental system of neighborhoods of $\xi \in \pi^{-1}(U)$ is given by the inverse images by Φ of the neighborhoods of $\Phi(\xi)$ in $\phi(U) \times \mathbf{R}^{2n}$. One must check that this topology is well defined (see the note following 1.8 and the reference), and that the topology is Hausdorff (exercise): TM has now a structure of topological (C^0) manifold. We now turn to the transition functions. If (U, ϕ) and (V, ψ) are two charts for M around m, the transition function for the associated charts $(\pi^{-1}(U), \Phi)$ and $(\pi^{-1}(V), \Psi)$ is

$$\Psi \circ \Phi^{-1} : \phi(U \cap V) \times \mathbf{R}^n \to \psi(U \cap V) \times \mathbf{R}^n$$

with
$$\Psi \circ \Phi^{-1}(p, u) = (\psi \circ \phi^{-1}(p), D_p(\psi \circ \phi^{-1}) \cdot u) :$$

it is C^{p-1}. $\blacksquare$

1.31 Exercise. Show that the manifold TM is always orientable.

1.B.3 Vector bundles

1.32 Definition. Let E and B be two smooth manifolds, and $\pi : E \to B$ be a smooth map. One says that (π, E, B) is a *vector bundle of rank n* (E is the *total space*, B the *basis*, and $\mathbf{R}^n$ the *fiber)* if

- π is surjective;

- there exists an open covering $(U_i)_{i \in I}$ of B, and diffeomorphisms
$$h_i : \pi^{-1}(U_i) \to U_i \times \mathbf{R}^n,$$
such that for any $x \in U_i$,
$$h_i\left(\pi^{-1}(x)\right) = \{x\} \times \mathbf{R}^n,$$

- and such that for $i, j \in J$, the diffeomorphisms
$$h_i \circ h_j^{-1} : (U_i \cap U_j) \times \mathbf{R}^n \to (U_i \cap U_j) \times \mathbf{R}^n$$
are of the form
$$h_i \circ h_j^{-1}(x, v) = (x, g_{ij}(x) \cdot v),$$
where $g_{ij} : U_i \cap U_j \to Gl(n, \mathbf{R})$ is smooth.

Each fiber $\pi^{-1}(x)$ is then equipped with a linear space structure. The maps h_i are called *local trivializations* for the vector bundle.

Example. For a smooth manifold M, the first projection $\pi : M \times \mathbf{R}^n \to M$ is a vector bundle of rank n: it is the product bundle.

1.33 Exercise. Show that $\pi : TM \to M$ is a vector bundle of rank n where $\dim M = n$ (imitate the proof of 1.30).

1.34 Definitions. a) A vector bundle of rank n, $\pi : E \to B$, is *trivial* if there exists a diffeomorphism $h : E \to B \times \mathbf{R}^n$ which preserves the fibers (that is with $h : \pi^{-1}(x) \to \{x\} \times \mathbf{R}^n$ vector spaces isomorphism).
b) A *section* of the bundle is a smooth map $s : B \to E$ such that $\pi \circ s = Id$ (smooth choice of one vector in each fiber).

1.35 Exercises. a) Show that a rank n vector bundle is trivial if and only if there exist n linearly independant sections $(s_i)_{(1 \le i \le n)}$ (that is such that for any point b of the basis, the vectors $(s_i(b))_{(1 \le i \le n)}$ are linearly independant in $\pi^{-1}(b) \simeq \mathbf{R}^n$).
b) With the notations of 1.11, show that the Möbius band is the total space of a non trivial vector bundle of rank one over the circle.

1.B.4 Tangent map

Let M and N be two manifolds of dimensions n and k respectively, and $f : M \to N$ a smooth map (in what follows, C^1 would be sufficient). If two smooth curves $c_k : I_k \to M$ ($k = 1, 2$), such that $c_k(0) = m$, define the same tangent vector at m, then, using the chain rule, one infers that the curves $f \circ c_k$ define the same tangent vector at $f(m)$.

1.36 **Definition.** The *tangent map* of f at m is the map

$$T_m f : T_m M \to T_{f(m)} N$$

defined in that way.

If M and N are open set of Euclidean spaces, $T_m f$ is of course the differential of f at m. If (U, φ) is a chart of M such that $\varphi(m) = 0$, then, using the alternative description 1.27 of tangent vectors, we see that $T_m \varphi$ sends the tangent vector class(U, φ, u) to u, so that it is linear. Moreover, it follows from the very definition of $T_m f$ that the tangent map satisfies the chain rule, and using definition 1.14, we conclude that $T_m f$ is linear. Now, we can define a map $Tf : TM \to TN$ in a natural way: denoting by v_m a tangent vector at m, we set

$$Tf(v_m) = T_m f(v_m).$$

With a little (easy) extra work, we get the following.

1.37 **Theorem.** *Let M, N be smooth manifolds, and $f : M \to N$ a C^p map ($p \geq 1$). Then $Tf : TM \to TN$ is a C^{p-1} bundle map. Moreover, if $g : N \to P$ is a C^p map in another manifold P, for $m \in M$ we have*

$$T_m(g \circ f) = T_{f(m)} g \circ T_m f.$$

In other words, $T(g \circ f) = Tg \circ Tf$.

1.C Vector fields

1.C.1 Definitions

1.38 Definition. A *(smooth) vector field* on a manifold M is a smooth section of the bundle TM (that is a smooth map X from M to TM such that, for any $m \in M$, $X(m) \in T_m M$). We will denote by $\Gamma(TM)$ the vector space of vector fields on M.

Example. The vector fields on $S^n \subset \mathbf{R}^{n+1}$ are the maps $X : S^n \to \mathbf{R}^{n+1}$ such that for $x \in S^n$, $\langle X(x), x \rangle = 0$.

1.39 **Exercises.** a) Let $\phi_1 : U_1 \to \Omega_1$ and $\phi_2 : U_2 \to \Omega_2$ be the stereographic charts for S^n. Let X_1 and X_2 be vector fields on Ω_1 and Ω_2 respectively. Which condition must X_1 and X_2 satisfy to represent in the charts the same vector field X on S^n?

b) Let $\Phi : \mathbf{R}^2 \to T^2$ be the parametrization defined by $\Phi(\theta, \theta') = (e^{i\theta}, e^{i\theta'})$, and Y be a vector field on $\mathbf{R}^2$. On what condition does Y represent, with respect to the parametrization Φ, a vector field X on T^2? Deduce then that the tangent bundle to T^2 is diffeomorphic to $T^2 \times \mathbf{R}^2$.

1.40 Theorem. *The tangent bundle TM is trivial if and only if there exist n linearly independant vector fields* $(\dim M = n)$.
Proof. It is a restatement of exercise 1.35 a). ∎

The study of the manifolds which tangent bundle is trivial (parallelizable) is taken into account by algebraic topology and K-theory (see for example [K]). One can prove for example that any 3-dimensional manifold is parallelizable, and that the only parallelizable spheres are S^1, S^3 and S^7.
The following result, which is much weaker, is however interesting.

1.41 Theorem. *Assume that n is even. Then any vector field X on S^n has at least one zero.*
Proof. (after J. Milnor). We consider the canonical embedding of S^n in $\mathbf{R}^{n+1}$. If X never vanishes, we can renormalize and assume that $\| X \| \equiv 1$. For ϵ small enough, the map $f : x \to x + \epsilon X(x)$ is a diffeomorphism from the sphere of radius one onto the sphere of radius $r = (1 + \epsilon^2)^{1/2}$ (check it, or see [H] for details). Let ω be the differential n−form defined by

$$\omega = \sum_{i=0}^{n} (-1)^i x^i dx^0 \wedge \cdots \wedge d\hat{x}^i \wedge \cdots \wedge dx^n.$$

The volume of $S^n(r)$ is $c_n r^n$ (no need to know c_n), but it is also, by change of variables in the integral, equal to

$$\frac{1}{r} \int_{S^n(r)} \omega = \frac{1}{r} \int_{S^n} f^* \omega.$$

The expression of f proves that $\int_{S^n} f^* \omega$ is polynomial in ϵ. But

$$\int_{S^n(r)} f^* \omega = c_n r^{n+1} = c_n (1 + \epsilon^2)^{\frac{(n+1)}{2}}.$$

This leads to a contradiction if n is even. ∎

Remark. We used differential forms, which will be introduced later (I.G). In fact, the theory of differential forms on $\mathbf{R}^n$ is sufficient to understand this proof.

1.42 Corollary. *If n is even, the tangent bundle TS^n is non trivial.*
Proof. Obvious with 1.40 and 1.41. ∎

1.43 Exercises. a) Give an example of a nowhere vanishing vector field on S^{2n+1}.
b) Show with elementary means that $TSO(n)$ is trivial.

1.C.2 Another definition for the tangent space

We now give a less geometrical but convenient definition for vector fields. We first come back to $T_m M$ itself.

We introduce on the set of (continuous, C^p, smooth, or analytic) functions defined in a neighborhood of $m \in M$ (the neighborhood does depend on the function) an equivalence relation, by identifying $(f : U \to \mathbf{R})$ and $(g : V \to \mathbf{R})$ if there exists a neighborhood $W \subset U \cap V$ of m such that $f_{|W} = g_{|W}$.

1.44 Definition. The equivalence classes for this relation are the *germs* at m of (continuous, C^p, smooth or analytic) functions. We will denote these spaces by $\mathcal{C}_m^0(M)$, $\mathcal{C}_m^p(M)$, $\mathcal{C}_m^\infty(M)$, and $\mathcal{C}_m^\omega(M)$.

Example. The space $\mathcal{C}_0^\omega(\mathbf{C})$ of germs at 0 of holomorphic functions can be identified with the space of power series with non zero convergence radius.

Remark. If M is a C^p manifold, the space $\mathcal{C}_m^p(M)$ $(0 \leq p \leq \infty)$ is isomorphic to $\mathcal{C}_0^p(\mathbf{R}^n)$: choose a chart (U, ϕ) for M around m, and associate to a function f defined in a neighborhood V of m the function $f \circ \phi^{-1}$ defined on $\phi(U \cap V)$. The map $f \to f \circ \phi^{-1}$ goes to the quotient and yields an isomorphism between $\mathcal{C}_m^p(M)$ and $\mathcal{C}_0^p(\mathbf{R}^n)$. Of course, this isomorphism is not canonical!

1.45 Definition. A *derivation* on $\mathcal{C}_m^p(M)$ is a linear map $\delta : \mathcal{C}_m^p(M) \to \mathbf{R}$ such that for $f, g \in \mathcal{C}_m^p(M)$:

$$\delta(f.g) = f(m)\delta(g) + g(m)\delta(f).$$

We denote by $\mathcal{D}_m^p(M)$ -or $\mathcal{D}_m(M)$ if there is no ambiguity- the set of derivations: their name is due to the following result.

1.46 Theorem. *Any derivation of $\mathcal{C}_0^\infty(\mathbf{R}^n)$ can be written as:*

$$\delta(f) = \sum_{j=1}^n \delta(x^j) \left(\frac{\partial f}{\partial x^j} \right)_{|0}.$$

1.47 Lemma. *Let f be C^p around 0 in $\mathbf{R}^n$. Then f can be written as*

$$f(x) - f(0) = \sum_{j=1}^n x^j h_j(x),$$

where h_j is C^{p-1}.

Proof. It is the integral Taylor formula: just write

$$f(x) - f(0) = \int_0^1 \frac{d}{dt} f(tx) dt = \sum_{j=1}^n x^j \int_0^1 \left(\frac{\partial f}{\partial x^j} \right)(tx) dt. \blacksquare$$

Proof of the Theorem. Let $\dot{f} \in C_0^\infty(\mathbf{R}^n)$, and f be a representative of $\dot{f}$. We have $\delta(f) = \delta(f - f(0))$, since from the definition $\delta(1.1) = \delta(1) + \delta(1)$ and hence $\delta(\text{constant}) = 0$. From the lemma,

$$\delta(f) = \sum_{j=1}^n \delta(x^j)h_j(0) \quad \text{with} \quad h_j(0) = \left(\frac{\partial f}{\partial x^j}\right)_{|0}.$$

The map $f \to \frac{\partial f}{\partial x^j}(0)$ clearly depends only on $\dot{f}$, and is itself a derivation. ∎

1.48 Exercise. Show there is no derivation on $C_0^0(\mathbf{R}^n)$.

We now turn to an analogous theorem for manifolds. We associate to $\xi \in T_m M$ the derivation L_ξ defined by $L_\xi(f) = T_m f.\xi$ (Lie derivative in the direction ξ).

1.49 Theorem. *The map $\xi \to L_\xi$ is a linear isomorphism between $T_m M$ and $\mathcal{D}_m(M)$.*

Proof. Let (U, ϕ) be a chart for M with $\phi(m) = 0 \in \mathbf{R}^n$, and (U, ϕ, v) be the corresponding representative for ξ. Then

$$L_\xi(f) = d(f \circ \phi^{-1})_{|0} \cdot v = \sum_{i=1}^n v^j \left(\frac{\partial f}{\partial x^j}\right)(0).$$

Just use then the isomorphism mentioned in the note following 1.44 and apply 1.46 ∎

If (U, ϕ) is a chart around m, and $(x^1, ..., x^n)$ are the corresponding coordinate functions, we will set

$$\left(\frac{\partial}{\partial x^j}\right)_{|m}(f) = \left(\frac{\partial}{\partial x^j}\right)(f \circ \phi^{-1})(\phi(m)).$$

For $m \in U$, we get derivations of $C_m(M)$, or equivalently by 1.49, vectors in $T_m M$. Check the maps so defined from U to TU are C^{p-1} if f is C^p: hence they define vector fields.

More generally if $X^1, \cdots, X^n$ are (smooth) functions on U, we introduce the vector field

$$X = \sum_{i=1}^n X^i \left(\frac{\partial}{\partial x^j}\right)$$

and the map L_X from $C^\infty(U)$ to itself defined by

$$L_X f(m) = \sum_{i=1}^n X^i(m) \left(\frac{\partial f}{\partial x^j}\right)$$

(L_X is a differential operator, see 1.113).

We now want to "globalize" the previous theorem (everything is smooth):

1.50 Definition. A *derivation* on M is a linear map δ from $C^\infty(M)$ to itself such that:

$$\text{for} \quad f, g \in C^\infty(M), \quad \delta(f.g) = f.\delta(g) + g.\delta(f).$$

1.51 Theorem. *The map* $X \to L_X$ *is an isomorphism between the vector space* $\Gamma(TM)$ *of vector fields on* M, *and the vector space* $\mathcal{D}(M)$ *of derivations.*
Proof. First L is injective: if X does not vanish at $a \in M$, there exists a smooth function f defined on all M and such that $L_X f(a) \neq 0$. Such a function exists obviously on a neighborhood of a (use a chart), then globalize with the help of a test function (see 1.132).

Now we prove that L is surjective. Use the same way as above test functions to prove that any germ at a of a smooth function defined on a neighborhood of a is also the germ of a smooth function defined on all M. Hence if δ is a derivation on M, the map $\delta : f \to \delta(f)(a)$ induces a derivation on the germs at a, and 1.49 yields a vector $X(a)$ such that $L_{X(a)} = \delta_a$. The map $a \to X(a)$ is indeed a smooth vector field: using local coordinates we see that

$$\delta_{|U} = \sum_{i=1}^{n} \delta(x^i) \left(\frac{\partial}{\partial x^i} \right),$$

and the functions $\delta(x^i)$ are smooth as the components of X in the chart (U, ϕ). ∎
Be careful: the existence of test functions was crucial in the proof. *The corresponding statement is false for holomorphic manifolds. For example, there are holomorphic vector fields on $P^1\mathbf{C}$ but there is no derivation on $C^\omega(P^1\mathbf{C})$, since any holomorphic function on a compact manifold is constant.*

The set of derivations has a natural structure of linear space. But the composition of two derivations is not automatically a derivation, since

$$\delta\delta'(fg) = f.(\delta\delta'g) + (\delta\delta'f).g + (\delta f).(\delta'g) + (\delta'f).(\delta g).$$

From this formula, we deduce immediately the following algebraic property:
1.52 Lemma. *If* δ *and* δ' *are two derivations, then*

$$\delta \circ \delta' - \delta' \circ \delta$$

is itself a derivation.

1.52 bis **Definition.** The *bracket* of two vector fields X and Y, denoted by $[X, Y]$, is the vector field corresponding to the derivation $L_X L_Y - L_Y L_X$.
It is possible to prove directly the preceding lemma. Say that in local coordinates

$$X = \sum_{i=1}^{n} X^i \frac{\partial}{\partial x^i} \quad \text{and} \quad Y = \sum_{i=1}^{n} Y^i \frac{\partial}{\partial x^i}.$$

Schwarz lemma yields for a smooth function f,

$$(L_X L_Y - L_Y L_X)f = \sum_{i=1}^{n} Z^i \frac{\partial}{\partial x^i}, \quad Z^i = \sum_{i=1}^{n} \left(X^j \frac{\partial Y^i}{\partial x^j} - Y^j \frac{\partial X^i}{\partial x^j} \right).$$

In fact, Schwarz lemma is a direct consequence (not under the minimal hypothesis!) of the preceding lemma and of 1.45: since $[X, Y]$ yields a derivation, for any smooth function f, the quantity

$$\sum_{i=1}^{n} Z^i \left(\frac{\partial f}{\partial x^i} \right) + \sum_{i,j=1}^{n} X^i Y^j \left(\frac{\partial^2 f}{\partial x^i \partial x^j} - \frac{\partial^2 f}{\partial x^i \partial x^j} \right)$$

depends only on the derivatives of first order of f, which is only the case if

$$\frac{\partial^2 f}{\partial x^i \partial x^j} = \frac{\partial^2 f}{\partial x^j \partial x^i}.$$

1.53 Lemma (Jacobi identity). *For any vector fields X, Y, Z on M*

$$[X, [Y, Z]] + [Y, [Z, X]] + [Z, [X, Y]] = 0.$$

Proof. Algebraic computation for the corresponding derivations. ∎

1.54 Exercise. Let X be a vector field on S^n. Extend X to $\mathbf{R}^{n+1}$ by $\tilde{X}(x) = \| x \| \cdot X \left(\frac{x}{\|x\|} \right)$ and let Y be the vector field on $\mathbf{R}^{n+1} \setminus \{0\}$ defined by $Y(x) = \frac{x}{\|x\|}$. Compute $[\tilde{X}, Y]$ and $[X_1, Y]$, where $X_1(x) = X \left(\frac{x}{\|x\|} \right)$.

1.C.3 Integral curves and flow of a vector field

We now turn to the differential equation $c' = X \circ c$ induced by a vector field X on M, where $c : I \subset \mathbf{R} \to M$ is a curve on M.

1.55 Proposition. *For any $x \in M$, there exist an interval I_x containing 0, and a unique smooth curve $c_x : I_x \to M$ such that $c_x(0) = x$ and for $t \in I_x$: $c'_x(t) = X(c_x(t))$.*
Proof. Working with a local chart for M around x, this result is a reformulation of the local existence and uniqueness for first order differential equations. Recall that the uniqueness means that c_x and $\tilde{c}_x$ coincide on the intersection of their intervals of definition. For more details, see [Wa]. ∎
We now vary x.

1.56 Proposition. *For $m \in M$, there exists an open neighborhood V of m, and an interval I containing 0, such that for $x \in V$ the map c_x is defined on I. Moreover, the map $(t, x) \to c_x(t)$ from $I \times V$ to M is smooth.*
Proof. The first part is a direct consequence of the proof of the preceding proposition. The second part is the regularity of solutions of differential equations with respect to initial conditions. ∎
1.57 Definition. The map $(t, x) \to c_x(t)$ is the *local flow* of the vector field X. Let us change our point of view, and set $\theta_t(x) = c_x(t)$. The following property is trivial, but fundamental.

1.58 **Proposition.** *The map θ_t is a local diffeomorphism. If the two members of the equality are defined, then*

$$\theta_t \circ \theta_{t'} = \theta_{t+t'}. \qquad (*)$$

Proof. Since

$$c_x'(t_1 + t_2) = X(c_x(t_1 + t_2))$$

we can give two interpretations for $c_x(t_1 + t_2)$. It is the value at $t_1 + t_2$ of the solution of the differential equation we consider, which satisfies to the initial condition x at 0. It is also the value at t_1 of the solution of the same differential equation with initial condition $c_x(t_2)$ at 0. ∎

1.59 **Definition.** A family of local diffeomorphisms (θ_t), $(t \in I)$ where I is an interval containing 0, and satisfying $(*)$ for $t, t', t + t' \in I$, is a *local one parameter group.*

For $x \in M$, there exists a maximal interval of definition for c_x. The first part of the following exercise shows that this interval is not in general equal to $\mathbf{R}$, and depends on x: the local group associated to a vector field is generally not a group.

1.60 **Exercises.** a) Give the flow of the vector field defined on $\mathbf{R}$ by $x \to x^2 \left(\frac{d}{dx}\right)$.

b) Show that the flow of the vector field defined on $\mathbf{R}^n$ by $\sum_{i=1}^{n} x^i \left(\frac{\partial}{\partial x^i}\right)$ is the group of dilations: $\theta_t(x) = e^t x$. Deduce from this result the Euler identity for homogeneous functions.

Be careful: the example of 1.60 a) shows also that the length of the maximal interval of definition of the function c_x can go to zero when x goes to infinity. In the sequel of this chapter, we shall use systematically identities involving local groups for vector fields. There will be no problem, since we shall be concerned with *local identities,* that are supposed to be written on open sets with compact closures.

However, we have the important following result.

1.61 **Theorem.** *The local group associated to a vector field X on M, with compact support K, is in fact a one parameter group of diffeomorphisms of M.*
Proof. For $p \in M$, there exists an open neighborhood U of p, and $\epsilon > 0$ such that for $m \in U$ the function $t \to c_m(t)$, solution of the differential equation $c' = X \circ c$ with initial condition $c_m(0) = 0$, is defined for $|t| < \epsilon$.

The compact K can be covered by a finite number of such open sets U and we denote by ϵ_0 the smaller of the corresponding ϵ. Hence $\theta_t(m) = c_m(t)$ is defined on $]-\epsilon_0, \epsilon_0[\times M$ and also, from 1.58, $\theta_{t+s} = \theta_t \circ \theta_s$ as soon as $|t|, |s| < \frac{\epsilon_0}{2}$. Then θ_t is indeed a diffeomorphism.

For $|t| > \frac{\epsilon_0}{2}$, we write $t = k\frac{\epsilon_0}{2} + r$, where k is an integer, and $|r| < \frac{\epsilon_0}{2}$, and set

$$\theta_t = \left(\theta_{\frac{\epsilon_0}{2}}\right)^k \circ \theta_r.$$

Check the relation $\theta_t \circ \theta_s = \theta_{t+s}$. ∎

1.62 Exercise. Let X, Y, Z be the vector fields defined on $\mathbf{R}^3$ by

$$X = z\frac{\partial}{\partial y} - y\frac{\partial}{\partial z}, \quad Y = x\frac{\partial}{\partial z} - z\frac{\partial}{\partial x},$$

$$Z = y\frac{\partial}{\partial x} - x\frac{\partial}{\partial y}.$$

Show that the map $(a, b, c) \to aX + bY + cZ$ is an isomorphism from $\mathbf{R}^3$ onto a subspace of the space of vector fields on $\mathbf{R}^3$, and that to the bracket of vector fields on $\mathbf{R}^3$ corresponds the cross product of vectors in $\mathbf{R}^3$. Compute the flow of the vector field $aX + bY + cZ$.

1.C.4 Image of a vector field by a diffeomorphism

1.63 Theorem. *Let $\phi : M \to N$ be a diffeomorphism, and X be a vector field on M. The map*

$$n \to Y(n) = T_{\phi^{-1}(n)}\phi \cdot X(\phi^{-1}(n))$$

defines a vector field Y on N. If L_X and L_Y are the associated derivations (on $C^\infty(M)$ and $C^\infty(N)$ respectively), then

$$\text{for} \quad f \in C^\infty(N), \qquad L_Y f = L_X(f \circ \phi) \circ \phi^{-1}.$$

Proof. The first part is a consequence of the expression of Y in local coordinates. Note that the hypothesis that $T\phi$ is invertible is crucial to prove that $n \to Y(n)$ is smooth.

As for the second part, note that

$$L_X(f \circ \phi)(m) = T_m(f \circ \phi) \cdot X(m),$$

and apply the chain rule. ∎

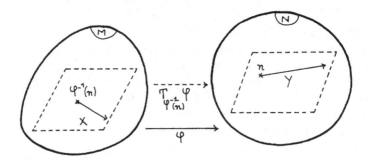

Fig. 1.9. Image of a vector field

1.64 Definition. The field Y is the *image* of X by ϕ, let $Y = \phi_* X$. If ϕ and ψ are two successive diffeomorphisms, we have

$$(\psi \circ \phi)_* = \psi_* \circ \phi_*.$$

1.65 Exercises. a) Let ϕ be the map from $\mathbf{R}$ to itself defined by $t \rightarrow t^3$, and X be the vector field $\frac{\partial}{\partial t}$. We would like to define $\phi_* X$ as the vector field associated to the derivation defined on $C^\infty(\mathbf{R})$ by $f \rightarrow (f \circ \phi)' \circ \phi^{-1}$. What happens in fact?
b) Compute the image under the map $x \rightarrow e^x$ of the vector field $\frac{\partial}{\partial x}$ on $\mathbf{R}$.

1.66 Proposition. *Let $\phi : M \rightarrow N$ be a diffeomorphism, and X and Y be two vector fields on M. Then*

$$[\phi_* X, \phi_* Y] = \phi_* [X, Y].$$

Proof. Immediate when considering the corresponding derivations. ∎

Now we give the relation between the local groups of X and $\phi_* X$.
1.67 Proposition. *Let X be a vector field on M, with local group θ_t, and $\phi : M \rightarrow N$ be a diffeomorphism: the local group associated to $\phi_* X$ is $\phi \circ \theta_t \circ \phi^{-1}$.*
Proof. Since $\phi \circ \theta_t \circ \phi^{-1}$ is a local one parameter group, it is sufficient to check the result around 0. But

$$\left(\frac{d}{dt} \right) [\phi(\theta_t(\phi^{-1}(n)))]_{|t=0} = T_{\phi^{-1}(n)}\phi \cdot \left(\frac{d}{dt} \right) (\theta_t(\phi^{-1}(n)))_{|t=0}$$

$$= T_{\phi^{-1}(n)}\phi \cdot X(\phi^{-1}(n)) = \phi_* X(n). \blacksquare$$

Let us see now what happens when ϕ is itself induced by a vector field.
1.68 Theorem. *Let X and Y be two vector fields on M, and θ_t be the local group of Y. Then*

$$\left(\frac{d}{dt} \right) (\theta_{t*} X)_{|t=0} = [X, Y].$$

Proof. We shall use the following fact, which is just a version with parameters of Lemma 1.47.
Let $f : [-\epsilon, \epsilon] \times M \rightarrow \mathbf{R}$ be a smooth map such that $f(0, p) = 0$. Then f can be written as

$$f(t, p) = t g(t, p)$$

where g is smooth and $\frac{\partial f}{\partial t}(0, p) = g(0, p)$.
Apply this remark to $\tilde{f}(t, p) = f(\theta_t(p)) - f(p)$, and set $g_t = g(t, .)$ –with $g_0 = Y.f$. Then

$$(\theta_{t*} X)_p.(f) = \left(X.(f \circ \theta_t) \right) \circ \theta_t^{-1} = X_{\theta_{-t}(p)}.(f + t g_t),$$

so that

$$\frac{1}{t}\big(X_p - (\theta_{t*}X)_p\big)(f) = \frac{1}{t}\big[(X.f)(p) - (X.f)(\theta_{-t}(p))\big] - X_{\theta_{-t}(p)}.g_t.$$

In the limit we get the claimed equality. ∎

Remark. Let X, Y, Z be three vector fields, and θ_t be the local group of Z. Derive with respect to t the identity

$$\theta_{t*}[X, Y] = [\theta_{t*}X, \theta_{t*}Y]$$

and apply the previous result to get another proof of Jacobi identity.

1.69 Exercise. Use theorem 1.68 and find another proof for the results of exercise 1.54.

1.D Baby Lie groups

Lie Groups theory is one of the cornerstones of Mathematics. However, we shall need only very basic results. A more detailed introduction, in the same spirit as ours, can be found in [Sp]. We also recommend [G].

1.D.1 Definitions

1.70 Definition. A Lie group is a group G equipped with a structure of a (smooth) manifold such that the map $(g, h) \rightarrow gh^{-1}$, defined from $G \times G$ to G, is smooth.

Examples. From 1.2, $Gl(n, \mathbf{R})$, $Gl(n, \mathbf{C})$ and $O(n)$ are Lie groups. We will also encounter in the sequel the following Lie groups:
- the unitary group $U(n)$, group of automorphisms for the hermitian form defined on $\mathbf{C}^n$ by $\sum_{i=1}^{n} \bar{z}_i z_i$,
- the special unitary group $SU(n)$, group of unitary automorphisms of determinant 1,
- the Lorentz group $O(1, n)$, group of automorphisms for the quadratic form defined in $\mathbf{R}^n$ by $\left(-x_0^2 + \sum_{i=1}^{n} x_i^2\right)$.

All of them are *linear groups*, i.e. subgroups of $Gl(n, \mathbf{R})$ or $Gl(n, \mathbf{C})$, which makes the theory is simpler.

Be careful: the definitions of $U(n)$ and $SU(n)$ involve complex numbers, but these groups *are not* complex manifolds.

For $g \in G$, we will denote the left and right translations, $x \rightarrow gx$ and $x \rightarrow xg$, by L_g and R_g respectively. They are diffeomorphisms.

The following notion yields the "infinitesimal" version of a Lie group.

1.71 Definition. A (real) *Lie algebra* is a (real) vector space $\underline{L}$, equipped with a bilinear map $[,] : \underline{L} \times \underline{L} \to \underline{L}$ such that
i) $[X, X] = 0$ for $X \in \underline{L}$;
ii) $[X, [Y, Z]] + [Y, [Z, X]] + [Z, [X, Y]] = 0$ for $X, Y, Z \in \underline{L}$ (Jacobi identity).

Example. We have seen in 1.53 that the vector space of vector fields on a manifold is equipped with a Lie algebra structure, with the bracket as bilinear map.

1.72 Proposition. *Let G be a Lie group. For $x \in T_e G$, the expression $X(g) = T_e L_g.x$ defines a left invariant vector field on G, (that is such that $L_{h*} X = X$ for any $h \in G$). The map $x \to X$ is an isomorphism between $T_e G$ and the vector space of left invariant vector fields. The later has a structure of Lie subalgebra of the Lie algebra of vector fields.*
Proof. By construction, X is left invariant, and is determined by its value at e. From 1.66, the left invariance is preserved under bracket, since

$$L_{g*}[X, Y] = [L_{g*} X, L_{g*} Y].$$

This isomorphism allows us to carry onto $T_e G$ the Lie algebra structure of the left invariant vector fields. ∎

1.73 Definition. The *Lie algebra* of G, denoted by $\underline{G}$, is the vector space $T_e G$, equipped with the bracket defined in the preceding section.
In the following, we will use freely both points of view, each of them is useful.
Example. The left invariant vector fields on the additive group $\mathbf{R}^n$ are the vector fields $\sum a^i \frac{\partial}{\partial x^i}$, where the a_i are constants, and their brackets are zero: hence $\mathbf{R}^n$ is the trivial n-dimensional Lie algebra.

1.74 Exercise. Show that the bracket in $\underline{Gl(n, \mathbf{R})}$ is just $[A, B] = AB - BA$.
We now turn to the flow of a left invariant vector field.

1.75 Theorem. *The flow θ_t of a left invariant vector field X on G is a global flow, and*

$$\theta_t(g) = g\theta_t(e)$$

for $g \in G$ and $t \in \mathbf{R}$.
Proof. From 1.67, and since X is left invariant, $\theta_t(gx)$ will be defined as soon as $\theta_t(x)$ is defined, and we will have $g\theta_t(x) = \theta_t(gx)$. From the existence theorem for a local flow (1.56), there exists an $\epsilon > 0$ such that $\theta_t(1)$, and hence $\theta_t(g)$, is defined for $| t | < \epsilon$. Setting, for $k \in \mathbf{Z}$, $\theta_{kt}(g) = g(\theta_t(e))^k$, we see that θ_t is defined on $\mathbf{R}$ (check the details). ∎

1.76 Definition. The map $X \to \theta_1(e)$, defined from $\underline{G}$ to G, is the *exponential map*, denoted by exp.
It is clear that $\theta_t(e) = \exp(tX)$, and that the map $t \to \exp(tX)$ is a (smooth) group homomorphism from $\mathbf{R}$ to G, such that $\frac{d}{dt}(\exp(tX))_{|t=0} = X$. Such an homomorphism is a *one parameter subgroup* of G.

1.77 Example. Let G be $Gl_n K$, with $K = \mathbf{R}$ or $\mathbf{C}$. The flow of the left invariant vector field defined by $A \in \underline{Gl_n K} = M_n(K)$ is obtained by integrating the differential equation $\frac{dX}{dt} = XA$, with initial condition $X(0) = Id$. If K^n is equipped with a norm, and $M_n(K)$ with the associated endomorphism norm, we have: $\| A^n \| \leq \| A \|^n$, and hence the series

$$\sum_{p=0}^{\infty} \left(\frac{t^p}{p!}\right) A^p,$$

and all its derivatives, converges normally. Hence $Y(t) = \exp(tA)$ equals the sum of this series: this justifies a posteriori the label for the exponential map.

1.78 Theorem. *The exponential map is a diffeomorphism from a neighborhood of 0 in $\underline{G}$ onto a neighborhood of Id in G, and its differential map at 0 is the identity.*

Proof. Once we know that exp is smooth (cf. 1.56), it is a tautology. Indeed, for $X \in \underline{G}$,

$$T_0 \exp \cdot X = \frac{d}{du} \exp uX_{|u=0} = X.$$

Be careful: In general the map exp is neither a diffeomorphism (take for G a compact group), nor a local homomorphism (if the connected component of Id in G is not commutative), nor a surjection. We will prove in 2.108, by using Riemannian geometry techniques, that if G is compact and connected, exp is surjective.

1.79 Exercise. Use 1.77 and describe, as a matrix algebra, the Lie algebras of the groups $O(n)$, $U(n)$, $SU(n)$ and $O(1, n)$.

1.D.2 Adjoint representation

Note that for a left invariant vector field X, $R_{h*}X$ is also left invariant, since the diffeomorphisms L_g and R_h of G commute (associativity in G).

1.80 Definition. The *adjoint representation* of G is the map

$$h \rightarrow \mathrm{Ad}h = (R_{h^{-1}})_*$$

defined from G to $\mathrm{Aut}(\underline{G})$.

Remarks: a) $\mathrm{Ad}h$ is an automorphism of Lie algebras (use 1.66).
b) We put h^{-1} to get, with the usual conventions, a representation: recall that $R_h \circ R_g = R_{gh}$.

1.81 Proposition. *For $h \in G$ and $X \in \underline{G}$,*

$$\mathrm{Ad}h.X = \left(\frac{d}{dt}\right) \left(h.\exp(tX).h^{-1}\right)_{|t=0}.$$

Proof. Apply 1.67 to the local group θ_t of X: the flow of $(R_{h^{-1}})_* X$ is $R_{h^{-1}} \circ \theta_t \circ R_h$ (recall that $\theta_t(g) = g.\exp(tX)$). ∎

Remark. Usually, that is when studying group theory rather than geometry, one defines $\mathrm{Ad}h.X$ as the infinitesimal generator for the one parameter group

$t \rightarrow h.\exp(tX).h^{-1}$. The previous proposition sets the equivalence of the two definitions.

Example. The adjoint representation of $Gl_n\mathbf{R}$ is the conjugation $\mathrm{Ad}h.X = hXh^{-1}$.

1.82 Proposition. *For $X, Y \in \underline{G}$,*

$$\left(\frac{d}{dt}\right)(\mathrm{Ad}(\exp(tY)).X)_{|t=0} = [X, Y].$$

Therefore, if $f : G \rightarrow H$ is a smooth group morphism, $T_e f$ is a Lie-algebra morphism, that is

$$T_e f \cdot [X, Y] = [T_e f \cdot X, T_e f \cdot Y].$$

Proof. For the first part, apply 1.68 to the local group of $-Y$, let $\phi_t(g) = g.\exp(-tY)$. For the second one, just remark that Ad and exp commute. ∎

We shall use repeatedly the following principle. If T is a bilinear or multilinear form on the Lie algebra $\underline{G}$ of G, there is a (unique) left invariant tensor (see F.) $\mathbf{T}$ on G such that $\mathbf{T}_e = T$. Morevoer, $\mathbf{T}$ is both left and right invariant if and only if T is $\mathrm{Ad}(G)$-invariant.

1.E Covering maps and fibrations

1.E.1 Covering maps and quotients by a discrete group

1.83 Definition. Let M and $\tilde{M}$ be two manifolds. A map $p : \tilde{M} \rightarrow M$ is a *covering map* if
i) p is smooth and surjective;
ii) for any $m \in M$, there exists a neighborhood U of m in M with $p^{-1}(U) = \bigcup_{i \in I} U_i$, where the U_i are disjoint open subsets of $\tilde{M}$ such that, for $i \in I$, $p : U_i \rightarrow U$ is a diffeomorphism.

1.84 Exercises. a) The map that sends $m \in M$ to $\mathrm{card}(p^{-1}(m))$ is locally constant. When this number is finite, it is called the number of leaves of the covering map.
b) Is the map $p : z \rightarrow z^3$, from $\mathbf{C}$ to $\mathbf{C}$, a covering map?
c) Show that the map

$$p : t \rightarrow (\cos(2\pi t), \sin(2\pi t))$$

is a covering map from $\mathbf{R}$ to S^1.
d) If $\tilde{M}$ and M are compact and if M is connected, show that a smooth map $f : \tilde{M} \rightarrow M$ is a covering map if and only if f has maximum rank at each point of M. Give a counter-example with $\tilde{M}$ non-compact but connected, and M compact.
Remark. A local diffeomorphism is not necessarily a covering map (1.84 d).

1.85 Definition. Let G be a discrete group (that is, equipped with the discrete topology), acting continuously on the left on a locally compact topological space E. On says that G acts *properly* on E if, for $x, y \in E$ where y does not belong to the orbit of x under G, there exist two neighborhoods V and W of x and y respectively such that, for any $g \in G$, $gV \cap W = \emptyset$ holds.

1.86 Proposition. *If a discrete group G acts continuously and properly on a locally compact topological space E, then the quotient E/G is Hausdorff.* The proof is an easy exercise in topology. See [Bo].

1.87 Definition. The group G acts *freely* if, for $x \in M$ and $g \in G$ with $g \neq Id$, then $gx \neq x$.

1.88 Theorem. *Let G be a discrete group of diffeomorphisms of a manifold M. If G acts properly and freely on M, there exists on the topological quotient space M/G a unique structure of smooth manifold such that the canonical projection $\pi : M \to M/G$ is a covering map.*

Proof. The topological quotient is Hausdorff since G acts properly.

Moreover, since G acts freely, every x in M in contained in an open set U such that all the open sets gU are pairwise disjoints when g runs through G. Therefore π yields an homeomorphism of U onto $\pi(U)$. Take the open sets U small enough, so that they are contained in chart domains for the differential stucture of M. Then the open sets $\pi(U)$ can be taken as chart domains for a differential structure on M/G with the required properties. We leave the details (and the proof of uniqueness) to the reader. See also [B-G]. For a counter-example if we do not assume that G acts freely, consider the diffeomorphism of order 2 of $\mathbf{R}$ given by $x \to -x$. ∎

1.89 Exercises. a) Take $\tilde{M} = \mathbf{R}^n$ and $G = \mathbf{Z}^n$ identified to the integer translations, and show that G is properly discontinuous without fixed points. Hence we get a quotient manifold $M = \tilde{M}/G$: show that M is diffeomorphic to the torus T^n (1.10 d).

b) Show that for $\tilde{M} = S^n$ and $G = \{Id, -Id\}$, we get as quotient space the real projective space $P^n\mathbf{R}$ (1.10 g). Deduce that $P^n\mathbf{R}$ is orientable for n odd, and non orientable for n even.

c) For $\tilde{M} = \mathbf{R}^n$ and G the group generated by the maps g_1 and g_2 from $\mathbf{R}^2$ to itself, defined by $g_1(x, y) = (x + \frac{1}{2}, -y)$ and $g_2(x, y) = (x, y + 1)$, we get as quotient manifold the *Klein bottle*. This manifold is built from the rectangle, with the identifications indicated on the figure. It is a compact and non-orientable manifold (exercise), which cannot be embedded in $\mathbf{R}^3$. Indeed Alexander theorem ([Gr]) says that a compact embedded surface in $\mathbf{R}^3$ divides the space in two connected components, one bounded, the other not: hence we can distinguish the interior and the exterior of the surface, and build on it a continuous field of normal vectors (outer normal for example), and the surface is orientable.

But the Klein bottle can be immersed in $\mathbf{R}^3$ (the picture 1.10 gives a rough idea of this property).

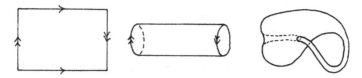

Fig. 1.10. The Klein bottle

1.E.2 Submersions and fibrations

1.90 Proposition. *For $f : M \to N$ a submersion and $y \in f(M)$, the set $f^{-1}(y)$ is an $(n - k)$ submanifold of M, where $\dim M = n$ and $\dim N = k$.*
Proof. Use 1.17 and 1.9. ■

1.91 Definition. Let E, B, and F be smooth manifolds and $\pi : E \to B$ be a smooth map. The triple (π, E, B) is a *fiber bundle* with *fiber* F, *basis* B, and *total space* E if:
i) the map π is surjective,
ii) there exists an open covering (U_i) $(i \in I)$ of B, and diffeomorphisms

$$h_i : \pi^{-1}(U_i) \to U_i \times F$$

such that $h_i(\pi^{-1}(x)) = \{x\} \times F$ for $x \in U_i$ (local triviality of the bundle).

1.92 Remarks. The map π is then a submersion since $\pi = pr_1 \circ h_i$. But a submersion does not necessarily come from a fibration (remove a point from the total space of a fiber bundle). Note that any submersion is locally a fibration (use 1.17).
Examples. a) The coverings, where the fiber is discrete.
b) The vector bundles defined in (1.32).

1.93 Exercises. a) If $\pi : E \to B$ is a fibration with basis and fiber F connected, show that the total space E is also connected.
b) Build a fibration $\pi : S^3 \to S^2$ with fiber S^1. Recall that S^2 is diffeomorphic to $P^1\mathbf{C}$, and use the canonical action of the unit circle of $\mathbf{C}$ on $\mathbf{C}^2$. This fibration is the *Hopf fibration*.
Similarly, use the action of the unit quaternions on $\mathbf{H}^2$ and prove that $P^1\mathbf{H}$ is diffeomorphic to S^4 to build a fibration $\pi : S^7 \to S^4$ with fiber S^3.

1.94 Definition. Recall that a *left (resp. right) action* of a group G on a set M is a map γ from $G \times M$ to M such that

$$\gamma(g_1\gamma(g_2, m)) = \gamma(g_1g_2, m) \qquad \text{resp.} \qquad \gamma(g_1\gamma(g_2, m)) = \gamma(g_2g_1, m).$$

If G is a Lie group G and M a manifold the action is said to be *smooth* if γ is smooth.
We shall write $\gamma(g, m) = g \cdot m$ (resp. $\gamma(g, m) = m \cdot g$ for a left (resp. right) action.

Remark. If not specified explicitly, all the actions we will consider in the following will be smooth left actions. For g fixed in G, the map $m \to g.m$ is then a diffeomorphism of M.

1.95 Theorem. *Let M be a manifold, and G be a Lie group acting smoothly on M. There exists on the topological quotient M/G a structure of manifold such that the canonical projection $\pi : M \to M/G$ is a submersion, if and only if G acts freely and properly on M.*
Then $\pi : M \to M/G$ is a smooth fibration with fiber G (cf. [Bo]).

Remark. Recall that G acts freely if it acts without fixed points, and that G acts properly if, for any compact $K \subset M$, the set

$$G_K = \{g \in G/gK \cap K \neq \emptyset\}$$

is relatively compact in G (compare with 1.85). In particular, a compact group always acts properly.
A map f from M/G into a manifold N is smooth if and only if $f \circ \pi$ is smooth. When working with M/G, nothing more that this simple criterion is necessary.

1.96 Exercise. Let the group of complex numbers of modulus 1 (identified with S^1) act on the unit sphere of $\mathbf{C}^{n+1}$, and prove that we get a manifold S^{2n+1}/S^1 which is diffeomorphic to $P^n\mathbf{C}$ (1.10 h)).

1.E.3 Homogeneous spaces

We deduce in a particular case the following
1.97 Theorem. *Let G be Lie group, and H a closed subgroup of G (H is then a Lie subgroup of G). Then there exists on the quotient space G/H a unique structure of manifold such that*
i) $\pi : G \to G/H$ *is a smooth fibration;*
ii) *the action of G on G/H given by $\gamma(g,p(x)) = p(g.x)$ is smooth.*
One says that G/H is an *homogeneous space*.

Remark. The property that a closed subgroup of a Lie group is a Lie subgroup, that is both a subgroup and a submanifold, is a deep result, known as Cartan-Von Neumann theorem (cf. [G], 4.22 for linear groups, and 6.12 for the general case). However, in all the examples we shall meet, it will be very easy to check directly.
We can also define homogeneous spaces in the following way.

1.98 Definition. Let G be a Lie group and M be a smooth manifold. M is a *G-homogeneous space* if G acts smoothly and transitively on M.
Example. A Lie group G is an homogeneous space, and this in several ways:

$$G = G \times G/G = G/\{e\}.$$

See 2.47 for more details.

1.99 Definition. Let $m \in M$. The group $H = \{g \in G/g(m) = m\}$ is a closed subgroup of G, called the *isotropy group* of m.

The previous definitions 1.97 and 1.98 are equivalent. More precisely:

1.100 Theorem. *The following properties are equivalent:*
i) *The Lie group G acts transitively and smoothly on M, with H as isotropy group of $m \in M$;*
ii) *M is diffeomorphic to the quotient manifold G/H.*
Actually, if $m \in M$, and if we denote by F the map from G to M defined by $F(g) = g.m$, then F yields by quotient by p a diffeomorphism $\tilde{F}$ between G/H and M, defined by $\tilde{F}(p(g)) = g.m$.
For a proof, see ([Bo], p.166).

1.101 Examples. a) The group $SO(n+1)$ acts transitively on S^n, the isotropy group of a point being isomorphic to $SO(n)$: if $x, y \in S^n$, and if $(x, a_1, ..., a_n)$ and $(y, b_1, ..., b_n)$ are two orthonormal basis of $\mathbf{R}^{n+1}$ inducing the same orientation, the transition matrix lies in $SO(n+1)$. The matrices that fix the first vector of basis are of the form

$$\begin{pmatrix} 1 & 0 \\ 0 & A \end{pmatrix}$$

with $A \in SO(n)$ (in an orthonormal basis!).
Since the action of $SO(n+1)$ on S^n is free and proper, we get

$$S^n = SO(n+1)/SO(n).$$

b) The group $Gl_n \mathbf{R}$ acts transitively on $\mathbf{R}^n \setminus \{0\}$, the isotropy group of a point being the set of matrices of the form

$$\begin{pmatrix} 1 & b \\ 0 & A \end{pmatrix}$$

with $A \in Gl_{n-1}\mathbf{R}$ and $b \in \mathbf{R}^n$. Since the action of $Gl_n\mathbf{R}$ is free and proper, we get $\mathbf{R}^n \setminus \{0\}$ as the quotient of $Gl_n\mathbf{R}$ by this group.
c) The unitary group $U(n+1)$ acts transitively on $P^n\mathbf{C}$: $U(n+1)$ acts transitively on S^{2n+1} (proof as in a)), and this action yields by quotient a transitive action on $P^n\mathbf{C}$. The isotropy group of a point is isomorphic to $U(n) \times U(1)$. We prove again that $U(n+1)$ acts freely and properly on $P^n\mathbf{C}$, and that

$$P^n\mathbf{C} = U(n+1)/U(n) \times U(1).$$

d) The homogeneous space $O(n)/O(n-p)$ is diffeomorphic to the submanifold of $\mathbf{R}^{np}$ which points are the sequences $(x_k)_{(k=1,...,p)}$ of orthonormal vectors in $\mathbf{R}^n$. It is the *Stiefel manifold $S_{n,p}$.*
e) The orthogonal group $O(n)$ acts smoothly on the right on $S_{n,p}$. The action is free and proper, and the quotient space

$$S_{n,p}/O(p) = O(n)/O(p) \times O(n-p)$$

is the manifold of the p-planes in $\mathbf{R}^n$: it is the *Grassman manifold $G_{n,p}$.*

f) Let q be the symmetric bilinear form defined on $\mathbf{R}^{n+1}$ by

$$q(x,y) = \left(-x_0 y_0 + \sum_{i=1}^{n} x_i y_i\right),$$

and set

$$O(1,n) = \{A \in M_n(\mathbf{R})/\forall x, y \in \mathbf{R}^{n+1}, q(Ax, Ay) = q(x,y)\}$$
$$H^n = \{x \in \mathbf{R}^{n+1}/q(x,x) = -1 \quad \text{and} \quad x_0 > 0\}$$
$$O_0(1,n) = \{A \in O(1,n)/A(H^n) = H^n\}.$$

We want to show that $O_0(1,n)$ acts transitively on H^n. First we show that the orthogonal symmetries (for q) with respect to an hyperplane of $\mathbf{R}^{n+1}$ belong to $O(1,n)$: if D is a non isotropic straight line, and if P is the orthogonal hyperplane to D for q, any vector $u \in \mathbf{R}^{n+1}$ can be decomposed as the sum of two vectors $u_1 \in D$ and $u_2 \in P$. The map

$$s : u_1 + u_2 \rightarrow u_1 - u_2$$

belongs to $O(1,n)$ since

$$q\big(s(u_1 + u_2), s(v_1 + v_2)\big) = q(u_1, v_1) + q(u_2, v_2)$$

$$= q(u_1 + u_2, v_1 + v_2).$$

Let now $x, y \in H^n$. To show that there exists an hyperplane containing 0 and $\frac{x+y}{2}$, and orthogonal to $y - x$, it is sufficient to prove that $q(x + y, x - y) = 0$, which is the case since $q(x,x) = q(y,y) = 1$.

The symmetry s with respect to P sends x to y. We still have to prove that $s \in O_0(1,n)$. Let

$$\overline{H}^n = \{x \in \mathbf{R}^{n+1}/q(x,x) = -1\}.$$

$\overline{H}^n$ has two connected components ($x_0 > 0$ or $x_0 < 0$) and is preserved under s. Since s was built to send x to y for a pair of points of H^n, s preserves also H^n and hence $s \in O_0(1,n)$. One can prove similarly that $SO_0(1,n)$ (matrices of $O_0(1,n)$ with positive determinant) acts transitively on H^n.

Now the isotropy group of e_0 is the set of matrices

$$\begin{pmatrix} 1 & 0 \\ 0 & A \end{pmatrix}$$

with $A \in O(n)$. The action of $O_0(1,n)$ on H^n is free and proper and hence $H^n = O_0(1,n)/O(n)$ (similarly we could prove that $H^n = SO_0(1,n)/SO(n)$). This result shows that $O(1,n)$ has four connected components ($\det = \pm 1$, preserving H^n or not): each of these components being isomorphic to $SO_0(1,n)$, we only have to show that $SO_0(1, n)$ is connected. Just apply the exercise 1.93 a) to the fibration

$$\pi : SO_0(1, n) \to SO_0(1, n)/SO(n) = H^n :$$

since the basis H^n and the fiber $SO(n)$ are connected, the total space $SO_0(1, n)$ is also connected.

1.F Tensors

It is essential, when studying differential or Riemannian geometry, to consider not only vector fields, but also sections of bundles canonically built from the tangent bundle. It is one of the tasks of the differential geometers to define them. The following principle (see [H] for an explanation) is implicitly used: every "canonical" construction concerning vector spaces can be made also for vector bundles, in particular for the tangent bundle to a manifold.

1.F.1 Tensor product (a digest)

If E and F are two finite dimensional vector spaces, a vector space $E \otimes F$, unique up to isomorphism and such that for any vector space G, $L(E \otimes F, G)$ is isomorphic to $L_2(E \times F, G)$ (we denote by $L_2(E \times F, G)$ the vector space of bilinear maps from $E \times F$ to G): $E \otimes F$ is the *tensor product* of E and F. Moreover, there exists a bilinear map from $E \times F$ to $E \otimes F$, denoted by $\otimes$, and such that if (e_i) and (f_j) are basis for E and F respectively, $(e_i \otimes f_j)$ is a basis for $E \otimes F$. It is easier to understand what happens if the vector spaces E and F are themselves vector spaces of linear forms. Indeed, if E^* and F^* denote the dual spaces of E and F, it is easy to check that

$$E^* \otimes F^* \simeq L_2(E \times F, \mathbf{R}),$$

and that $\alpha \otimes \beta$ is the bilinear form $(x, y) \to \alpha(x)\beta(y)$.

Finally, if E' and F' are two other vector spaces and if $u \in L(E, E')$, $v \in L(F, F')$, we define

$$u \otimes v \in L(E \otimes E', F \otimes F')$$

by deciding that

$$(u \otimes v)(x \otimes y) = u(x) \otimes v(y).$$

We define also the tensor product of a finite number of vector spaces: it is associative.

1.F.2 Tensor bundles

1.102 Be given a vector bundle E on M, we can define the vector bundles E^*, $\otimes^p E$, $\Lambda^p E$,... whose respective fibers at m are E_m^*, $\otimes^p E_m$, $\Lambda^p E_m$ (see [H] for details). Rather than developping a general theory we do not need here, let us see how it works in the case $E = TM$.

First we define $\otimes^p T_m M$ and $\otimes^p TM$ by imitating the construction of $T_m M$ and TM (1.30). If (U, ϕ) and (V, ψ) are two charts for M such that $m \in U \cap V$ and $\phi(m) = \psi(m) = 0$, and if $u, v \in (\mathbf{R}^n)^{\otimes p}$, one says that the triples (U, ϕ, u) and (V, ψ, v) are equivalent if

$$\left[\psi \circ \phi^{-1}{}'(0) \right]^{\otimes p} \cdot u = v.$$

By definition, $\otimes^p T_m M$ is the set of equivalence classes for the previous relation and, as in 1.27, the map $u \to \mathrm{class}(U, \phi, u)$ denoted by $\theta_{U,\phi,m}^{\otimes p}$, is bijective. It carries on $\otimes^p T_m M$ a linear space structure, which does not depend on the chart.

Then we set $\otimes^p TM = \bigcup_{m \in M} \otimes^p T_m M$. Imitate the construction of 1.30 to check the structure of fiber bundle over M: just replace $T_m M$ by $\otimes^p T_m M$, and $\theta_{U,\phi,m}$ by $\theta_{U,\phi,m}^{\otimes p}$.

Replacing tensor powers by transpositions, we define in the same way T^*M: the triples (U, ϕ, ξ) and (V, ψ, η) will be equivalent if

$$\xi = \left[{}^t(\psi \circ \phi^{-1})'(0) \right] \cdot \eta.$$

To define $(\otimes^p TM) \otimes (\otimes^q T^*M)$, we begin of course with the equivalence relation defined on triples (U, ϕ, r) and (V, ψ, s) with

$$r, s \in (\otimes^p \mathbf{R}^n) \otimes (\otimes^q \mathbf{R}^{n*}),$$

given by

$$s = \left((\psi \circ \phi^{-1})'(0) \right)^{\otimes p} \otimes \left({}^t(\phi \circ \psi^{-1})'(0) \right)^{\otimes q} \cdot r.$$

1.103 **Definition.** The space $T_m^* M$ is the *cotangent space* to M at m, and T^*M is the *cotangent bundle*.

The bundle $(\otimes^p TM) \otimes (\otimes^q T^*M)$ will be denoted by $T_q^p M$.

1.104 **Definition.** A *(p,q)-tensor* on M is a smooth section of the bundle $T_q^p M$.

Remark. In this context, the term of tensor is a commonly used abbreviation: we should say "tensor field".

A (p, q)-tensor is also called a p (times)-contravariant and q (times)-covariant tensor. The set of tensors of type (p, q) is denoted by $\Gamma(T_q^p M)$. It is a real vector space and a $C^\infty(M)$-module.

1.F.3 Operations on tensors

There are two natural algebraic operations on (ordinary) tensors on a vector space:

i) the tensor product, that is for each (p, q, p', q'), the natural bilinear map from

$$E^{\otimes p} \otimes E^{*\otimes q} \otimes E^{\otimes p'} \otimes E^{*\otimes q'} \longrightarrow E^{\otimes(p+p')} \otimes E^{*\otimes(q+q')} \ ;$$

ii) the contractions $(c_{i,j})_{((i=1,\cdots,p),\,(j=1,\cdots,q))}$, which are the linear maps

$$c_{i,j} : E^{\otimes p} \otimes E^{*\otimes q} \longrightarrow E^{\otimes(p-1)} \otimes E^{*\otimes(q-1)}$$

defined on the decomposed tensors by the formula

$$c_{i,j}(x_1 \otimes \cdots \otimes x_p \otimes y_1^* \otimes \cdots \otimes y_q^*)$$

$$= y_j^*(x_i)\, x_1 \otimes \cdots \otimes \hat{x}_i \otimes \cdots \otimes x_p \otimes y_1^* \otimes \cdots \otimes \hat{y}_j \otimes \cdots \otimes y_q^*.$$

Example. On $E \otimes E^*$, which is canonically isomorphic to $\mathrm{End}E$, the unique contraction, which takes its values in $\mathbf{R}$, is the trace (exercise).

Considering tensors on a manifold, we naturally define tensor products and contractions by working fiber by fiber. For example, the contraction of a tensor $T \in \Gamma(TM \otimes T^*M)$ will be the function $\mathrm{tr}T$... The following example is most important:

1.105 Definition. The *interior product* of a vector field X with a covariant tensor of type $(0, p)$ is the tensor $c_{1,1}(X \otimes S)$ (of type $(0, p-1)$), defined as:

$$(i_X S)_m(x_1, \cdots, x_{p-1}) = S_m(X(m), x_1, \cdots, x_{p-1}).$$

To see how this notion is used, cf.1.121.

Another important operation on tensors is the *pull-back*.

1.106 Proposition, Definition. *Let $\phi : M \to N$ be a smooth map, and S be a (0,p) tensor on N. The formula*

$$S_m^1(x_1, \cdots, x_p) = S_{\phi(m)}(T_m\phi.x_1, \cdots, T_m\phi.x_p),$$

(where $m \in M$ and $x_i \in T_m M$) defines a (0,p)-tensor on M, called the pull-back *of S by ϕ, and denoted by $\phi^* S$.*

Proof. It is clear that $m \to S_m^1$ is a section of $T_p^0 M$, which is smooth as ϕ (compare with 1.63: it is not necessary to assume that ϕ is a diffeomorphism as it was the case for vector fields). ∎

1.107 Proposition. a) *Let $\phi : M \to N$ be smooth. Then for α and β in $\Gamma T_p^0 N$ and $\Gamma T_q^0 N$ respectively, $\phi^*(\alpha \otimes \beta) = \phi^*\alpha \otimes \phi^*\beta$.*
b) *If $\psi : N \to P$ is another smooth map, then $(\psi \circ \phi)^* = \phi^* \circ \psi^*$.*

Proof. The first part comes from a purely algebraic computation, the second is a direct consequence of the chain rule. ∎

If we assume that ϕ is a *diffeomorphism*, we can define ϕ^* on tensors of any type. First if $X \in \Gamma(TN)$ is a vector field, we set $\phi^*X = \phi_*^{-1}X$. Then we extend ϕ^* to $\Gamma(T_q^pM)$ by deciding that the property a) of the previous proposition is satisfied. The map ϕ^* so defined commutes with the contractions (algebraic computation).

1.108 Example. Let (U_i, ϕ_i) be an atlas for M. Consider for a (p,q)-tensor S on M, the tensors $S_i = (\phi^{-1})^*S$ defined on the open subsets $\phi_i(U_i)$ of $\mathbf{R}^n$. For any pair (i, j) with $U_i \cap U_j \neq \emptyset$, the following holds:

$$\left(\phi_j \circ \phi_i^{-1}\right)^* \left(S_{j \,|\, \phi_j(U_i \cap U_j)}\right)$$

$$= S_{i \,|\, \phi_i(U_i \cap U_j)}.$$

Conversely, be given a family of tensors S_i on $\phi_i(U_i)$ that satisfy the previous condition, there exists a unique tensor S on M such that

$$S_i = \left(\phi_i^{-1}\right)^* S.$$

Hence, as soon as tensors and pull-back are defined on open sets in $\mathbf{R}^n$, we have another way to define tensors on manifolds: this remark will be used in 1.119.

1.F.4 Lie derivatives

1.109 Definition. The *Lie derivative* associated to a vector field X on M is the linear map

$$L_X : \Gamma(T_q^pM) \to \Gamma(T_q^pM)$$

which sends S to the tensor

$$\left(\frac{d}{dt}\right)(\phi_t^*S)_{|t=0},$$

where ϕ_t is the local one parameter group associated to X.

Check that this formula actually defines a tensor, and that the expression of L_XS at m in local coordinates only depends on the values at m of S, X and of their derivatives of order one.

1.110 Proposition. *The operator L_X is actually defined by the following properties:*
i) *for $f \in C^\infty(M)$, $L_Xf = df(x) = X.f$,*
ii) *for $Y \in \Gamma(TM)$, $L_XY = [X, Y]$,*
iii) *for any tensors S, T: $L_X(S \otimes T) = L_XS \otimes T + S \otimes L_XT$ *(i.e. the map L_X is a derivation of the tensor algebra on M)*,*
iv) *for any (p,q)-tensor S (with $p, q > 0$), and for any contraction c,*

$$L_X(c(S)) = c(L_XS).$$

Proof. These properties are satisfied by L_X: i) is obvious, and ii) is a consequence of 1.68. One proves iii) and iv) by differentiating the identities

$$\phi_t^*(S \otimes T) = \phi_t^* S \otimes \phi_t^* T, \quad \text{and} \quad \phi_t^*(c(S)) = c(\phi_t^* S). \blacksquare$$

Conversely, Let P_X be a linear operator on tensors, type preserving, equal to L_X on $C^\infty(M)$ and $\Gamma(TM)$, and satisfying iii) and iv). Then

1.111 **Lemma.** *The operator P_X is a local operator, that is for any open subset U of M, and for two tensors S and S' which coincide on U, then $P_X S$ and $P_X S'$ also coincide on U.*

Proof of the lemma. For V a relatively compact open subset of U and f a smooth function with support in U and equal to 1 on V, we have $fS = fS'$. Apply P_X to both members to find that (with iii):

$$(X.f)S + fP_X S = (X.f)S' + fP_X S',$$

and hence $P_X S$ and $P_X S'$ coincide on V.

Now we can use local coordinates and check that $L_X = P_X$. $\blacksquare$

1.112 **Corollary.** *For two vector fields X and Y, we have*

$$L_X \circ L_Y - L_Y \circ L_X = L_{[X,Y]}.$$

Proof. Both members do coincide on $C^\infty(M)$ (definition of the bracket) and on $\Gamma(TM)$ (Jacobi identity). One checks directly that the operator defined on the left satisfies the properties iii) and iv) of 1.110. $\blacksquare$

1.F.5 Local operators, differential operators

1.113 **Definition.** A *local operator* on a manifold M is a linear map P from $\Gamma(E)$ to $\Gamma(F)$, where E and F are two vector bundles on M, which satisfies the property introduced in lemma 1.111: for an open subset U of M, and a section s of E with $s_{|U} = 0$, one has $Ps_{|U} = 0$.

If M is an open subset U of $\mathbf{R}^n$, and if $E = F = U \times \mathbf{R}^n$, in which case

$$\Gamma(E) = \Gamma(F) = C^\infty(U),$$

then any differential operator with smooth coefficients is a local operator. One can prove conversely, by using elementary theory of distributions, that any local operator from $C^\infty(M)$ to itself is a differential operator (Cf. [N]). From this result, it is easy to see what happens in the general case of two vector bundles E and F on a manifold M: if $U \subset M$ is an open set over which the bundles E and F are trivial, and if U is also the domain of a chart (U, ϕ) for M, the local expression of P with respect to this chart and to trivializations of E and F over U is given by a matrix of differential operators on $C^\infty(U)$. We leave the details to the reader. In any case, he or she will check them easily for the differential operators (that is local operators) which will be met in the sequel:

i) the exterior derivative (1.119),

ii) the covariant derivative associated with a Riemannian metric (2.58)

iii) the Laplace operator on functions, and on differential forms (4.7, 4.29).

1.F.6 A characterization for tensors

The following test will be used frequently in the sequel (see for example 3.2 and 5.1).

1.114 Proposition. *Let P be a $\mathbf{R}$-linear map between two spaces of tensors $\Gamma(T_q^p M)$ and $\Gamma(T_s^r M)$. The following are equivalent:*

a) *P is $C^\infty(M)$-linear,*

b) *for s,s' in $\Gamma(T_q^p M)$, and m in M with $s_m = s'_m$, then $(Ps)_m = (Ps')_m$.*

Proof. It is clear that b) implies a).

Now if a) is satisfied, P is a local operator (use a test function as we did in lemma 1.111). Assume then that $s_m = s'_m$. Use a local trivialization of $T_q^p M$ and lemma 1.111 to built an open subset U of M containing m, smooth functions $(f_i)_{(i=1,\cdots,N)}$ (where N is the dimension of the fiber), and sections $(\sigma_i)_{(i=1,\cdots,n)}$ of $T_q^p U$ such that $(s - s')_{|U} = \sum f_i \sigma_i$. Hence

$$P(s - s')_{|U} = P\left(\sum f_i \sigma_i\right) = \sum f_i P(\sigma_i),$$

therefore $P(s - s')_m = \sum f_i(m) P(\sigma_i)_m = 0$.

Simpleminded example. Let $P : \Gamma(T_0^1 M) \to C^\infty(M)$ be a $C^\infty(M)$-linear map. For any vector field X and any $m \in M$, the previous proposition ensures that the real $P(X)_m$ depends only on X_m in $T_m M$. It is clear that the map $X_m \to P(X)_m$ is a linear form on $T_m M$. Hence we got a section (with a priori no regularity) of the bundle $T^* M$. One can use local trivializations of $T_0^1 M$ and prove that this section is smooth. Conversely, it is clear that if ξ is a smooth section of $T^* M$, the map $X \to \xi(X)$ from $\Gamma(T_0^1 M)$ to $C^\infty M$ is $C^\infty(M)$-linear.

This is a general phenomenon. One can use the isomorphisms $E^* \otimes F = \mathrm{Hom}(E, F)$ and $(E \otimes F)^* = E^* \otimes F^*$ (for finite dimensional vector spaces), and the proposition 1.114, to prove that it is equivalent to give a $C^\infty(M)$-linear map from $\Gamma(T_q^p M)$ to $\Gamma(T_s^r M)$, or a section of the bundle $\Gamma((T_q^p M)^* \otimes (T_s^r M))$, that is a (q+r,p+s)-tensor. Do not be satisfied with these abstract considerations, but apply them to the curvature tensor (3.3).

Counter-example. The bracket, seen as a bilinear map from $\Gamma(TM) \times \Gamma(TM)$ to $\Gamma(TM)$, is not a tensor. For X, Y in $\Gamma(TM)$ and f, g in $C^\infty(M)$, we have indeed

$$[fX, gY] = f(X.g)Y - g(Y.f)X + fg[X, Y].$$

Another example. A q-covariant tensor can be seen as a q-linear form on the $C^\infty(M)$ module $\Gamma(TM)^{\otimes q}$. This point of view will be very useful in the following: we shall principally meet the bundle $S^2 M$ of bilinear symmetric forms, (which is clearly a subbundle of $T_2^0 M$), the bundles $\Lambda^k M$ of antisymmetric k-forms ($k = 1, \cdots, n = \dim M$), and the corresponding spaces of sections, that is the bilinear symmetric forms, and the exterior forms on M.

1.115 Exercise. Let $S \in \Gamma(T_q^0 M)$, and $X, X_1, ... X_q$ be (q+1) vector fields. Show that

$$(L_X S)(X_1, \cdots, X_q) = X.S(X_1, \cdots, X_q)$$

$$-\sum_{i=1}^{q} S(X_1, \cdots, X_{i-1}, [X, X_i], \cdots, X_{i+1}, X_q).$$

1.G Differential forms

The most important tensors are differential forms. The main reason for their importance is the fact that, under mild compactness assumptions, it is possible to define the integration of a form of degree k on a (sub)manifold of dimension k.

1.G.1 Definitions

1.116 Definition. A *differential form of degree k* on a manifold M is a smooth section of the bundle $\Lambda^k M$. We will set $\Gamma(\Lambda^k M) = \Omega^k M$.

1.117 Algebraic recall:
For a vector space E, there exists on $\otimes^k E$ an antisymmetrization operator, defined on the decomposed elements by:

$$\text{Ant}(x_1 \otimes \cdots \otimes x_k) = \sum_{s \in S_k} \text{sign}(s) x_{s(1)} \otimes \cdots \otimes x_{s(k)}.$$

This formula is more suggestive for k-forms: for $f \in \otimes^k E^*$ we have

$$(\text{Ant} f)(x_1, \cdots, x_k) = \sum_{s \in M} \text{sign}(s) f(x_{s(1)}, \cdots, x_{s(k)}).$$

We denote by $\bigwedge^k E^*$ the vector space of antisymmetric k-forms on E. The *exterior product* of $f \in \Lambda^k E^*$ and $g \in \Lambda^l E^*$ is the (k+l) antisymmetric form defined by

$$f \wedge g = \frac{1}{k!l!} \text{Ant}(f \otimes g).$$

Example. For f, g in E^*,

$$(f \wedge g)(x, y) = f(x)g(y) - f(y)g(x).$$

We will admit that the exterior product is associative and anticommutative, that is

$$f \wedge g = (-1)^{kl} g \wedge f, \quad \text{for } f \in \Lambda^k E^* \quad \text{and } g \in \Lambda^l E^*$$

(see [Sp] t.1 for a proof).

Recall finally that $\Lambda^{(\dim E)} E^* \simeq \mathbf{R}$ and that $\Lambda^k E^* = 0$ for $k > \dim E$.

It can also be useful to consider forms with complex values, and the corresponding bundles. The amateur of abstract nonsense can check easily that that $L_{\mathbf{R}}(E, \mathbf{C}) = E^* \otimes_{\mathbf{R}} \mathbf{C}$ and that, if $E_{\mathbf{C}} = E \otimes_{\mathbf{C}} \mathbf{R}$ is the complexified of the vector space E, then $\Lambda^k E \otimes_{\mathbf{R}} \mathbf{C}$ is $\mathbf{C}$-isomorphic to $(\Lambda^k E_{\mathbf{C}})$.

It is clear that all the previous definitions extend directly to the subbundle $\Lambda^k M$ of antisymmetric tensors of $T_k^0 M$.

As a consequence of the properties of the action of ϕ^* and L_X on tensors (1.107 and 1.110), we have

$$\phi^*(\alpha \wedge \beta) = \phi^*\alpha \wedge \phi^*\beta, \quad \text{and } L_X(\alpha \wedge \beta) = L_X\alpha \wedge \beta + \alpha \wedge L_X\beta.$$

1.118 Exercises. a) Let $\omega \in \Omega^1(S^2)$ be a differential 1-form on S^2 such that for any $\phi \in SO(3)$, $\phi^*\omega = \omega$ holds. Show that $\omega = 0$. State and prove an analogous result for differential forms on S^n.

b) Let p be the canonical projection from $\mathbf{C}^{n+1} \setminus \{0\}$ to $P^n\mathbf{C}$. Show that there exists a 2-form ω on $P^n\mathbf{C}$ such that

$$p^*\omega = \left(\sum_{k=0}^{n} dz^k \wedge d\bar{z}^k \right) \bigg/ \left(\sum_{k=0}^{n} |z_k|^2 \right),$$

where $\quad dz^k = dx^k + idy^k \quad$ and $\quad d\bar{z}^k = dx^k - idy^k.$

Show that the form ω is invariant under the action of $U(n+1)$ on $P^n\mathbf{C}$, and that for $k = 2, \cdots, n$ the $2k$-form ω^k is non zero and $U(n+1)$ invariant.

We will see that the forms ω^k generate the cohomology of $P^n\mathbf{C}$ (see 4.35).

Exterior forms are more interesting than tensors, for the following reason: we shall define on $\sum_{k=0}^{\dim M} \Omega^k M$ a "natural" differential operator (see 4.1) – that is depending only on the differential structure of M. This operator gives information on the topology of the manifold (cf. 1.125).

1.G.2 Exterior derivative

1.119 Theorem and definition. *Let M be a smooth manifold. For any $p \in \mathbf{N}$, there exists a unique local operator d from $\Omega^p M$ to $\Omega^{p+1} M$, called the exterior derivative and such that*

i) for $p=0$, $d : C^\infty(M) \to \Omega^1 M$ is the differential on functions,

ii) for $f \in C^\infty(M)$, we have $d(df) = 0$,

iii) for $\alpha \in \Omega^p M$ and $\beta \in \Omega^q M$, we have

$$d(\alpha \wedge \beta) = d\alpha \wedge \beta + (-1)^p \alpha \wedge d\beta.$$

Short proof. (for more details, see [Sp]). Let us first treat the case where M is an open subset U of $\mathbf{R}^n$. Then $\alpha \in \Omega^p U$ can be decomposed in a unique way as:

$$\alpha = \sum \alpha_{i_1 \cdots i_p} dx^{i_1} \wedge \cdots \wedge dx^{i_p},$$

where the sum is understood on all the strictly increasing sequences $i_1 < \cdots < i_p$ of $[1, n]$, and the $\alpha_{i_1, \cdots i_p}$ being smooth. We must have

$$d\alpha = \sum d\alpha_{i_1 \cdots i_p} \wedge dx^{i_1} \wedge \cdots \wedge dx^{i_p}.$$

One checks directly that the operator just defined (which satisfies i) and ii) by construction) also satisfies iii). To prove that d is a local operator, just imitate the argument of 1.111, using iii) for the product of a test function and a p-form.

To extend this result to manifolds, we will need the following:

1.120 Lemma. Let $\phi : U \to V$ be a diffeomorphism between two open subsets of $\mathbf{R}^n$. Then $\phi^* \circ d = d \circ \phi^*$, that is the diagram below commutes.

$$
\begin{array}{ccc}
\Omega^p(V) & \xrightarrow{\phi^*} & \Omega^p(U) \\
{\scriptstyle d}\downarrow & & \downarrow{\scriptstyle d} \\
\Omega^{p+}(V) & \xrightarrow{\phi^*} & \Omega^{p+1}(U)
\end{array}
$$

Proof. It is the chain rule in the case $p = 0$. Use 1.119 b) and the behaviour of ϕ^* with respect to the exterior product (1.117), and proceed by induction on p. ∎

End of the proof of the theorem. If M is equipped with an atlas (U_i, ϕ_i), we can define $d\omega_i$ for the local expressions of a p-form ω on M (begining of the proof) and, from the lemma:

$$\left(\phi_j \circ \phi_i^{-1}\right)^* \left(d\omega_{j|\phi_j(U_i \cap U_j)}\right) = d\omega_{i|\phi_i(U_i \cap U_j)}.$$

This proves (see 1.108) that the $d\omega_i$ are local expressions for a (p+1)-form on M, which is the form $d\omega$ we are looking for: use 1.120 and check that $d\omega$ doesn't depend on the atlas. ∎

1.121 Proposition. For any p-form ω on M, the following holds:
i) $d(d\omega) = 0$, that is $d \circ d = 0$;
ii) for a smooth map $\phi : M \to N$, $d(\phi^*\omega) = \phi^*(d\omega)$, that is $\phi^* \circ d = d \circ \phi^*$;
iii) for a vector field X on M, $L_X d\omega = d(L_X \omega)$, that is $L_X \circ d = d \circ L_X$, and $L_X \omega = d(i_X \omega) + i_X(d\omega)$, that is $L_X = d \circ i_X + i_X \circ d$.
Proof. i), ii) and the first part of iii) are clear. Let

$$P_X = d \circ i_X + i_X \circ d.$$

One checks directly that

$$P_X(\alpha \wedge \beta) = P_X \alpha \wedge \beta + \alpha \wedge P_X \beta,$$

i.e. that P_X is, as L_X, a derivation of the $C^\infty(M)$-algebra $\Omega(M)$. Look carefully at the existence theorem for Lie derivative on tensors (1.110), and note that a derivation is determined by its values on functions and 1-forms: hence we only need to check that $L_X = P_X$ on functions and 1-forms (even only on 1-forms which can be written as dx^i in local coordinates), which is immediate. ∎

1.122 Corollary. *For $\alpha \in \Omega^p(M)$ and $(X_0, \cdots, X_p)$ $p+1$ vector fields on M, we have*

$$d\alpha(X_0, \cdots, X_p) = \sum_{i=0}^{p} (-1)^i L_{X_i}\big(\alpha(X_0, \cdots, \hat{X}_i, \cdots, X_p)\big)$$

$$+ \sum_{0 \le i < j \le p} (-1)^{i+j} \alpha([X_i, X_j], X_0, \cdots, \hat{X}_i, \cdots, \hat{X}_j, \cdots, X_p).$$

Proof. It is the definition of d for $p = 0$. Proceed by induction on p (use 1.115). ∎

Example. For $\alpha \in \Omega^1(M)$, we have

$$d\alpha(X, Y) = X.\alpha(Y) - Y.\alpha(X) - \alpha([X, Y]).$$

Compare these formulas with those of 2.61, where d is computed in terms of the covariant derivative associated with a Riemannian metric.

1.123 Exercise. Check that the right member of the equation in 1.122 is $C^\infty(M)$-linear with respect to the X_i.

Remark. We can take the expression of d obtained at 1.122 as a definition for the exterior derivative. This point of view is technically less simple, (convince yourself by trying and prove that $d \circ d = 0$ using Jacobi identities!), but is coordinate-free: that is why it is useful in infinite dimension (see for example [Bt]).

Another application of the second formula on 1.121 iii) is the proof of the the following

1.124 Poincaré lemma. *Let U be an open star-shaped subset of $\mathbf{R}^n$, and α be a p-form on U such that $d\alpha = 0$. Then there exists a $(p-1)$-form $\beta \in \Omega^{(p-1)}(U)$ such that $d\beta = \alpha$.*

Proof. See [Wa] or [Sp] ∎

The analogous property is false on general manifolds. This leads to the definitions of important (smooth, *in fact topological*) of the manifold.

1.125 Definitions. Let M be a smooth manifold. The pth *de Rham group* of M, denoted by $H^p_{DR}(M)$, is the quotient

$$\{\alpha \in \Omega^p(M), d\alpha = 0\}/d\Omega^{p-1}(M).$$

The *p-the Betti number* $b_p(M)$ is the dimension of $H^p_{DR}(M)$.

* **Remark.** In Algebraic Toplogy, Betti numbers are defined for any coefficient field K: $b_k(M, K) = \dim H^k(M, K)$. Both definitions coincide for real numbers, in view of de Rham's theorem (cf. [Wa] or [B-T]).*
It is clear that $H^p_{DR}(M) = 0$ when $p > \dim M$, and that $H^0_{DR}(M) = \mathbf{R}^k$ if M has k connected components. It can be proved (see [W]) that the vector spaces $H_{DR}(M)$ are indeed topological invariants, and that for M compact, connected, orientable and n-dimensional, $H^n_{DR}(M) = \mathbf{R}$. Finally, $H^k_{DR}(S^n) = 0$ for $0 < k < n$ (ibidem). We will come back to this subject in chapter IV, and we will see how to compute de Rham cohomology by using analytic methods.

1.G.3 Volume forms

1.126 Definition. A *volume form* on an n-dimensional manifold is a never zero exterior form of degree n.
The volume forms are interesting in view of the following theorem.

1.127 Theorem. *Let M be a countable at infinity and connected manifold. The following are equivalent:*
i) there exists a volume form on M;
ii) the bundle $\Lambda^n T^* M$ is trivial;
iii) M is orientable.
Proof. i) and ii) are clearly equivalent: since $\Lambda^n T^* M$ is a bundle of rank one, there exists a non zero section of this bundle (that is a volume form on M) if and only if it is trivial (1.35 a).
Let us show now that i) implies iii). Let ω be a volume form on M and (U_i, ϕ_i) be an atlas for M. The local expression of ω in a chart is:

$$\phi_i^{-1*}(\omega_{|U_i}) = a_i dx^1 \wedge dx^2 \wedge \cdots \wedge dx^n,$$

where the function $a_i \in C^\infty(\phi_i(U_i))$ is never zero. One can assume that the a_i are positive (compose ϕ_i with an orientation reversing symmetry if necessary). From the very definition of exterior forms, we know that on $\phi_i(U_i \cap U_j)$:

$$\left(\phi_i \circ \phi_j^{-1}\right)^* (a_i dx^1 \wedge dx^2 \wedge \cdots \wedge dx^n) = a_j dx^1 \wedge dx^2 \wedge \cdots \wedge dx^n,$$

and hence the jacobian $J\left(\phi_i \circ \phi_j^{-1}\right) = a_j/a_i$ is always positive.
That iii) implies i) is more technical, and uses partitions of unity (see 1.H). Assume that M is orientable, and let (U_i, ϕ_i) be an orientation atlas for M (the jacobians of all the transition functions are positive). From the hypothesis of the theorem, we can assume that the family (U_i) is locally finite. Hence there exists a subordinate partition of unity (ρ_i). Let $\omega_0 = dx^1 \wedge dx^2 \wedge \cdots \wedge dx^n$: since $\mathrm{supp}(\rho_i) \subset U_i$, the forms

$$\omega = \rho_i\left((\phi_i^{-1})^* \omega_0\right)$$

and $\omega = \sum_i \omega_i$ (finite sum at each point) are defined on the whole M. We must now check that ω is a volume form: work on the domain of a chart U_k, and note that

$$\phi_k^* \omega = \sum_i \rho_i \left(\phi_k \circ \phi_i^{-1} \right)^* \omega_0$$

(this sum is actually finite) is of the form $(\sum_i \mu_i) \omega_0$, where, from the hypothesis, the $\mu_i(x)$ are nonnegative and one of them at least is strictly positive at each point of U_k. $\blacksquare$

Exercise. Using the previous theorem, prove that $P^n \mathbf{R}$ is orientable if and only if n is odd.

1.G.4 Integration on an oriented manifold

1.128 One says that two orientation atlases on an orientable manifold M are equivalent if their union is still an orientation atlas. An orientation for an orientable manifold is an equivalence class of orientation atlases. If M is connected, there are two distinct orientations.

It can be seen more easily using volume forms: from the proof of 1.127, a volume form defines an orientation. If ω and ω' are two volume forms, there exists a never zero function f such that $\omega = f\omega'$, and ω and ω' define the same orientation if and only f is positive.

Let now α be an exterior form of degree n, with compact support in M, and assume that M is oriented. We define the integral of α over M, denoted by $\int_M \alpha$ in the following way: let (U_i, ϕ_i) be an atlas for M such that the cover U_i is locally finite. If $\mathrm{supp}(\alpha)$ is included in U_i, then

$$\left(\phi_i^{-1} \right)^* \alpha = f_i dx^1 \wedge \cdots \wedge dx^n,$$

where $f_i \in C^\infty \left(\phi_i(U_i) \right)$, and we set

$$\int_M \alpha = \int_{\mathbf{R}^n} f_i dx^1 \wedge \cdots \wedge dx^n.$$

Using the change of variables formula, and the fact that the jacobians are positive, we see that this integral does not depend on the chart (if $\mathrm{supp}(\alpha) \subset U_i \cap U_j$).

In the general case, use a partition of unity (ρ_i) subordinate to the covering (U_i), and decompose α into a finite sum (since α has compact support) of n-forms with support in some U_i. The same use of the change of variables formula shows that the result does not depend on the orientation atlas. It can be checked that the result does not depend on the partition of unity either. But if we change of orientation, the integral is changed into its opposite. For more details, see [Wa], or the first volume of [Sp].

Now if ω is a volume form giving the orientation chosen for M, we associate to ω a positive measure μ on M by setting $\mu(f) = \int_M f\omega$. One checks easily that, if f is continuous and nonnegative, then $\mu(f) = 0$ if and only if $f \equiv 0$.

1.G.5 Haar measure on a Lie group

1.129 Theorem. *There exists on any Lie group G a non-trivial left-invariant measure (that is, if f is continuous with compact support, and if $g \in G$, $\mu(f \circ L_g) = \mu(f)$). This measure is unique up to a scalar factor.*
Proof. Let $n = \dim G$. For an n-exterior non zero form α on $\underline{G} = T_e G$, we define an L_g-invariant form on G by, for $g \in G$ and $x_i \in T_g G$:

$$\overline{\alpha}(g)(x_1, x_2, \cdots, x_n) = \alpha(e)(T_g L_{g^{-1}} x_1, \cdots, T_g L_{g^{-1}} x_n),$$

where $\alpha(e) = \alpha$ (compare to 2.90). Then, for $f \in C^0(G)$ with compact support, we have

$$\int_G f\overline{\alpha} = \int_G (f \circ L_g) L_g^* \overline{\alpha} = \int_G (f \circ L_g)\alpha.$$

The uniqueness, which will not be used in the sequel, is left to the reader. ∎

Remark. The previous theorem is true for any locally compact topological group, but the proof is more difficult.
Misusing the term, we will say that the measure defined above is "the" Haar measure on G.
1.130 Exercises. a) Explicit the Haar measure μ (that is the corresponding exterior form) in the case $G = Gl(n, \mathbf{R})$.
b) Let K be a compact Lie subgroup of $Gl(n, \mathbf{R})$ (actually, any closed subgroup of a Lie group is a Lie subgroup, see [G]). Show that there exists a K-invariant quadratic form q on $\mathbf{R}^n$, and deduce there exists $g \in Gl(n, \mathbf{R})$ such that $g.K.g^{-1} \subset O(n)$ *(this is the starting point for the proof that all the maximal compact subgroups of a Lie group are conjugate)*.
c) Check the uniqueness we claimed in 1.129. Hint: let μ be left-invariant and ν right-invariant, both non trivial. Set $\check{\nu}(f) = \nu(\check{f})$, where $\check{f}(g) = f(g^{-1})$. Then $\check{\nu}$ is left-invariant.
Pick a function f such that $\mu(f) \neq 0$, and set

$$D_f(s) = \frac{1}{\mu(f)} \int_G f(t^{-1}s) d\nu(t)$$

Show that, for any $\phi \in C^0(G)$, $\nu(\phi) = \mu(D_f \phi)$. Infer that D_f does not depend on f. The take $s = e$ and conclude.
Remark. We will introduce in 3.90 the notion of *density* on a manifold, and will then be able to compute the integral of functions (in place of maximal degree exterior forms) without any orientability assumption.

1.H Partitions of unity

Be given a smooth manifold M, we know what is a smooth "object" on the manifold (for example a function, a vector field, an exterior form...): we just check that this object, read through the charts defining the structure of manifold of M, is smooth.

To prove existence theorems (for Riemannian metrics for example), it is be useful to proceed backwards. We begin with an atlas $(U_i, \varphi_i)_{i \in I}$ for M, and smooth objects defined on the $\varphi_i(U_i)$. We transport them to the U_i, and want to "glue" them to obtain a similar object on the whole manifold. That is possible, thanks to *partitions of unity.*

1.131 Theorem. *Let M be smooth and countable at infinity (countable union of compact sets). Let $(U_i)_{(i \in I)}$, be an open covering of M. Then:*
i) *there exists a locally finite open covering $(V_k)_{(k \in K)}$, subordinate to (U_i). Namely:*
a) *each V_k is included in some U_i,*
b) *there exists around each point m of M an open set W which meets only a finite number of the V_k,*
ii) *there exists a family $(\alpha_k)_{(k \in K)}$ of real smooth functions on M such that:*
a) *the support of α_k is included in V_k,*
b) *α_k is nonnegative, and for any $m \in M$, we have*

$$\sum_{k \in K} \alpha_k(m) = 1,$$

(for m fixed, the sum is finite).

1.131 bis Definition. One says that (α_k) is *a partition of unity subordinate to the covering (V_k).*

1.132 Remarks. a) It is necessary to assume that M is countable at infinity. In practice this will not be a restriction, since all the manifolds we will be interested in (the "natural" manifolds) will have this property.
b) To build a partition of unity, we need first topological tools, and then test functions on $\mathbf{R}^n$ (that is smooth functions with compact support included in a ball of radius a, and equal to 1 on the ball of radius b). Since the test functions are crucial in several proofs of chapter I (1.114, 1.119), let us explain how they are built: begin with a smooth function with compact support in $\mathbf{R}$, for example

$$f_r(x) = e^{\frac{1}{x^2 - r^2}},$$

for $| x | < r$, and $f_r(x) = 0$ for $| x | > r$. Then for proper choices of primitives F_1 and F_2 of f, and of the constants b and c, the function $bF_1(x)F_2(c - x)$ is a test function on $\mathbf{R}$. In $\mathbf{R}^n$, take the corresponding radial function.
c) See [B-T] for explanations on some subtleties related to partitions of unity, and for numerous applications to the topology of manifolds.

Partitions of unity are used in this book, to show the existence of volume forms on orientable manifolds (1.126), and the existence of Riemannian metrics (2.2). *These two theorems are actually particular cases of a (folk) result of algebraic topology, which says that any (locally trivial) bundle with locally compact, countable at infinity basis, and contractible fiber has a section (see [St]).*

Let us turn to another application: any countable at infinity, abstract manifold, can be embedded in a numerical space $\mathbf{R}^N$ for N big enough, and hence is diffeomorphic to a submanifold of $\mathbf{R}^N$. But such an embedding is not canonical: the study of abstract manifolds cannot be reduced to the study of submanifolds of numerical spaces!

1.133 **Theorem.** (Whitney, 1935). *Any countable at infinity smooth manifold M of dimension n can be embedded in $\mathbf{R}^{2n+1}$.*

Sketch of proof. We give it in the case M is *compact*. Let (U_i, ϕ_i) be an atlas for M with a finite number of charts $(i = 1, \cdots, p)$, and (α_i) be a subordinate partition of unity. For $i = 1, \cdots, p$, we define from the map ϕ_i, which is only defined on U_i, a map ψ_i which is defined on the whole M and with values in $\mathbf{R}^n$, as follows:

for $x \in U_i$, $\psi_i(x) = \alpha_i(x)\phi_i(x)$; and for $x \notin U_i$, $\psi_i(x) = 0$.

The map ψ_i is smooth. Let then Φ be the map defined from M to $\mathbf{R}^{np+p}$ by

$$\Phi(x) = \big(\psi_1(x), \cdots, \psi_p(x); \alpha_1(x), \cdots, \alpha_p(x)\big).$$

One checks easily that the linear tangent map to Φ is never singular, hence Φ is an immersion. More Φ is injective, and hence bijective on its image and, since M is compact, Φ is an homeomorphism between M and $\Phi(M)$: the map Φ yields an embedding of M in $\mathbf{R}^N$ with $N = np + p$.

It is easy to build an embedding of M in $\mathbf{R}^{2n+1}$ from the previous construction (but we won't need this refinement). If indeed $N > 2n + 1$, Sard's lemma (see [H]) ensures that almost all the projections on the hyperplanes of $\mathbf{R}^N$ are non singular. This is an immediate consequence of Sard's theorem. Then we get an embedding of M in $\mathbf{R}^{N-1}$ by projecting Φ on a proper hyperplane. For more details, and the proof in the non-compact case, see [H] again.

Riemannian metrics

Pythagoras theorem says that the squared length of an infinitesimal vector, say in $\mathbf{R}^3$, whose components are dx, dy and dz, is $dx^2 + dy^2 + dz^2$. Thus, the length of a parametrized curve $c(t) = \big(x(t), y(t), z(t)\big)$ is given by the integral

$$\int ds = \int (x'^2 + y'^2 + z'^2)^{1/2} dt.$$

Formally, this means that we get the length by setting

$$ds^2 = c^*(dx^2 + dy^2 + dz^2).$$

Now, for a curve on the unit sphere of $\mathbf{R}^3$, which is parametrized (in spherical coordinates) by

$$c(t) = \big(\sin\theta(t)\cos\phi(t), \sin\theta(t)\sin\phi(t), \cos\theta(t)\big),$$

the infinitesimal length is

$$\big(\theta'^2 + \sin^2\theta\,\phi'^2\big)^{1/2} dt.$$

If we forget the 3-space and view the sphere as an abstract manifold equipped with the chart (θ, ϕ), we see that the differential expression

$$ds^2 = d\theta^2 + \sin^2\theta\, d\phi^2$$

provides a way for computing the length of curves on the sphere, i.e. a *Riemannian metric*. The factor $\sin^2\theta$ takes into account the geometric distorsion of the chart.Some difficulties occur at the poles, where the "spherical coordinates" are no longer coordinates. We already met this problem in Chapter I. It will be easily overcome when we give a formal definition of Riemannian manifolds.

In the Euclidean space, straight lines are length minimizing. The curves which (locally) minimize length in a Riemannian manifold are the *geodesics*. For example, the geodesics of the sphere are the great circles. But too long geodesics

— for example great circles which contain two antipodal points — may be no longer length minimizing.

We begin with a long series of examples of Riemannian manifolds. Before studying geodesics, we must know how to take directional derivatives in a Riemannian manifold (this is not possible in a differentiable manifold, without any extra structure). The *covariant derivative* does that, and permits to set up the differential equation of geodesics, to prove their existence and minimization properties. Why and how do too long geodesics stop minimizing is explained in detail in 2.C.

As by-product of the existence theorem for geodesics, we notice that a Riemannian metric is always tangent to the Euclidean metric (see 2.86 and 2.100 for a more precise statement). In other words, there are no order one differential Riemannian invariants. Invariants of order two, provided by the *curvature*, will be studied in the next chapter.

We finish the chapter with a few words about the geodesic flow, and an excursion in the realm of pseudo-Riemannian geometry.

2.A Existence theorems and first examples

2.A.1 Basic definitions

A Riemannian metric on an open set U of $\mathbf{R^n}$ is a family positive definite quadratic forms on $\mathbf{R^n}$, depending smoothly on $m \in U$. The general definition is not fundamentally different.

2.1 Definition. A *Riemannian metric* on M is a smooth and positive definite section g of the bundle $S^2 T^* M$ of the symmetric bilinear 2-forms on M.
Let $\left(\frac{\partial}{\partial x^i}\right)_{(1 \leq i \leq n)}$ be the coordinate vector fields in a local chart around m. Let $u, v \in T_m M$ with

$$u = \sum_{i=1}^n u^i \frac{\partial}{\partial x^i}_{|m} \quad \text{and} \quad v = \sum_{i=1}^n v^i \frac{\partial}{\partial x^i}_{|m}.$$

Then $g_m(u, v) = \sum_{i,j} g_{ij}(m) u^i v^j$, where

$$g_{ij}(m) = g\left(\frac{\partial}{\partial x^i}_{|m}, \frac{\partial}{\partial x^j}_{|m}\right).$$

We will denote $g = \sum_{i,j} g_{ij} dx^i \otimes dx^j$, or shortly

$$g = \sum_{i,j} g_{ij} dx^i dx^j.$$

Recall that all the manifolds we work with are supposed to be connected and countable at infinity. Under these assumptions, we have the following:

2.2 Theorem. *There exists at least one Riemannian metric on any manifold M.*

Proof. We first build Riemannian metrics on the domains of local charts for M and then globalize with a partition of unity (I.H).

Let $(U_k, f_k)_{(k \in K)}$ be an atlas such that (U_k) is a locally finite covering of M, and let (α_k) be a subordinate partition of unity. For a scalar product q on $\mathbf{R}^n$ $(n = \dim M)$ and $k \in K$, let $q_k = f_k^*(q)$ (q_k is a Riemannian metric on U_k) and $g = \sum_k \alpha_k p_k$: g is a smooth and positive definite section of $S^2 T^* M$. If indeed $m \in M$, there exists at least one $j \in K$ with $\alpha_j(m) > 0$ and if $u \in T_m M$ is non zero, then

$$g(u, u) = \sum_k \alpha_k(m) q_k(u, u) \geq a_j(m) q_j(u, u) > 0. \ \blacksquare$$

Remark. Since we know that M can be embedded in some $\mathbf{R}^N$, we could also have taken the restriction to M of the Euclidean metric $\sum_{i=1}^N dx_i^2$. Both arguments use partitions of unity, together with elementary properties of positive quadratic forms: the former argument uses the fact that the sum of two positive quadratic forms is positive, the latter that the restriction to a vector subspace of a positive definite quadratic form is positive definite. These trivial properties are not satisfied for pseudo-Riemannian metrics, and the situation becomes very different in that case (see 2.D below).

2.3 Definition. A *pseudo-riemannian metric* of signature (p, q) on M is a smooth section g of the bundle $S^2 T^* M$ such that, for any $m \in M$, the quadratic form g_m on $T_m M$ has signature (p, q). If p or q is 1, the pseudo-riemannian metric is called *Lorentzian*.

We confess that the term "metric" is rather missleading! Not surprisingly in view of 2.2, it is not always possible to build on a given manifold M a a Lorentzian or a pseudo-Riemannian metric. For instance, there is no Lorentzian metric on the sphere S^2 (check it as an exercise!).

Recall. Let (U, f) and (V, f') be two local charts for M. If g is given by (g_{kl}) (resp. (g'_{ij})) with respect to the related coordinate systems (x_k) (resp. (y_j)), then

$$g'\left(\frac{\partial}{\partial y^i}, \frac{\partial}{\partial y^j}\right) = \sum_{k,l} \frac{\partial x^k}{\partial y^i} \frac{\partial x^l}{\partial y^j} g\left(\frac{\partial}{\partial x^k}, \frac{\partial}{\partial x^l}\right),$$

that is

$$(g'_{ij}) = {}^t(\Phi^{-1})(g_{kl})(\Phi^{-1}),$$

where Φ is the jacobian matrix of the transition function $(f' \circ f^{-1})$ (1.108). This is just the familiar formula for quadratic forms.

2.4 Example. The Euclidean space $\mathbf{R}^2$ is canonically equipped with a Riemannian structure g: let us compute the local expression of this metric in polar coordinates. We have

$$\frac{\partial}{\partial r}(r, \theta) = (\cos \theta, \sin \theta) \quad \text{and} \quad \frac{\partial}{\partial \theta}(r, \theta) = (-r \sin \theta, r \cos \theta),$$

$$\text{hence} \quad g(\frac{\partial}{\partial r}, \frac{\partial}{\partial r}) = 1, \quad g(\frac{\partial}{\partial \theta}, \frac{\partial}{\partial \theta}) = r^2,$$

$$\text{and} \quad g(\frac{\partial}{\partial r}, \frac{\partial}{\partial \theta}) = 0,$$

that is $g = dr^2 + r^2 d\theta^2$.

2.5 Definition. Let (M, g) and (N, h) be two Riemannian manifolds. A map $f : M \to N$ is a *local isometry* if for any $m_i n M$, the tangent map $T_m f$ is an isometry between the Euclidean vector spaces $(T_m M, g_m)$ and $(T_{f(m)} N, h_{f(m)})$. If moreover f is a diffeomorphism, we say that f is an *isometry*.

2.6 Length of curves.
Let $c : [0, a] \to M$ be a curve, and $0 = a_0 < a_1 < ... < a_n = a$ be a partition of $[0, a]$ such that $c_{|[a_i, a_{i+1}]}$ is of class C^1 (c is piecewise C^1). The length of c is defined by

$$L(c) = \sum_{i=0}^{n-1} \int_{a_i}^{a_{i+1}} | c'(t) | \, dt,$$

where $| c'(t) | = \sqrt{g(c'(t), c'(t))}$.
The length of a curve does not depend on the choice of a regular parametrization.

In particular, *Riemannian geometry in dimension 1 is a void field*: a change of metric can be translated into a mere change of parametrization for the curve. Any two 1-dimensional Riemannian manifolds are locally isometric.

2.7 Volume element.
Suppose for simplicity that (M, g) is oriented. Then M carries a natural volume form v_g. It is defined by setting

$$(v_g)_m(e_1, \cdots, e_n) = 1$$

for any direct orthogonal basis $(e_1, \cdots, e_n)$ of $(T_m M, g_m)$. In a local oriented chart,

$$v_g = \sqrt{\det(g_{ij})} dx^1 \wedge dx^2 \cdots \wedge dx^n.$$

If M is not orientable, a natural measure can be defined in a similar way (cf. 3.H).

2.A.2 Submanifolds of Euclidean or Minkowski spaces

2.8 Submanifolds of $\mathbf{R}^n$
A submanifold M of an Euclidean space $\mathbf{R}^n$ is canonically equipped with the Riemannian metric g defined by restricting to each tangent space $T_m M$ the ambient scalar product of $\mathbf{R}^n$. The length of a curve in M is equal to its length in $\mathbf{R}^n$.

Example. Let $S^n \subset \mathbf{R}^{n+1}$ be equipped with the induced metric, and let us compute the length of an arc of great circle of angle α. Since the length does not depend on the parametrization, we can chose the curve $c : [0, \alpha] \to S^n$ defined by $c(t) = (\cos t)x + (\sin t)y$ (where $\mid x \mid = \mid y \mid = 1$ and $\langle x, y \rangle = 0$). Then

$$L(c) = \int\limits_0^\alpha \mid c'(t) \mid dt = \int\limits_0^\alpha dt = \alpha.$$

Remark. Any Riemannian manifold (M, g) can be isometrically embedded in $(\mathbf{R}^N, \mathrm{can})$ for N big enough (Nash theorem, cf. [G-R]). That is, there is an embedding $F : M \to \mathbf{R}^N$ such that $g = F^*(\mathrm{can})$. Once again, there will be no canonical embedding: we will have to be careful not to mistake the intrinsic properties of abstract Riemannian manifolds for the properties of their isometric embeddings (extrinsic). For instance, two regular curves in the Euclidean plane or the 3-space are locally isometric, but this local isometry is not, in general, the restriction of an isometry of the ambient space: it is well known for instance that plane curves are classified, up to Euclidean motions, by their curvature. This curvature has very little to do with the Riemannian curvature we shall meet in the next chapter. See also 2.12. for another example of extrinsic versus intrinsic geometry.

2.9 Revolution surfaces in the Euclidean space $\mathbf{R}^3$.

Let $c : [0, 1] \to \mathbf{R}^2$ be a plane curve, parametrized by arclength, set $c(u) = (r(u), z(u))$ and assume that $r > 0$. If the image of c is a submanifold of $\mathbf{R}^2$, the surface S (fig.a) obtained by rotation of c around the z-axis, and which can be "parametrized" by

$$S(u, \theta) = (r(u) \cos \theta, r(u) \sin \theta, z(u))$$

is a submanifold of $\mathbf{R}^3$ (use 1.3).

The surface S being equipped with the metric g induced by the Euclidean metric of $\mathbf{R}^3$, we write this metric in the local coordinate system (u, θ), and compute the length of a parallel ($u = \mathrm{constant}$). At the point $S(u, \theta)$, we have

a.   ь.

Fig. 2.1. A revolution surface

$$\frac{\partial}{\partial u} = \big(r'(u)\cos\theta, r'(u)\sin\theta, z'(u)\big) \tag{2.1}$$

$$\frac{\partial}{\partial \theta} = \big(-r(u)\sin\theta, r(u)\cos\theta, 0\big). \tag{2.2}$$

Hence $|\frac{\partial}{\partial u}| = \sqrt{r'(u)^2 + z'(u)^2}$, $|\frac{\partial}{\partial \theta}| = r(u)$ and $\langle \frac{\partial}{\partial u}, \frac{\partial}{\partial \theta} \rangle = 0$ that is, since we assumed that c is parametrized by arclength,

$$g = (du)^2 + r(u)^2 (d\theta)^2.$$

If we want to get such manifolds as the sphere, we have to accept points on the curve c with $r(u) = 0$: to make sure that the revolution surface generated by c is regular, we must demand that in these points, $z'(u) = 0$. For details concerning conditions at the poles, see [B1]. In the case of the sphere, the curve c is given by $c(u) = (\sin u, \cos u)$ and $r(u) = \sin u$. Outside the poles, the induced metric is given in local coordinate (u, θ) by

$$g = du^2 + (\sin^2 u)d\theta^2.$$

2.10 The hyperbolic space.

We introduce first the *Minkowski space* of dimension $n + 1$, that is $\mathbf{R}^{n+1}$ be equipped with the quadratic form

$$\langle x, x \rangle = -x_0^2 + x_1^2 + \cdots + x_n^2.$$

Let H^n be the submanifold of $\mathbf{R}^{n+1}$ defined by

$$H^n = \{x \in \mathbf{R}^{n+1} / \langle x, x \rangle = -1, x_0 > 0\}$$

(the second condition ensures that H^n is connected, see 1.1). The quadratic form

$$-dx_0^2 + dx_1^2 + \ldots + dx_n^2 \in \Gamma(S^2 T^* \mathbf{R}^{n+1})$$

induces on H^n a positive definite symmectric 2-tensor g. If indeed $a \in H^n$, $T_a H^n$ can be identified with the orthogonal of a for the quadratic form $\langle , \rangle$, and g_a with the restriction of $\langle , \rangle$ to this subspace. Since $\langle a, a \rangle < 0$, g_a is positive definite from the Sylvester theorem, or an easy computation. The hyperbolic space of dimension n is just H^n, equipped with this Riemannian metric. Using 1.101 f), we see that $O_o(1, n)$ acts *isometrically* on H^n. There exist other presentations for this space. They are conceptually easier, but technically more complicated. They are treated in the following exercise.

2.11 Exercise: Poincaré models for (H^n, g).

Let f be the "pseudo-inversion" with pole $s = (-1, 0, \cdots, 0)$ defined by

$$f(x) = s - \frac{2(x - s)}{\langle x - s, x - s \rangle},$$

where $\langle , \rangle$ is the quadratic form defined in 2.10. For $X = (0, X_1, .., X_n)$ in the hyperplane $x_0 = 0$, we note that

$$\langle X, X \rangle = \sum_{i=1}^{n} X_i^2 = |\, X\,|^2,$$

(square of the Euclidean norm in $\mathbf{R}^n$).

a) Show that f is a diffeomorphism from H^n onto the unit disk

$$\{x \in \mathbf{R}^n,\, |\, x\,| < 1\}, \quad \text{and that} \quad (f^{-1})^* g = 4 \sum_{i=1}^{n} \frac{dx_i^2}{(1 - |\, x\,|^2)^2}.$$

b) What happens if we replace H^n by the unit sphere of the Euclidean space $\mathbf{R}^{n+1}$, and f by the usual inversion of modulus 2 from the South pole?

c) Let f_1 be the inversion of $\mathbf{R}^n$ with pole $t = (-1, 0, \cdots, 0)$, given by

$$f_1(X) = t + \frac{2(x - t)}{|\, x - t\,|^2}.$$

Show that f_1 is a diffeomorphism from the unit disk onto the half space $x_1 > 0$. If $g_1 = (f^{-1})^* g$ (see b)), show that

$$(f_1)^* g_1 = \sum_{i=1}^{n} \frac{dx_i^2}{x_1^2}.$$

In that way, we have obtained two Riemannian manifolds which are isometric to (H^n, g), namely the *Poincaré disk* in a), and the *Poincaré half-plane* in c).

2.11 bis Definition Two Riemannian metrics g_0 and g_1 on a manifold M are *(pointwise) conformal* if there is a nowhere vanishing smooth function f on M such that $g_1 = f g_0$.

Both Poincaré models are conformal to the Euclidean metric (on the unit ball or the half-plane). For two conformal metrics, the angles are the same.

Let us now give other examples to show that isometric Riemannian metrics may look quite different.

2.12 Exercise. Let $C \subset \mathbf{R}^3$ be a *catenoid: C* is the revolution surface generated by rotation around the z-axis of the curve of equation $x = \cosh z$. Let $H \subset \mathbf{R}^3$ be an *helicoid: H* is generated by the straight lines which are parallel to the xOy plane and meet both the z-axis and the helix $t \to (\cos t, \sin t, t)$.

a) Show that H and C are submanifolds of $\mathbf{R}^3$, and give a "natural" parametrization for both.

b) If g is the Euclidean metric $dx^2 + dy^2 + dz^2$ of $\mathbf{R}^3$, give the expressions of $g_{|C}$ and $g_{|H}$ in the parametrizations defined in a), and show that C and H are locally isometric. Are they globally isometric?

It is not possible to guess from the embeddings that C and H are locally the same from the Riemannian point of view. For example, C does not contain any straight line, even no segment: the local isometries between C and H do not come from isometries of the ambient space.

2.13 Exercises. a) Show that the curves with "constant tangent" (i.e. such that in the figure 2.2 below $mm' \equiv 1$) are obtained by horizontal translation from the curve $t \to (t - \tanh t, \frac{1}{\cosh t})$.
b) The *pseudo-sphere* is the surface generated by rotation around the x-axis of the curve defined in a). Show that the pseudo-sphere, with the circle generated by the cusps taken away, is locally isometric to $(\mathbf{R}^2, \frac{dx^2 + dy^2}{y^2})$.

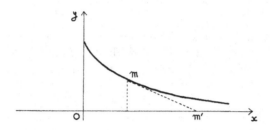

Fig. 2.2. A meridian of the pseudo-sphere

2.A.3 Riemannian submanifolds, Riemannian products

2.14 Definition. Let (M, g) be a Riemannian manifold. (N, h) is a *Riemannian submanifold* of (M, g) if:
i) N is a submanifold of M,
ii) for any $m \in N$, h_m is the restriction of g_m to $T_m N$.
Exercise. Show that the length of the curves drawn on N is the same when computed in N or M.
Be careful: We will define in 2.91 the distance associated to a Riemannian metric by $d(m, n) = \mathrm{Inf}\, L(c)$, for all piecewise C^1 curves c from m to n: the distance associated with the metric h induced on N will generally be different from the distance associated with the metric g of M. Give counter-examples.
2.15 Definition. Let (M, g) and (N, h) be two Riemannian manifolds. The *product metric* $g \times h$, defined on $M \times N$ is defined by:

$$(g \times h)_{(m,n)}(u, v) = g_m(u_1, v_1) + h_n(u_2, v_2),$$

when indexing by 1 and 2 respectively the components of u and v on $T_m M$ and $T_n N$.

2.16 Exercise. Denote by g_1 and dr^2 the canonical metrics of S^{n-1} and $I =]0, \infty[$ respectively. Define on $S^{n-1} \times I$ a metric g by:

$$g_{(m,r)} = r^2 g_1 + dr^2.$$

Is g the product metric?

Show that $(S^{n-1} \times I, g)$ is isometric to $(\mathbf{R}^n \setminus \{0\}, \text{can})$, and that $(S^{n-1} \times I, g_1 \times dr^2)$ is isometric to the cylinder

$$C = \{x \in \mathbf{R}^{n+1} / x_1^2 + \ldots + x_n^2 = 1 \quad \text{and} \quad x_0 > 0\},$$

equipped with the Riemannian metric induced by the Euclidean structure of $\mathbf{R}^{n+1}$.

Let X be a unit vector tangent to S^{n-1}, and Y be the same vector considered as tangent to $S^{n-1} \times \{r\}$. Compute the norm of Y at (m, r) for the two metrics.

2.A.4 Riemannian covering maps, flat tori

2.17 Definition. Let (M, g) and (N, h) be two Riemannian manifolds. A map $p : N \to M$ is a *Riemannian covering map* if:
i) p is a smooth covering map,
ii) p is a local isometry.

2.18 Proposition. *Let $p : N \to M$ be a smooth covering map. For any Riemannian metric g on M, there exists a unique Riemannian metric h on N, such that p is a Riemannian covering map.*
Proof. If such a metric h exists, it has to satisfy for $n \in N$ and $X, Y \in T_n N$:

$$h_n(X, Y) = g_{p(n)}(T_n p(X), T_{p(n)} p(Y)).$$

Conversely, this formula defines on each tangent space to N a scalar product h_n (since $T_n p$ is a vector spaces isomorphism). This scalar product depends smoothly on n: since p is a local diffeomorphism, h and g have the same expression in the local charts (U, ϕ) around n, and $(p(U), \phi \circ p^{-1})$ around $p(n)$. ∎
On the contrary, a Riemannian metric h on N does not automatically yield a metric g on M such that p is a Riemannian covering map.

2.19 Example. Let us equip the sphere S^2 with a Riemannian metric which can be written in "spherical coordinates" as:

$$h = d\theta^2 + a^2(\theta)d\phi^2.$$

Under which condition does this metric induce on $P^2\mathbf{R}$ a metric g such that the canonical projection from (S^2, h) to $(P^2\mathbf{R}, g)$ is a Riemannian covering map?
Such a metric must satisfy $h_n = g_{p(n)}(T_n p \cdot, T_n p \cdot)$. Since $p(x) = p(y)$ if and only if $x = y$ or $x = -y$, it is necessary and sufficient for g to exist that $a(\pi - \theta) \equiv a(\theta)$: g is then unique.
More generally, we have the following:

2.20 Proposition. *Let (N, h) be a Riemannian manifold and G be a free and proper group of isometries of (N, h). Then there exists on the quotient manifold $M = N/G$ a unique Riemannian metric g such that the canonical projection $p : N \to M$ is a Riemannian covering map.*

Proof. Let n and n' be two points of N in the same fiber $p^{-1}(m)$ ($m \in M$). There exists an isometry $f \in G$ such that $f(n) = n'$.

Since $p \circ f = p$ and p is a local diffeomorphism, we can define a scalar product g_m on $T_m M$ by, for $u, v \in T_m M$:

$$g_m(u, v) = h_n \left((T_n p)^{-1} \cdot u, (T_n p)^{-1} \cdot v \right) \qquad (*)$$

for $n \in p^{-1}(m)$. This quantity does not depend on n in the fiber since

$$(T_{n'} p)^{-1} = T_n f \circ (T_n p)^{-1}$$

and $T_n f$ is an isometry of Euclidean vector spaces between $T_n N$ and $T_{n'} N$. The same proof as above yields that g is smooth. Hence we have built a metric g on M such that p is Riemannian covering map. Since any such g must satisfy (*), this metric is the only possible. ∎

2.21 **Example.** We can use 2.20 and equip $P^n \mathbf{R}$ with a canonical metric such that the canonical projection $p : S^n \to P^n \mathbf{R}$ is a Riemannian covering map, with $G = \{Id, -Id\}$ as isometry group of the sphere.

The map p is an isometry from the open hemisphere U onto $P^n \mathbf{R}$ with a submanifold isometric to $P^{n-1} \mathbf{R}$ taken away. The map $p_{|U}$ is indeed a diffeomorphism on its image, and hence an isometry, and $(P^n \mathbf{R} \setminus p(U)) = p(S^{n-1})$ where S^{n-1} is the equator which bounds U. The metric induced by the metric of S^n on S^{n-1} is actually the canonical one, and the restriction of p to S^{n-1} is still the quotient by the antipody relation.

2.22 **Flat tori.**

Those are quotients of the Euclidean space by a lattice.

Let $(a_1, \cdots, a_n)$ be a basis of $\mathbf{R}^n$. The lattice Γ associated to this basis is the set of all vectors $\sum_{j=1}^{n} k_j a_j$, where $k_j \in \mathbf{Z}$. Identifying Γ with a group of translations of $\mathbf{R}^n$, we can equip canonically the quotient $\mathbf{R}^n / \Gamma$ with a structure of smooth manifold diffeomorphic to T^n. Indeed, the map $p : \mathbf{R}^n \to T^n$ defined by

$$p(\sum_j x_j a_j) = \left(e^{2i\pi x_j} \right)_{(j=1, \cdots, n)}$$

is constant on Γ and yields by quotient by p a continuous and bijective map $\hat{p} : \mathbf{R}^n / \Gamma \to T^n$. This map is an homeomorphism since $\mathbf{R}^n / \Gamma$ is compact. A local chart for T^n around

$$\dot{x} = \left(e^{2i\pi x_j} \right)_{(j=1, \cdots, n)}$$

is given by

$$\Phi :]x_1 - 1/2, x_1 + 1/2[\times \cdots \times]x_n - 1/2, x_n + 1/2[\longrightarrow T^n$$

where $\Phi(y_1, \cdots, y_n) = $ class in T^n of $\left(e^{2i\pi y_j} \right)_{(j=1, \cdots, n)}$. In this chart $\hat{p}$ is just

$$\Phi^{-1} \circ \hat{p}(\sum x_j a_j) = (x_1, \cdots, x_n),$$

and hence $\hat{p}$ and $\hat{p}^{-1}$ are smooth.

Fig. 2.3. Lattices corresponding to non isometric tori

We can now equip $\mathbf{R}^n/\Gamma$ with the canonical metric for which $\pi : \mathbf{R}^n \to \mathbf{R}^n/\Gamma$ is a Riemannian covering map. Then, using $\hat{p}$, we get a metric g_Γ on T^n such that $\hat{p}$ is an isometry. If $\langle , \rangle$ is the Euclidean scalar product on $\mathbf{R}^n$ and $\left(\frac{\partial}{\partial x^j}\right)$ are the coordinate vector fields on T^n associated with the chart Φ, the metric is given in this chart by

$$g_\Gamma = \sum_{i,j} \langle a_i, a_j \rangle dx_i\, dx_j.$$

Such metrics on the torus are called *flat metrics* on T^n. The Riemannian manifolds (T^n, g_Γ) can be classified up to isometries by considering the lattices Γ.

2.23 Theorem. *The metrics g_Γ and $g_{\Gamma'}$ defined on T^n are isometric if and only if there exists an isometry of $\mathbf{R}^n$ which sends the lattice Γ on the lattice Γ'.*

Proof. If there exists an isometry $F : \mathbf{R}^n \to \mathbf{R}^n$ with $F(\Gamma) = \Gamma'$, then F goes to the quotient and gives an isometry f between $\mathbf{R}^n/\Gamma$ and $\mathbf{R}^n/\Gamma'$ (clear, check it).

$$
\begin{array}{ccc}
\mathbf{R}^n & \xrightarrow{\ F\ } & \mathbf{R}^n \\
f \downarrow & & \downarrow p' \\
\mathbf{R}^n/\Gamma & \xrightarrow{\ f\ } & \mathbf{R}^n/\Gamma'
\end{array}
$$

Conversely, if f is an isometry from $\mathbf{R}^n/\Gamma$ to $\mathbf{R}^n/\Gamma'$, since $\mathbf{R}^n$ is simply connected, there is a map F from $\mathbf{R}^n$ to itself such that $f \circ p = p' \circ f$. The map f is an isometry (same type of computation as for the first step), and by construction, $f(\Gamma) = \Gamma'$. ∎

2.24 Classification of 2-dimensional flat tori.

We want to classify the flat metrics on $\mathbf{T}^2$ up to isometries and homotheties (two metrics g_1 and g_2 on a manifold are *homothetic* if there exists a real $\lambda > 0$ such that $g_1 = \lambda g_2$). This classification is equivalent to the classification of the lattices of $\mathbf{R}^2$ up to isometries and dilations.

Let a_1 be the shortest non zero vector of Γ. After using if necessary a dilation and a rotation, we can assume that $a_1 = e_1$ (first vector of the canonical orthonormal basis of $\mathbf{R}^2$) Let a_2 be the shortest vector of $\Gamma \setminus \mathbf{Z} \cdot a_1$. The

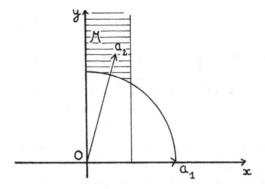

Fig. 2.4. (One half of) the modular domain

vectors a_1 and a_2 generate Γ. If not there would exist a vector $z \in \Gamma$ with $z = \lambda_1 a_1 + \lambda_2 a_2$, and $\mid \lambda_i \mid < 1/2$. Then

$$\| z \|^2 = \lambda_1^2 + 2\lambda_1\lambda_2\langle a_1, a_2\rangle + \lambda_2^2 \| a_2 \|^2$$
$$< 1/4 \left(1 + 2 \| a_2 \| + \| a_2 \|^2\right),$$

so that

$$\|z\|^2 < 1/4 \left(1 + \| a_2 \|\right)^2 \leq \| a_2 \|^2,$$

a contradiction.

Now we can assume (eventually use symmetries with respect to the coordinate axis), that a_2 lies in the first quadrant: the x-coordinate of a_2 is then smaller than $1/2$ (if not $a_2 - a_1$ would be shorter than a_2). The class of the lattice Γ is hence determined by the position of a_2 in the hachured domain

$$\mathcal{M} = \{(x, y)/ x^2 + y^2 \geq 1, 0 \leq x \leq 1/2, y > 0\},$$

and two lattices corresponding to two different points of $\mathcal{M}$ belong to two different classes (see figure 2.4).

The flat tori coming from a square lattice $\left(a_2 = (0, 1)\right)$, a rectangle one $\left(a_2 = (0, x)\right)$ or an hexagonal one $\left(a_2 = (1/2, \sqrt{3/2})\right)$ are called square, rectangle or hexagonal tori.

Remark. The reader may have recognized (one half) of the famous *modular domain*, which, from our point of view, classifies oriented 2-lattices. It can be viewed as the double quotient $SO(2)\backslash Sl(2, \mathbf{R})/Sl(2, \mathbf{Z})$, while $\mathcal{M} = O(2)\backslash Gl(2, \mathbf{R})/Gl(2, \mathbf{Z})$. * By the way, the classification of oriented flat tori (up to isometries and homotheties) is equivalent to that of genus 1 Riemann surfaces (compare with [J2], 2.7).*

Although flat tori are locally isometric to $(\mathbf{R}^n, \mathrm{can})$, their global geometry is very rich. Their study belongs to Arithmetic as well as to Geometry (see [Ma]).

2.25 Exercises. a) Let T^2 be embedded in $\mathbf{R}^3$ as image of $\mathbf{R}^2$ by the map Φ defined by

$$\Phi(\theta, \phi) = \left[(2 + \cos \theta) \cos \phi, (2 + \cos \theta) \sin \phi, \sin \theta\right].$$

Let g be the metric induced on T^2 by the Euclidean metric of $\mathbf{R}^3$. Since the map Φ yields by quotient a diffeomorphism between $S^1 \times S^1$ and T^2, we can also define the product metric

$$(T^2, g_0) = (S^1, \text{can}) \times (S^1, \text{can}).$$

Write g_0 in the (θ, ϕ) coordinate patch, and prove that g is not the product metric (one can prove more generally that g is not a flat metric (see 3.11)).
b) Let M be the submanifold of $\mathbf{R}^4$, image of $\mathbf{R}^2$ under the map $\Phi(\theta, \phi) = (e^{i\theta}, e^{i\phi})$. We equip M (which is diffeomorphic to T^2) with the induced metric g. Show that (M, g) is a square torus (Clifford torus).
c) Let (e_1, e_2) be a basis of $(\mathbf{R}^2, \text{can})$, and let G be the group of diffeomorphisms generated by γ_i ($i = 1, 2$), where

$$\gamma_1(x, y) = (x + 1, -y) \quad \text{and} \quad \gamma_2(x, y) = (x, y + 1).$$

Show that $\mathbf{R}^2$ is diffeomorphic to the Klein bottle K (1.89), but that the Euclidean metric can only go to the quotient in case the basis (e_1, e_2) is orthogonal. These Riemannian Klein bottles (endowed with the quotient metrics) are the *flat Klein bottles*. Show that a flat Klein bottle has a 2-folded Riemannian covering, which is a rectangle flat torus.
d) Compute the isometry group of a flat 2-dimensional torus.
e) Let $\theta \in \mathbf{R}$ and $n \in \mathbf{Z}$ be given. For $(x, z) \in \mathbf{R} \times S^1$, we set

$$T_n(x, z) = (x + n, e^{in\theta} z),$$

(S^1 is identified with the set of complex numbers of modulus 1). Let us equip S^1 with the metric induced by the metric of $\mathbf{R}^2$ ($\text{length}(S^1) = 2\pi$), $\mathbf{R}$ with its canonical metric, and $\mathbf{R} \times S^1$ with the product metric. Show that the T_n are isometries and that the quotient $\mathbf{R} \times S^1/T_n$ is a flat torus. Compute the lattice Λ of $\mathbf{R}^2$ such that $\mathbf{R} \times S^1/T_n$ is isometric to $\mathbf{R}^2/\Lambda$.
Is it possible to build the Klein bottle by an analogous procedure?

2.A.5 Riemannian submersions, complex projective space.

2.26 Let $p : M \to N$ be a submersion: note that any complement in $T_m M$ of $(T_m p)^{-1}(0)$ is isomorphic to $T_{p(m)} N$, but that there exists no canonical choice for such a complement. However, in the case M is equipped with a Riemannian metric g, we can chose the orthogonal complement of $(T_m p)^{-1}(0)$ in $T_m M$, let H_m: it is the *horizontal subspace* of $T_m M$.

2.27 Definition. A map p from (M, g) to (N, h) is a *Riemannian submersion* if

i) p is a smooth submersion,

ii) for any $m \in M$, $T_m p$ is an isometry between H_m and $T_{p(m)} N$.

Examples: i) The projections defined from a Riemannian product, onto the factor spaces, are plainly Riemannian submersions.

ii) The projection from $S^{n-1} \times]0, \infty[$ equipped with the metric $r^2 g_1 + dr^2$ (2.16), onto $(]0, \infty[, dr^2)$ is a Riemannian submersion. Note that the fibers are not isometric one to each other.

iii) More generally if (M, g) and (N, h) are Riemannian manifolds, and if f is a never zero smooth function on N, the projection from $(M \times N, f^2 g + h)$ onto (N, h) is a Riemannian submersion.

2.28 Proposition. *Let $(\tilde{M}, \tilde{g})$ be a Riemannian manifold, and G be a Lie group of isometries of $(\tilde{M}, \tilde{g})$ acting properly and freely on $\tilde{M}$ (hence $M = \tilde{M}/G$ is a manifold, and $p : \tilde{M} \to \tilde{M}/G$ is a fibration). Then there exists on M a unique Riemannian metric g such that p is a Riemannian submersion.*

Proof. Let $m \in M$, and $u, v \in T_m M$. For $\tilde{m} \in p^{-1}(m)$, there exist unique vectors $\tilde{u}, \tilde{v} \in H_{\tilde{m}}$ such that $T_{\tilde{m}} p \cdot \tilde{u} = u$ and $T_{\tilde{m}} p \cdot \tilde{v} = v$. Set $g_m(u, v) = g_{\tilde{m}}(\tilde{u}, \tilde{v})$ (i.e. decide that $(T_m M, g_m)$ is isometric to $H_{\tilde{m}}$ with the induced metric). The metric g_m does not depend on the choice of $\tilde{m}$ in the fiber: indeed if $p(\tilde{m}) = p(\tilde{m}') = m$, there exists $\gamma \in G$ such that $\overline{\gamma}(\tilde{m}) = \tilde{m}'$ (where $\overline{\gamma}$ is the isometry of $\tilde{M}$ associated to γ), and $T_{\tilde{m}} \overline{\gamma}$ is an isometry between $H_{\tilde{m}}$ and $H'_{\tilde{m}}$ which commutes with p.

Now we must check that $m \to g_m$ is a smooth section of $S^2 M$. Note first that, if $\pi_{\tilde{m}}$ is the orthogonal projection from $T_{\tilde{m}} \tilde{M}$ onto $H_{\tilde{m}}$, then the section $\tilde{m} \to \pi_{\tilde{m}}$ of $\mathrm{End}(TM)$ is smooth. Moreover, the fibration p has local sections: hence if $U \subset M$ is an open subset over which such a section s exists, $g_{|U}$ is given by

$$g_m(u, v) = g_{s(m)}(\pi_{s(m)} \tilde{u}, \pi_{s(m)} \tilde{v}). \blacksquare$$

2.29 A basic example: the complex projective space.

Recall that $P^n \mathbf{C}$ is the quotient of $\mathbf{C}^{n+1} \setminus \{0\}$ by the group $\mathbf{C}^*$ of dilations of $\mathbf{C}^{n+1}$. Set, for $z = (z^0, ..., z^n)$ in $\mathbf{C}^{n+1}$, $|z|^2 = \sum_{j=0}^n |z^j|^2$. Then $P^n \mathbf{C}$ can also be seen as the quotient S^{2n+1}/S^1 (1.96) where S^{2n+1} is the unit sphere in $\mathbf{C}^{n+1} \sim \mathbf{R}^{2n+2}$, and S^1 is the unit circle in $\mathbf{C}$, with the action given by $(u, z) \to uz$. The canonical map $p : S^{2n+1} \to P^n \mathbf{C}$ is a submersion from 1.96. We now give a direct proof of this fact, to make ourselves familiar with this fundamental example.

Let us consider the vector space $\mathbf{C}^{n+1}$ as a real vector space, equipped with the Euclidean metric given by $|dz|^2$: for $z \in S^{2n+1}$, $\mathbf{C}^{n+1}$ is decomposed into the orthogonal sum $\mathbf{R}.z \oplus T_z S^{2n+1}$. More, since the tangent space to the orbit of z under the S^1 action is the real subspace $\mathbf{R}.iz$ of $T_z S^{2n+1}$, we have also the orthogonal decomposition

$$\mathbf{C}^{n+1} = \mathbf{R}.z \oplus \mathbf{R}.iz \oplus H_z.$$

This decomposition is preserved under the action of the unitary group $U(n+1)$, seen under our identifications as the subgroup of $SO(2n+2)$ of the **C**-linear maps. We want to show that p is a submersion: we are done if we check that p is an isomorphism between H_z and $T_{p(z)}P^n\mathbf{C}$. Using the action of $U(n+1)$, it is sufficient to prove the result for $z_0 = (1, 0, ..., 0)$. We can use in a neighborhood of $p(1, 0, ..., 0)$ the chart ϕ given by

$$\phi\big(p(z)\big) = \left(\frac{z^i}{z^0}\right)_{(1 \le i \le n)}.$$

Then,

$$T_{z_0}S^{2n+1} = (i\eta^0, \xi^1, ..., \xi^n),$$

where $\eta^0 \in \mathbf{R}$ and $\xi^j \in \mathbf{C}$, and $T_{z_0}(\phi \circ p)$ is plainly the map

$$(i\eta^0, \xi^1,, \xi^n) \to (\xi^1, ..., \xi^n).$$

We denote in the following by $x = p(z)$ the generic point of $P^n\mathbf{C}$.

2.30 Exercise. Prove that H_z is a complex vector subspace of $\mathbf{C}^{n+1}$. Using the isomorphism $T_z p$ from H_z onto $T_x P^n\mathbf{C}$, the multiplication by i becomes an isomorphism of $T_x P^n\mathbf{C}$. Prove that this isomorphism does not depend on z in the fiber above x: it will be denoted by J_x. Prove that $x \to J_x$ is a smooth section of the bundle $\mathrm{End}(TP^n\mathbf{C})$ *(J is the almost complex structure associated with the holomorphic structure of $P^n\mathbf{C}$: we will not use this result, which is easy to check).*

Now we can equip $P^n\mathbf{C}$ with a "natural" Riemannian metric: the sphere S^{2n+1} being equipped with the metric induced by the canonical metric of $\mathbf{C}^{n+1} = \mathbf{R}^{2n+2}$, there exists from 2.28 a unique metric on $P^n\mathbf{C}$ such that p is a Riemannian submersion, the horizontal subspace at z being H_z. This metric is often called the *Fubini-Study metric*.

2.31 Example. Let $z \in S^{2n+1}$, and $v \in H_z$. The vector v, seen as a vector in $\mathbf{C}^{n+1}$, is normal to z and iz. What is the length of the curve of $P^n\mathbf{C}$ parametrized on $[0, \pi]$ by $c(t) = p(\cos t.z + \sin t.v)$?
Answer: Let γ be the curve of S^{2n+1} given by $\gamma(t) = \cos t.z + \sin t.v$. The vector $\gamma'(t)$ is everywhere normal to $\gamma(t)$ and $i\gamma(t)$, hence $\gamma'(t) \in H_{\gamma(t)}$,

$$g_{P^n\mathbf{C}}\big(c'(t), c'(t)\big) = g_{S^{2n+1}}\big(\gamma'(t), \gamma'(t)\big),$$

and $L(c) = \pi$. Note that c is a smooth closed curve.

2.32 Exercise. Prove that $(P^1\mathbf{C}, \mathrm{can}) = (S^2, \frac{1}{4}\mathrm{can})$, making 1.19 b) more precise. (Using 2.43, one can prove that these metrics are homothetic with the least computations as possible).

2.A.6 Homogeneous Riemannian spaces

2.33 Definition. An *homogeneous Riemannian space* is a Riemannian manifold (M, g) whose isometry group $\mathrm{Isom}(M, g)$ acts transitively on M.

The following results are related with the differential theory of homogeneous spaces (chap.I).

2.34 Theorem. (Myers-Steenrod, see [M-Z] 6.3, [Hn] th 2.5 p.204). *The isometry group of a Riemannian manifold is a Lie group.*

Furthermore, it is proved by the way that the topology as a Lie group of $\text{Isom}(M, g)$ coincide with the topology of uniform convergence on compact sets.

2.35 Theorem. *Let (M, g) be a Riemannian manifold. The isotropy subgroup of a given point is a compact subgroup of $\text{Isom}(M, g)$. If $\text{Isom}(M, g)$ acts transitively on M, then $\text{Isom}(M, g)$ is compact if and only if M is compact.*

Proof. Using the remark following 2.34, it is a straightforward consequence of Ascoli's theorem. But be careful: we use the fact that the Riemannian isometries of (M, g) are also the isometries for the distance on M associated to g (see 2.91). ■

Hence, a Riemannian homogeneous space is diffeomorphic to a smooth homogeneous space G/H, where H is a compact subgroup of the Lie group G: just take $G = \text{Isom}(M, g)$, and H will be the isotropy group of a point. This naive approach is inadequate since:

i) there can exist distinct pairs (G, H) and (G', H') with $G/H = G'/H'$. For example $\mathbf{R}^n$ can be seen as the homogeneous space $\text{Isom}(\mathbf{R}^n)/SO(n)$ (we will see that the group of Riemannian isometries of $\mathbf{R}^n$ coincides with the group of isometries for the Euclidean distance). But the group of translations is also transitive on $\mathbf{R}^n$, and $\mathbf{R}^n = \mathbf{R}^n/\{0\}$. A Riemannian homogeneous space can have transitive isometry groups smaller than $\text{Isom}(M, g)$;

ii) there are "natural" examples where the canonical map from G to $\text{Diff}(G/H)$ is not injective. Take for example the homogeneous space $U(n + 1)/(U(1) \times U(n))$ (1.101).

2.36 Definition. An homogeneous space is *effective* if the map which sends $\gamma \in G$ to the diffeomorphism of G/H defined by $L_\gamma : gH \to \gamma gH$ is injective, and *almost effective* if the kernel of this map is discrete.

2.37 Exercise. $SU(n + 1)/S\big(U(1) \times U(n)\big) = P^n\mathbf{C}$ is almost effective, but non effective.

To take into consideration the previous remarks i) and ii), the problem of knowing wether an homogeneous space can be considered as a Riemannian homogeneous space, must be set in the following way: does there exist on the manifold G/H one or several G-invariant Riemannian metrics, such that G acts isometrically? We do not ask to the natural map from G to $\text{Isom}(G/H)$ to be surjective (i), nor injective (ii). We now give an important example, where such a metric exists, and is unique up to a scalar factor.

If G/H is effective, 2.35 yields that a necessary condition for the existence of a G-invariant metric on G/H is the *compactness* of H. In the general case, we will need the following:

2.38 Definition. The (linear) *isotropy representation* of the homogeneous space $M = G/H$ is the homomorphism from H to $Gl(T_{[e]}M)$ defined by $h \to T_{[e]}L_h$.

2.39 Example (and exercise): the isotropy representations of

$$SO(n+1)/SO(n) = S^n \quad \text{and} \quad SO_0(n,1)/SO(n) = H^n$$

are identified with the natural representation of $SO(n)$ in $\mathbf{R}^n$.

2.40 Proposition. *The isotropy representation of G/H is equivalent to the adjoint representation of H in $\underline{G/H}$.*

Proof. Since H is a Lie subgroup of G then, for $h \in H$, $\mathrm{Ad}_H h = \mathrm{Ad}_G h_{|\underline{H}}$. Hence $\mathrm{Ad}_G h$ yields by quotient a linear isomorphism of $\underline{G/H}$. This isomorphism will be denoted by $\mathrm{Ad}_{G/H} h$, or $\mathrm{Ad} h$ if there is no ambiguity. The tangent space $T_{[e]} G/H$ is identified with $\underline{G/H}$. On the other hand, for $h \in H$ and $X \in \underline{G}$, the sets $h.\exp(tX)H$ and $h.\exp(tX).h^{-1}H$ are identical. Going to the quotient and differentiating, we get

$$T_{[e]}L_h = T_e \pi \circ \mathrm{Ad}_G h = \mathrm{Ad}_{G/H} h,$$

where we denoted by π the natural projection from G onto G/H. ∎

2.41 Exercise. What is the isotropy representation of $U(n+1)/U(n) \times U(1)$?

2.42 Theorem. i) *There exists a G-invariant Riemannian metric on the homogeneous space G/H if and only if $\overline{\mathrm{Ad}_{G/H}H}$ is compact in $Gl(\underline{G/H})$.*
ii) *If we assume that the isotropy representation has no trivial invariant subspaces (that is different from 0 and G/H), this metric is unique up to a scalar factor.*

Proof. i) For such a metric $\langle , \rangle$, $\mathrm{Ad}_{G/H}H$ must act isometrically on the Euclidean vector space $(T_{[e]}G/H, \langle , \rangle_{[e]})$, and hence $\overline{\mathrm{Ad}_{G/H}H}$ must be a compact subgroup of $Gl(\underline{G/H})$.
Conversely, if $H_1 = \overline{\mathrm{Ad}_{G/H}H}$ is compact, one first build on $T_{[e]}G/H$ an Euclidean metric for which $\mathrm{Ad}_{G/H}H$ acts isometrically: let γ be any Euclidean metric on $T_{[e]}G/H$, and set

$$\langle x, y \rangle_{[e]} = \overline{\gamma}(x, y) = \int_{H_1} \gamma(\mathrm{Ad}h_1.x, \mathrm{Ad}h_1.y)v_{H_1},$$

where v_{H_1} is a Haar measure on H_1. This integral is finite since H_1 is compact, and $\langle , \rangle_{[e]}$ is obviously positive definite.
Now, if $m \in G/H$ is equal to $L_g[e]$, then

$$\langle x, y \rangle_m = \langle T_m L_{g^{-1}}x, T_m L_{g^{-1}}y \rangle_{[e]}.$$

This quadratic form does not depend on g: if $m = L_g[e] = L_{g'}[e]$, then $g'^{-1}g = h \in H$, and we get:

$$T_m L_{g^{-1}}x = T_{[e]} L_h(T_m L_{g'_{-1}}x)$$

and hence

$$\langle T_m L_{g^{-1}} x, T_m L_{g^{-1}} y \rangle_{[e]} = \langle T_m L_{g'^{-1}} x, T_m L_{g'^{-1}} y \rangle_{[e]}.$$

Eventually, we must check that the map $m \to \langle , \rangle_m$ from M to $S^2 M$ we have just defined is smooth. It works because the canonical map $p : G \to G/H$ is a fibration (1.97) and hence has local sections (compare with 2.28).

ii) Let $\langle , \rangle$ and $\langle , \rangle'$ be two such metrics and diagonalize $\langle , \rangle'_{[e]}$ with respect to $\langle , \rangle_{[e]}$: we get a decomposition of $\underline{G/H}$ into a direct sum of $\mathrm{Ad}_{G/H} H$-invariant subspaces. Just use the irreducibility, and the fact that if the metrics $\langle , \rangle_{[e]}$ and $\langle , \rangle'_{[e]}$ are proportional, then so are $\langle , \rangle$ and $\langle , \rangle'$. ∎

2.43 Definition. An homogeneous space G/H is *isotropy irreducible* if the isotropy representation is irreducible -in other words if it has no non trivial invariant subspaces.

Be careful: This property depends on the pair (G, H), and not only on the manifold $M = G/H$.

2.44 Example, exercise. The homogeneous space $SO(n + 1)/SO(n)$ is isotropy irreducible from 2.39. $U(n + 1)/U(n)$ is diffeomorphic to S^{2n+1}, but is not isotropy irreducible.

An isotropy irreducible homogeneous space G/H has, from 2.43, a "natural" Riemannian metric, which is the unique (up to a scalar factor) G-invariant metric. Basic examples are

$$SO(n + 1)/SO(n) \simeq S^n$$
$$SO(n + 1)/S(O(n) \times O(1)) \simeq P^n \mathbf{R}$$
$$SU(n + 1)/S(U(1) \times U(n)) \simeq P^n \mathbf{C}$$

In particular, the $U(n+1)$-invariant metric on $P^n \mathbf{C}$ we built in 2.30 by using Riemannian submersions is forced to be this natural metric.

2.45 Exercise: the Veronese surface.

Let $\mathbf{R}^3$ be equipped, in the canonical basis, with the standard Euclidean metric.

a) Let $v = (x, y, z)$ be a unit vector, and p_v be the orthogonal projection on $\mathbf{R}.v$. Prove that the matrix of p_v is

$$\begin{pmatrix} x^2 & xy & xz \\ xy & y^2 & yz \\ xz & yz & z^2 \end{pmatrix}.$$

b) Use the map $v \to p_v$ from S^2 to $\mathrm{End}(\mathbf{R}^3)$, and define an embedding V of $P^2 \mathbf{R}$ into $\mathrm{EndSym}(\mathbf{R}^3) \sim \mathbf{R}^6$ (this method yields in fact an embedding into $\mathbf{R}^5$, even in $\mathbf{R}^4$).

c) Prove that $P^2 \mathbf{R}$, seen as the quotient of S^2 by $\{Id, -Id\}$, is the homogeneous space $SO(3)/O(2)$. Explain what it means.

d) Let $A, B \in \text{End}(\mathbf{R}^3)$, $Q \in Gl(\mathbf{R}^3)$, set $\langle A, B \rangle = \text{tr}(^t AB)$ and $c_Q(A) = QAQ^{-1}$. Prove shortly that $\langle , \rangle$ is an Euclidean metric on $\text{End}(\mathbf{R}^3)$, and that if $Q \in SO(3)$, then c_Q is an isometry.

e) Prove that the Riemannian metric induced by this metric on $V(P^2\mathbf{R}) \sim P^2\mathbf{R}$ is proportional to the canonical one (coming from S^2). A very few computations are needed. What is the proportionality coefficient?

2.46 Back with Lie groups.

A Lie group can be viewed as an homogeneous space. There are two extremal cases: one can consider $G = G/\{e\}$, G acting on itself by left translations. Any scalar product on $\underline{G}$ gives rise to a left invariant metric on G. For example, there is on $G = SO(3)$ a 3-parameters family of such metrics, which are most important in solid mechanics (see [Ar] p.318).

One can also consider $G = (G \times G)/G$, where $G \times G$ acts on G by left and right translations. G is diagonally embedded in $G \times G$ as isotropy group, and the isotropy representation can be identified with the adjoint representation of G in $\underline{G}$. The bi-invariant metrics on G are then given by the $\text{Ad}G$-invariant scalar products on $\underline{G}$ (see I.D).

In particular, *any compact Lie group has at least one bi-invariant metric.*

2.47 Exercise.
Prove that for $G = O(n)$, $U(n)$ or $SU(n)$, the bilinear form $(X, Y) \to -\text{tr}(XY)$ on $\underline{G}$ (where X and Y are seen as endomorphisms of $\mathbf{R}^n$ or $\mathbf{C}^n$), is real, positive definite, and $\text{Ad}G$-invariant.

From 2.42, if the adjoint representation of G is irreducible, and if G is compact there exists, up to a scalar factor, a unique bi-invariant metric on G. *This can only be the case if G is compact, and if $\underline{G}$ is a simple Lie algebra (see [Hn]). One proves also that this is the case for $O(n)$ ($n \neq 4$), and $SU(n)$. Of course, $\text{tr}(XY)$ is in these two cases, proportional to the Killing form.*

For more details on the Riemannian metrics on homogeneous spaces and Lie groups, see [C-E], [d'A-Z] and [B2].

2.48 Geometric structures.

This has become the name for the compact quotients by a discrete isometry group of an homogeneous simply connected Riemannian manifold * (or, equivalently, of a locally homogeneous complete Riemannian manifold).* In dimension 2, geometric structures are isometric quotients of S^2, $\mathbf{R}^2$ and H^2, equipped with their (essentially unique, cf. 2.43) Riemannian metric. This follows (for example) from 3.82. It is more delicate (cf. III.L) to prove that any compact 2-manifold carries a geometric structure.

In dimension 3, besides S^3, $\mathbf{R}^3$ and H^3, the Riemannian products $S^2 \times \mathbf{R}$ and $H^2 \times \mathbf{R}$ define of course geometric structures. But there are three more models spaces, provided by some 3-dimensional Lie groups. This is beautifully explained in P. Scott's paper [Sc]. Lastly, let us mention Thurston's conjecture: any compact 3-manifold can be decomposed into pieces which carry a geometric structure (cf. [Sc] again for a precise statement, and of course [Th]).

2.B Covariant derivative

2.B.1 Connections

A vector field Y on $\mathbf{R}^n$ can be seen as a smooth map from $\mathbf{R}^n$ to itself. The derivative of Y at $m \in \mathbf{R}^n$ in the direction $v \in T_m\mathbf{R}^n \simeq \mathbf{R}^n$ is then the vector $T_mY \cdot v = d_mY \cdot v$. If X is another vector field, the derivative of Y at m in the direction X_m will be $T_mY \cdot X_m$. This yields, when m goes through $\mathbf{R}^n$, the vector field $dY(X)$.

It is clear that, for any $f \in C^\infty(\mathbf{R}^n)$, $dY(fX) = fdY(X)$, and $d(fY)X = df(X)Y + fd(Y)X$. Furthermore, an explicit computation (using coordinates) shows that $dY(X) - dX(Y) = [X, Y]$.

On a manifold, there is no natural way to define the directional derivative of a vector field. The job is done by connections.

2.49 Definition. If M is a smooth manifold, a *connection* on TM is an **R**-bilinear map from $\Gamma(TM) \times \Gamma(TM)$ to $\Gamma(TM)$, such that, for $X, Y \in \Gamma(TM)$ and $f \in C^\infty(M)$, we have

$$D_{fX}Y = fD_XY \quad \text{and} \quad D_X(fY) = (X.f)Y + fD_XY.$$

The first condition just says that, for any $m \in M$, $(D_XY)_m$ only depends on X_m (cf. 1.114). The second one is a Leibniz type property. This definition makes sense for vector bundle $E \to M$: a connection will be defined as a map from $\Gamma(TM) \times \Gamma(E)$ to $\Gamma(E)$ with the above properties. Coming back to the case when $E = TM$, we introduce the following symmetry property.

2.50 Definition. A connection on TM is *torsion-free* if, for $X, Y \in \Gamma(TM)$,

$$D_XY - D_YX = [X, Y].$$

Remark. For any connection on TM, it can be easily checked that the map

$$T : (X, Y) \to D_XY - D_YX - [X, Y]$$

defines a tensor. This tensor is called the *torsion* of the connection D.

2.51 Theorem. *There exists on any Riemannian manifold (M, g) a unique torsion-free connection consistent with the metric, i.e. such that, if X, Y, Z are vector fields on M,*

$$X.g(Y, Z) = g(D_XY, Z) + g(Y, D_XZ).$$

Proof. Recall that $X.g(Y, Z) = L_X(g(Y, Z))$ is the derivative of the function $m \to g_m(Y_m, Z_m)$ in the direction of the vector field X. Then

$$X.g(Y, Z) = g(D_XY, Z) + g(Y, D_XZ)$$
$$Y.g(Z, X) = g(D_YZ, X) + g(Z, D_YX)$$
$$Z.g(X, Y) = g(D_ZX, Y) + g(X, D_ZY)$$

Add the first two equalities, subtract the last one, and use the torsion-free condition. This yields

$$2g(D_X Y, Z) = X.g(Y, Z) + Y.g(Z, X) - Z.g(X, Y)$$

$$+g([X, Y], Z) - g([X, Z], Y) - g([Y, Z], X) \qquad (2.52),$$

and hence the *uniqueness* of the map D.

To prove the existence of D, check first (using the properties of the bracket) that the right side of (2.52) is $C^\infty(M)$-linear with respect to Z. From 1.114, the vector fields X and Y being given, this right side defines a $(0,1)$-tensor, and $D_X Y$ will be the vector field associated to this tensor by the metric g. The properties of a connection are easily checked, using 2.52. ∎

2.53 Definition. This connection is the *Levi-Civita connection,* or the *canonical connection,* of the metric g.

Example. The canonical connection of the Euclidean space is just $D_X Y = dY(X)$.

Remark. In the proof of the existence of the Levi-Civita connection, we only used the property for g to be non-degenerate. Therefore, this result works for any pseudo-riemannian metric. For example, the connection $D_X Y = dY(X)$ of $\mathbf{R}^n$ is the Levi-Civita connection of any pseudo-riemannian metric $\sum_{i,j} a_{ij} dx_i dx_j$ with constant coefficients, whatever the signature may be.

2.54 Proposition. *Let g be a Riemannian metric, and X, Y be two vector fields on a manifold M, defined respectively in local coordinates by*

$$g = \sum_{i,j} g_{ij} dx^i \otimes dx^j, \ \sum_{i=1}^n X^i \frac{\partial}{\partial x^i} \text{ and } \sum_{i=1}^n Y^i \frac{\partial}{\partial y^i}.$$

In this coordinate patch, we have

$$D_X Y = \sum_{i=1}^n \left(\sum_{j=1}^n X^j \frac{\partial Y^i}{\partial x^j} + \sum_{j,k=1}^n \Gamma_{jk}^i X^j Y^k \right) \frac{\partial}{\partial x^i},$$

where the functions Γ_{jk}^i are defined by the relation $D_{\frac{\partial}{\partial x^j}} \frac{\partial}{\partial x^k} = \sum_{i=1}^n \Gamma_{jk}^i \frac{\partial}{\partial x^i}$. The Γ_{jk}^i are the Christoffel symbols. *Setting $\partial_k g_{ij} = \frac{\partial}{\partial x^k} g_{ij}$, we have:*

$$\Gamma_{jk}^i = \frac{1}{2} \sum_{l=1}^n g^{il} \left(\partial_j g_{kl} + \partial_k g_{lj} - \partial_l g_{jk} \right).$$

Proof. Note first that $\Gamma_{jk}^i = \Gamma_{kj}^i$, since $[\frac{\partial}{\partial x^j}, \frac{\partial}{\partial x^k}] = 0$. The first part is a consequence of the general properties of connections. To get the explicit value of the Γ_{jk}^i, remark that (using 2.52),

$$2g \left(D_{\frac{\partial}{\partial x^j}} \frac{\partial}{\partial x^k}, \frac{\partial}{\partial x^l} \right) = \partial_j g_{kl} + \partial_k g_{lj} - \partial_l g_{jk}$$

The formula giving Γ_{jk}^i is then straightforward. ∎

This formula can sometimes be useful, but not so often. In this book, it is use only once, cf. 2.124. When an explicit computation of a connection is needed, the main technique is to compute it in terms of another one.

2.55 Exercise. a) Prove that the difference of two connections on a manifold is a *tensor*.

b) Let g be a Riemannian metric on M and $\tilde{g} = f^2 g$, where f is a never zero smooth function on M. Give the relation between the canonical connections $\tilde{D}$ of $\tilde{g}$, and D of g.

2.B.2 Canonical connection of a Riemannian submanifold

Let (M, g) be a Riemannian submanifold of $(\tilde{M}, \tilde{g})$, X be a vector field on M, and $m \in M$. Then there exists an open subset U of $\tilde{M}$ containing m, and a vector field $\tilde{X}$ defined on U, which restriction to $M \cap U$ is X: assume that M is n-dimensional, and is of codimension p in $\tilde{M}$. Let then (U, ϕ) be a chart for $\tilde{M}$ around m such that $\phi(U \cap M) = \mathbf{R}^n \times \{0\} \subset \mathbf{R}^{n+p}$. In this chart, let $X = \sum_{i=1}^n X^i \frac{\partial}{\partial x^i}$, where X^i is a smooth function on $U \cap M$. Just set

$$\tilde{X}\big(\phi^{-1}(x, y)\big) = \sum_{i=1}^n X^i\big(\phi^{-1}(x, 0)\big) \frac{\partial}{\partial x^i}.$$

2.56 Proposition. *Let D and $\tilde{D}$ be the canonical connections of (M, g) and $(\tilde{M}, \tilde{g})$ respectively. Let X, Y be two vector fields on M, and $\tilde{X}$, $\tilde{Y}$ be extensions of X and Y to an open subset $m \in U \subset \tilde{M}$. Then*

$$(D_X Y)_m = (\tilde{D}_{\tilde{X}} \tilde{Y})_m^\top,$$

where we denoted by $v^\top$ the orthogonal projection of $v \in T_m \tilde{M}$ on $T_m M$.
Proof. Let $\tilde{X}$, $\tilde{Y}$ and $\tilde{Z}$ be local extensions for the vector fields X, Y and Z on M. Then $\tilde{X} = \sum_{i=1}^{n+p} \tilde{X}^i \frac{\partial}{\partial x^i}$, with $\tilde{X}^i_{|U \cap M} = X^i$ for $i \leq n$, $\tilde{X}^i_{|U \cap M} = 0$ for $i > n$, and likewise for Y and Z. This shows that $([\tilde{X}, \tilde{Y}])_m = ([X, Y])_m$, and $\tilde{g}(\tilde{X}, \tilde{Y}) = g(X, Y)$ for any $m \in U \cap M$. Just apply 2.52 to both connections D and $\tilde{D}$ to find that

$$\tilde{g}(\tilde{D}_{\tilde{X}} \tilde{Y}, \tilde{Z}) = g(D_X Y, Z)$$

on $U \cap M$, and we are done. ∎

2.57 Exercises. a) Spherical coordinates on S^2 being defined by

$$(\theta, \phi) \to (\cos\theta \cos\phi, \cos\theta \sin\phi, \sin\theta),$$

compute $D_{\frac{\partial}{\partial\theta}} \frac{\partial}{\partial\theta}$, $D_{\frac{\partial}{\partial\phi}} \frac{\partial}{\partial\phi}$ and $D_{\frac{\partial}{\partial\theta}} \frac{\partial}{\partial\phi}$.

b) Consider the parametrizations of T^2 defined on $S^1 \times S^1$ by

$$\phi_1(\theta, \phi) = (e^{i\theta}, e^{i\phi}) \quad \text{and}$$

$$\phi_2(\theta, \phi) = \big((2 + \cos\theta)\cos\phi, (2 + \cos\theta)\sin\phi, \sin\theta\big).$$

The torus T^2 is hence equipped by two Riemannian submanifold structures (of $\mathbf{R}^4$ and $\mathbf{R}^3$ respectively). Compare successively in these two cases $\left[\frac{\partial}{\partial\theta}, \frac{\partial}{\partial\phi}\right]$ and $D_{\frac{\partial}{\partial\theta}}\frac{\partial}{\partial\phi}$.

2.B.3 Extension of the covariant derivative to tensors

The relation $D_X fY = (X.f)Y + fD_X Y$ shows a similarity between D_X and L_X, viewed as endomorphisms of the vector space $\Gamma(TM)$. And since D_X is something like a directional derivative, it is natural to set, for any smooth function on M, $D_X f = X.f$. The following statement is analogous to the characterization of the Lie derivative by its values on the functions and vector fields, and can be proved in the same way.

2.58 **Proposition.** *Let X be a vector field on M. The endomorphism D_X of $\Gamma(TM)$ has a unique extension as an endomorphism of the space of tensors, still denoted by D_X, which is type-preserving and satisfies the following conditions:*
i) *For any tensor $S \in \Gamma(T_q^p M)$ (with $p > 0$ and $q > 0$), and any contraction c on $T_q^p M$, then $D_X(c(S)) = c(D_X S)$.*
ii) *For any tensors S and T,*

$$D_X(S \otimes T) = D_X S \otimes T + S \otimes D_X T.$$

If for example, $S \in \Gamma(T_q^0 M)$, and the X_i are vector fields on M, then

$$(D_X S)(X_1, ..., X_q) = X.\big(S(X_1, ..., X_q)\big)$$

$$- \sum_{i=1}^{q} S(X_1, ..., X_{i-1}, D_X X_i, ..., X_q).$$

The reader should be aware that, in general,

$$(D_X S)(X_1, \dots, X_n) \neq D_X\left(S(X_1, \dots, X_n)\right)$$

(exercise). For convenience, we will often set $D_X S(\dots) = (D_X S)(\dots)$: be careful not to be confused!
In the case $S = g$, we get from the very definition of the Levi-Civita connection the following obvious, but fundamental, property:

2.59 **Proposition.** *Let D be the canonical connection of a Riemannian manifold (M, g). Then, for any vector field X on M, $D_X g = 0$.*

Now, since $D_{fX} = fD_X$, the quantity $(D_X S)_m$ depends only, for any tensor S, on the value of X at m. Hence, we get for any $m \in M$ a linear map from $T_m M$ to $(\otimes^p T_m M) \otimes (\otimes^q T_m^* M)$, hence a (smooth) section of the bundle $T_{q+1}^p M$, that is a $(p, q+1)$-tensor.

2.60 Definition. The *covariant derivative*, denoted by D, is the family of linear maps from $\Gamma(T_q^p M)$ to $\Gamma(T_{q+1}^p M)$ defined just above $(0 \leq p, q < \infty)$, namely

$$DS(X; X_1, \ldots, X_n) = D_X S(X_1, \ldots, X_n).$$

The label for D comes from the fact that, for any tensor S, DS has a covariance degree one more than S. It is important to compare the covariant derivative to the operators we defined previously, only using the differential structure of M: the Lie derivative and the exterior derivative.

2.61 Proposition. *Let (M, g) be a Riemannian manifold, and $(X, X_1, ..., X_p)$ be vector fields on M. Then:*
i) *For $S \in \Gamma(T_p^0 M)$,*

$$(L_X S)(X_1, ..., X_p) = (D_X S)(X_1, ..., X_p)$$

$$+ \sum_{i=1}^{p} S(X_1, ..., X_{i-1}, D_{X_i} X, ..., X_p).$$

ii) *For any exterior form α of degree p,*

$$d\alpha(X_0, X_1, ..., X_p) = \sum_{i=0}^{p+1} (-1)^i D_{X_i} \alpha(X_1, ..., X_{i-1}, X_0, X_{i+1}, ..., X_p).$$

Proof. Just come back to formulas 1.115 and 1.122, and replace the brackets $[U, V]$ by $D_U V - D_V U$. Note that the Riemannian structure is not directly used in that proof. ∎

2.62 Example: Killing fields.
If we apply to g the previous proposition, we get

$$(L_X g)(Y, Z) = g(D_Y X, Z) + g(Y, D_Z X).$$

A *Killing field* is a vector field X on (M, g) such that $L_X g = 0$. It amounts to the same to say that, for any $m \in M$, the endomorphism $(D_* X)_m$ of the Euclidean space $(T_m M, g_m)$ is skew-symmetric, or that the $(0, 2)$-tensor $DX^\flat$ (see 2.66) is antisymmetric.

2.63 Exercises. a) Compute $L_X g$ in local coordinates.
b) Show that a vector field X is a Killing field if and only if the local group associated to X consists in local isometries of (M, g), and that the bracket of two Killing fields is still a Killing field.
c) Show that, on $S^n \subset \mathbf{R}^{n+1}$, with the induced metric, the vector fields

$$x^i \frac{\partial}{\partial x^j} - x^j \frac{\partial}{\partial x^i}$$

$(0 \leq i, j \leq n)$ are Killing fields.

2.64 **Another example: the Hessian of a function.**
If $f \in C^\infty(M)$, 2.50 yields that the $(0,2)$-tensor Ddf is *symmetric*. In fact, this property is one of the main motivations for the property to be torsion-free. The tensor Ddf is called the *Hessian* of the function f. We insist on the fact that it is a Riemannian notion, and not merely a differential one.

2.65 **Exercises.** a) Give the expression of Ddf in a local chart.
b) Show that if $S^n \subset \mathbf{R}^{n+1}$ is equipped with its canonical metric g, and if f is the restriction to S^n of a linear form, then $Ddf = -fg$. State and prove a similar property for the hyperbolic space H^n (*hint:* give a geometrical proof, by noticing that for any Riemannian manifold, $2Ddf = L_{\nabla f}g$).

2.66 **Musical isomorphisms.**
In a Riemannian manifold (M, g), one can associate to any tangent vector $v \in T_m M$ the linear form on $T_m M$ defined by $x \to g_m(x, v)$. In the same way, working fiber by fiber, we can associate to any vector field $X \in \Gamma(TM)$ the 1-form ω defined by $\omega(Y) = g(X, Y)$. Let $\sum_{i=1}^n X^i \frac{\partial}{\partial x^i}$ be the expression of X in a local coordinate system $(x^1, ..., x^n)$: then the local expression for ω is $\sum_{i=1}^n \omega_i dx^i$, where $\omega_i = \sum_{j=1}^n g_{ij} X^j$. Therefore, we will denote by $\flat$ (flat) the map $X \to \omega$ so defined (the indices are lowered!). It is clearly an isomorphism. The inverse isomorphism will of course be denoted by $\sharp$ (sharp), and is given in local coordinates by

$$(\omega^\sharp)^i = \sum_{i=1}^n g^{ij} \omega_j,$$

where (g^{ij}) $(1 \le i, j \le n)$ denotes the inverse matrix of (g_{ij}).
For instance the vector $df^\sharp$ is the Riemannian analogue of the familiar *gradient*, and denoted in the same way by ∇f.
The same procedure can be applied to (p, q)-tensors, and shows that all the spaces $\Gamma(T_p^q M)$, where $p+q = r$ is given, are isomorphic. These isomorphisms will be used frequently in the following, very often implicitely. There will be no problem, since 2.59 clearly implies that the covariant derivative commutes with the musical isomorphisms $\flat$ and $\sharp$, and with their tensor powers.
Exercise. Show that the scalar product carried by $\flat$ on $T_m^* M$ is given in local coordinates by:

$$\langle \omega, \omega' \rangle = \sum_{i,j} g^{ij} \omega_i \omega_j'.$$

2.B.4 Covariant derivative along a curve

2.67 **Definition.** A *vector field along a curve* $c : I \subset \mathbf{R} \to M$ is a curve $X : I \to TM$ such that $X(t) \in T_{c(t)}M$ for any $t \in I$.

For example, $c'(t)$ is a vector field along c. A vector field along a curve is the formalized notion of infinitesimal deformation of the curve.

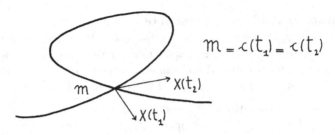

Fig. 2.5. A vector field along a curve

2.68 Theorem. *Let (M, g) be a Riemannian manifold, D be its canonical connection, and c be a curve of M. There exists a unique operator $\frac{D}{dt}$ defined on the vector space of vector fields along c, which satisfies to the following conditions:*

i) for any real function f on I,

$$\frac{D}{dt}(fY)(t) = f'(t)Y(t) + f(t)\frac{D}{dt}Y(t);$$

ii) if there exists a neighborhood of t_0 in I such that Y is the restriction to c of a vector field X defined on a neighborhood of $c(t_0)$ in M, then

$$\frac{D}{dt}Y(t_0) = (D_{c'(t_0)}X)_{c(t_0)}.$$

Proof. It would be pleasant to define $\frac{D}{dt}Y$ directly from the connection, using ii). Unfortunately, a vector field along a curve cannot always be extended to a neighborhood of the curve in M (for example when the curve has a dense image in M -compare to 1.4 e). But the local coordinate expression gives

$$Y(t) = \sum_{i=1}^{n} Y^i(t)\left(\frac{\partial}{\partial x^i}\right)_{c(t)},$$

and shows that a vector field along c is always a linear combination (with coefficients depending on t) of vector fields along c which can be extended. Hence, if $\frac{D}{dt}$ exists, condition ii) gives

$$\frac{D}{dt}Y(t) = \sum_j Y^j(t)\left(\frac{D}{dt}\frac{\partial}{\partial x^j}\right)(t) + \sum_i (Y^i)'(t)\left(\frac{\partial}{\partial x^i}\right).$$

If $c(t) = (x^1(t), ..., x^n(t))$ in local coordinates, we have

$$c'(t) = \sum_i (x^i)'(t)\left(\frac{\partial}{\partial x^i}\right)_{c(t)} , \text{ therefore}$$

$$\frac{D}{dt}\left(\frac{\partial}{\partial x^j}\right)_{c(t)} = D_{c'(t)}\frac{\partial}{\partial x^j} = \sum_k (x^k)'(t)D_{\frac{\partial}{\partial x^k}}\frac{\partial}{\partial x^i}$$

and
$$\frac{D}{dt}Y(t) = \sum_i \left[(Y^i)'(t) + \sum_{j,k}\Gamma^i_{jk}(x^j)'(t)Y^k(t)\right]\frac{\partial}{\partial x^i}. \qquad (2.69)$$

This proves the uniqueness of $\frac{D}{dt}$. Conversely, the operator defined by 2.69 satisfies

$$\frac{D}{dt}Y = \sum_j Y^j D_{c'}\frac{\partial}{\partial x^j} + \sum_i (Y^i)'(t)\left(\frac{\partial}{\partial x^i}\right).$$

Property i) is then immediate. To prove ii), it is sufficient to note that if f is a function defined on a neighborhood of $c(t)$ in M, then $c'.f = (f \circ c)'(t)$. Hence if Y is the restriction along c of a vector field X defined on a neighborhood of $c(t_0)$, and if $X = \sum_i X^i(\frac{\partial}{\partial x^i})$, we have:

$$\frac{D}{dt}Y(t) = \sum_j X^j(c(t))D_{c'}\left(\frac{\partial}{\partial x^j}\right) + \sum_i (c'(t).X^i)\left(\frac{\partial}{\partial x^i}\right) = (D_{c'(t)}X)_{c(t)}. \blacksquare$$

2.70 **Examples.** Note, using 2.56 and the previous theorem, that if (M, g) is a Riemannian submanifold of $(\tilde{M}, \tilde{g})$ and if X is a vector field tangent to M along a curve c drawn on M, then $\frac{D}{dt}X = \left(\frac{\tilde{D}}{dt}X\right)^{\top}$. Let us study the case of a great circle $c(t) = \cos t.u + \sin t.v$ of the sphere $S^n \subset \mathbf{R}^{n+1}$, equipped with the induced metric (we denoted by u a unit vector normal to S^n, and by v a unit tangent vector). For $X(t) = c'(t)$, we have

$$\frac{D}{dt}c'(t) = \left(\frac{d}{dt}c'(t)\right)^{\top} = 0.$$

In the general case, $X(t)$ can be written as

$$X^0(t)c'(t) + \sum_{i=1}^{n-1} X^i(t)w_i,$$

where (w_i) is a basis of the vector space of normal vectors to u and v, and

$$\frac{D}{dt}X(t) = (X^0)'(t)c'(t) + \sum_{i=1}^{n-1}(X^i)'(t)w_i.$$

The covariant derivative along a curve is much easier to compute than the connection. On the other hand, it determines entirely the connection (use property ii) of the theorem. if c is a curve through m such that $c'(0) = x \in T_m M$, then $(D_x Y)_m = \frac{D}{dt}(Y \circ c)(0)$.

2.B.5 Parallel transport

2.71 Definition. A vector field X along a curve c is *parallel* if $\frac{D}{dt}X = 0$.

2.72 Proposition. *Let c be C^1 curve defined on an interval I, and $t_0 \in I$. For any $v \in T_{t_0}M$, there exists a unique parallel vector field X along c such that $X(t_0) = v$.*

Proof. It is sufficient to prove the result for any compact subinterval $[t_0, T] \subset I$. Now there exists a finite subdivision of $[t_0, T]$, let $t_0 < t_1 < ... < t_n = T$, such that $c([t_i, t_{i+1}])$ is included in the domain of a chart (U_i, ϕ_i). The condition of being parallel is expressed in each chart by a first order linear differential system. Apply successively to the intervals $[t_i, t_{i+1}]$ the classical result of existence of uniqueness for such differential systems (see [Sp], t.1). Recall that solutions of *linear* equations can be extended to their whole interval of definition. ∎

Remark. The previous proposition can of course be extended to piecewise C^1 curves.

2.73 Definition. The *parallel transport* from $c(0)$ to $c(t)$ along a curve c in (M, g) is the linear map P_t from $T_{c(0)}M$ to $T_{c(t)}M$, which associates to $v \in T_{c(0)}M$ the vector $X_v(t)$, where X_v is the parallel vector field along c such that $X_v(0) = v$.

2.74 Proposition. *The parallel transport defines for any t an isometry from $T_{c(0)}M$ onto $T_{c(t)}M$. More generally, if X and Y are vector fields along c, then*

$$\frac{d}{dt}g(X(t), Y(t)) = g\left(\frac{D}{dt}X(t), Y(t)\right) + g\left(X(t), \frac{D}{dt}Y(t)\right).$$

Proof. The general properties of linear differential equations show that the maps P_t are vector spaces isomorphisms. The formula stated in the proposition, from which we deduce the fact that the P_t are isometries, is proved by the method employed in theorem 2.68. ∎

2.75 Example: the Euclidean space.

For any vector field along a curve c, $\frac{D}{dt}X(t) = \frac{d}{dt}(X(t))$ (obvious). The map P_t is just the translation from $c(0)$ to $c(t)$.

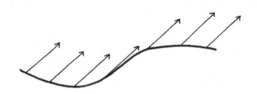

Fig. 2.6. Parallel transport in $\mathbf{R}^n$

It depends only on the extreme points of c. We will see later on that this property is very peculiar: it implies that the manifold is locally isometric to $\mathbf{R}^n$. The following example shows that the converse is false (see also 2.82).

2.76 Example: the revolution cone.

Assume that the vertex angle of the cone is 2α. The map ϕ which consists in developping the cone is an isometry, and hence preserves the covariant derivative. The parallel transport along $\gamma = \phi \circ c$ is the translation in $\mathbf{R}^2$.

Let $X(t)$ be a parallel vector field along a curve c turning around the cone. With respect to its initial condition $X(0)$, the vector $X(\alpha)$ has rotated of an angle β equal to $2\pi(1 - \sin\alpha)$, in the same direction as c.

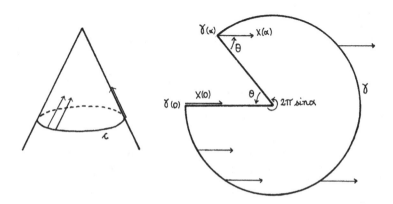

Fig. 2.7. Parallel transport on a revolution cone

Application. We are going to compute the result of the parallel transport along a small circle of the sphere S^2 (intersection of S^2 with a cone of revolution whose vertex at the center of the sphere, the vertex angle being 2θ). The vertex angle of the tangent cone to S^2 along c is $\pi - 2\theta$. The tangent planes to S^2 and this cone being the same along the curve, if $\frac{D}{dt}$ and $\frac{\nabla}{dt}$ are the covariant derivatives of S^2 and of this cone along c respectively, we have from 2.56,

$$\frac{D}{dt}X(t) = \frac{\nabla}{dt}X(t) = [X'(t)]^\top,$$

and hence $X(t)$ is parallel on S^2 if and only if it is parallel on the cone. We have just seen that this parallel transport is a rotation of angle $2\pi(1 - \cos\theta)$ in the same direction as c'.

2.B.6 A natural metric on the tangent bundle[1]

Let $p : TM \to M$ the canonical projection. It is a submersion, and the kernel of the tangent map at $x_m \in TM$ is an n-dimensional vector subspace of $T_{x_m}TM$, namely the subspace of *vertical vectors*. Putting theses subspaces alltogether, we get a vector subbundle of TTM, which is called the *vertical bundle* $\mathcal{V}M$. The inclusion $T_m M \subset TM$ defines, by taking its tangent map at x_m, a natural isomorphism between $T_m M$ and $\mathcal{V}_{x_m} M$.

The data of a connection gives a transverse subbundle to the vertical bundle. Indeed, a vector field along a curve in M is just a curve in TM. Let (c, X) be such a vector field, definined on $]-\epsilon, \epsilon[$. Suppose that $c(0) = m, X(0) = x_m$. The velocity at 0 of (c, X) defines a vector in $T_{x_m}TM$, whose projection on $T_m M$ is $c'(0)$. This vector will be called *horizontal* if this X is parallel along c. One infers from 2.72 that for any v_m and $x_m \in T_m M$, there is a unique horizontal vector in $T_{x_m}TM$ whose projection is v_m. We are going to us describe what happens using coordinates.

Let $(u_1, \cdots, u_n)$ be local coordinates for M, and $(u_1, \cdots, u_n, X_1, \cdot X_n)$ the corresponding local coordinates for TM. Lastly, let $(v_1, \cdots, v_n, \varXi_1, \cdot \varXi_n)$ the coordinates of a tangent vector at (u, X) with respect to the $\frac{\partial}{\partial u_k}$ and the $\frac{\partial}{\partial X_k}$. Such a vector will be vertical if the v_k are zero, and horizontal if

$$\forall i, \ \varXi^i + \sum_{j,k} \varGamma^i_{jk}(u_1, \cdots, u_n)v^j X^k = 0$$

It is clear that horizontal vectors form a rank n subbundle $\mathcal{H}$ of TM, which is transverse to the vertical bundle, so that $T_{x_m}p : T_{x_m}TM \to T_m M$ is an isomorphism.

Now, we can endow TM with a Riemannian metric g_T, in the following way.

- If $\varXi, \varXi' \in T_{x_m}TM$ are both vertical, their scalar product is the scalar product of the corresponding vectors in $(T_m M, g_m)$.
- If they are both horizontal, their scalar product is the scalar product of their projections onto $T_m M$.
- If one is vertical and the other horizontal, we decide they are orthogonal.

This metric is called the *Sasaki metric* on TM

Remark. The word horizontal has not been used by accident. Indeed, from its very definition, $p : (TM, g_T) \to (M, g)$, is a Riemannian submersion. For an explicit example, cf. 2.120.

2.C Geodesics

2.C.1 Definition, first examples

Let us consider a surface M embedded in $\mathbf{R}^3$, and equipped with the induced metric g. The curves drawn on this surface, provided with an arc-

---

[1] This passage may be skipped on first reading

length parametrization, and which are locally the shortest way between two points, are the curves with normal acceleration vector field (this comes from the Euler-Lagrange equations for the problem ([Sp], t.1, p.370,380)). Let $c : \mathbf{R} \to M \subset \mathbf{R}^3$ be such a curve, and D be the covariant derivative associated to g: this condition is just $D_{c'}c' = 0$.

We now imitate this property and introduce the notion of geodesic of an abstract Riemannian manifold. We will prove in 2.94 that they are actually locally shortest ways between two points.

2.77 Definition. Let (M, g) be a Riemannian manifold, and D be the covariant derivative associated to g. A parametrized curve on M is a *geodesic* if $D_{c'}c' = 0$ (or $\frac{D}{dt}c' = 0$).

In particular, the norm of $c'(t)$ is constant, since

$$\frac{1}{2}\frac{d}{dt}g\big(c'(t), c'(t)\big) = g\big(D_{c'}c', c'\big) = 0.$$

In local coordinates, the geodesics are the solutions of the differential system

$$(1 \le i \le n) \quad \frac{d^2x^i}{dt^2} + \sum_{j,k} \Gamma^i_{jk}(x(t))\frac{dx^k}{dt}\frac{dx^j}{dt} = 0, \qquad (2.78)$$

where $x(t) = (x^1(t), ..., x^n(t))$. The proof is obvious (2.69).

2.79 Theorem (local existence and uniqueness). *Let $m_0 \in M$. There exist an open set $m_0 \in U \subset M$, and $\epsilon > 0$ such that, for $m \in U$ and $v \in T_m M$ with $\mid v \mid < \epsilon$, there is a unique geodesic $c_v :] - 1, 1[\to M$ with $c_v(0) = m$ and $c'_v(0) = v$.*

We postpone the proof of this theorem to the next section, and proceed with a few examples.

2.80 Examples. a) The geodesics of $\mathbf{R}^n$ are the straight lines parametrized with constant velocity: indeed,

$$\frac{D}{dt}c'(t) = 0 \quad \text{iff} \quad c''(t) = 0 \quad \text{iff} \quad c(t) = x_0 + tv.$$

b) The geodesics of a Riemannian submanifold $M \subset \mathbf{R}^{n+k}$ are the curves with normal acceleration vector field. If indeed D is the connection of M and $\overline{D}$ is the connection of $\mathbf{R}^{n+k}$, for a curve c on M we have

$$c''(t) = \frac{\overline{D}}{dt}c'(t) = \overline{D}_{c'(t)}c'(t).$$

This yields the result since $D_{c'(t)}c'(t) = \big(\overline{D}_{c'(t)}c'(t)\big)^{\perp}$

c) The great circles of the sphere (S^n, can), parametrized proportionally to arclength, are geodesics. Let us consider (S^n, can) isometrically embedded in $\mathbf{R}^{n+1}$: for $p \in S^n$ and $u \in T_p S^n$, let c be the geodesic of (S^n, can) with initial conditions $c(0) = p$ and $c'(0) = u$. If P is the 2-dimensional vector space defined by the vector u and the point p, let s be the orthogonal symmetry with respect to P. The uniqueness theorem ensures that c is invariant under s, since $s \circ c$ and c are two geodesics (s is an isometry), satisfying the same initial conditions: hence c is the great circle $S^n \cap P$, parametrized with constant velocity. The uniqueness theorem ensures that all the geodesics of S^n are obtained in this manner.

d) One can prove similarly that the geodesics of H^n (2.10) are the curves obtained by intersection of a 2-dimensional subspace of $\mathbf{R}^{n+1}$. If a 2-plane P intersects H^n, the restriction of the quadratic form

$$\langle x, x \rangle = -x_0^2 + x_1^2 + \cdots x_n^2$$

is non degenerate, so that, denoting by $P^\perp$ the orthogonal of P for $\langle , \rangle$, we have

$$\mathbf{R}^{n+1} = P \bigoplus P^\perp$$

The linear transform s such that $s(x) = x$ for $x \in P$ and $s(x) = -x$ for $x \in P^\perp$ is an isometry of the Minkowski space which leaves H^n stable, so that its restriction to H^n is an isometry. Up from now, the arguments are the same as for the sphere.

2.80 bis Definition. A submanifold M of $(\tilde{M}, \tilde{g})$ is *totally geodesic* if, for any $m \in M$ and $v \in T_m M$, the geodesic c of $(\tilde{M}, \tilde{g})$ such that $c(0) = m$ and $c'(0) = v$ is contained in M.

The previous discussion shows that, if M is the fixed points set of some isometry of $(\tilde{M}, \tilde{g})$, it is totally geodesic.

2.81 Proposition. *Let $p : (N, h) \to (M, g)$ be a Riemannian covering map. The geodesics of (M, g) are the projections of the geodesics of (N, h), and the geodesics of (N, h) are the liftings of those of (M, g).*

Proof. The map p is a local isometry. Hence if γ is a geodesic of N, the curve $c = p \circ \gamma$ is also a geodesic of M. The uniqueness theorem 2.79 shows that they are the only geodesics on M. Conversely, if $p \circ \gamma$ is a geodesic of M, then γ is a geodesic of N. ■

2.82 Geodesics of flat tori, flat Klein bottles, real projective spaces.
a) Let $p : \mathbf{R}^n \to \mathbf{R}^n / \mathbf{Z}^n \sim T^n$ be the covering map. We have seen in 2.23 that two quadratic forms q_1 and q_2 define two isometric Riemannian structures on $\mathbf{R}^n$ which can induce on the quotient T^n two non isometric flat metrics.

However, the geodesics of any flat torus T^n are the projections of the straight lines in $\mathbf{R}^n$, parametrized proportional to length. If Γ and Γ' are two lattices in $\mathbf{R}^n$, there exists an affine map A which exchanges these lattices and gives by quotient a diffeomorphism $\bar{A}$ between $\mathbf{R}^n / \Gamma$ and $\mathbf{R}^n / \Gamma'$: $\bar{A}$ is generally not an isometry, but exchanges the geodesics of the corresponding flat tori.

b) The geodesics of a flat Klein bottle are the images of the straight lines in $\mathbf{R}^2$ by the covering map p (1.89). If the fundamental domain in $(\mathbf{R}^2, \text{can})$ of a flat Klein bottle (K, g) is given by the rectangle with side lengths a and b (the sides being identified as shown on the diagram 2.9), we can mention:

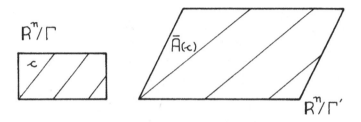

Fig. 2.8. Two flat tori

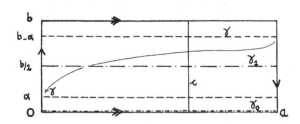

Fig. 2.9. Geodesics of a flat Klein bottle

i) the images under p of the vertical straight lines in $\mathbf{R}^2$ which give in (K, g) periodic geodesics of length $2b$ (geodesic c of figure 2.9).

ii) the images under p of the horizontal straight lines in $\mathbf{R}^2$ yield two exceptional periodic geodesics of length a (geodesics γ_0 and γ_1), or periodic geodesics of length $2a$ (geodesic γ).

Remark. If we build the Klein bottle by gluing together along their boundaries two rectangular Moebius bands, the two exceptional geodesics are the souls of these Moebius bands.

c) We denote the canonical metrics by can (2.21). The map $p : (S^n, \text{can}) \to (P^n\mathbf{R}, \text{can})$ is a Riemannian covering map (quotient by the antipodal map): the geodesics of $P^n\mathbf{R}$ are the projections of the geodesics of the sphere. If c is a geodesic of $P^n\mathbf{R}$, parametrized by arclength, we note that $c(t + \pi) = c(t)$, since $p(x) = p(-x)$: the geodesics of $P^n\mathbf{R}$ are periodic with period π.

2.83 Exercises. a) Describe the geodesics of the revolution cone in $\mathbf{R}^3$ whose vertex angle is α. Catch this cone with a lasso whose loop has length L. From which value of α will the lasso slip? (we assume the rope to be weightless, the cone without sliding friction, and that we pull downwards).

b) Describe the parallel transport along the geodesics γ, γ_1 and c of the flat Klein bottle (2.82 b)).

c) Describe the parallel transport along a geodesic on a 2-dimensional Riemannian manifold.

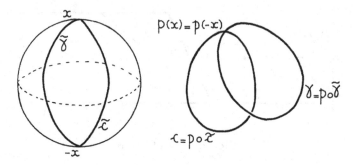

Fig. 2.10. Geodesics of a projective space

d) Let $P = \{z \in \mathbf{C}/\Im z > 0\}$ be equipped with the metric $g_z = \frac{dx^2+dy^2}{(\Im z)^2}$.

i) Compute the Christoffel symbols and write down the differential system satisfied by the geodesics.

ii) Prove that the curves $c : t \to (x_0, e^{at})$ are geodesics. Give a proof which does not use i). Hint: go on reading this section!

iii) Use the transitivity of the isometry group of (P, g) on the unit tangent bundle and prove that the geodesics of (P, g) are the circles and the straight lines normal to the x-axis.

e) Let M be a revolution surface in $\mathbf{R}^3$, endowed with the induced metric g. The meridian line being parametrized by the length u, and the angle of rotation being denoted by θ, the metric is given by

$$g = du^2 + a^2(u)d\theta^2,$$

where $a(u)$ is the distance to the axis (2.9). Show that the geodesics are:

i) the meridian lines,

ii) the parallels ($u = $ constant) for which $a'(u) = 0$,

iii) the curves which, when parametrized by length, satisfy

$$\left(\frac{du}{dt}\right)^2 + a^2(u(t)))\left(\frac{d\theta}{dt}\right)^2 = 1$$

$$a^2((u(t))\frac{d\theta}{dt} = C,$$

where C is a parameter associated to the geodesic.

The geodesics oscillate between two consecutive parallels satisfying $a(u) = C$ -except in the case one of these parallels is extremal (that is $a'(u) = 0$): our geodesic is then asymptotic to this parallel, which is itself a geodesic.

2.C.2 Local existence and uniqueness for geodesics, exponential map

Since we now turn to a local result, we can work in a local chart around m_0 where the geodesics are the solutions of the differential system 2.78. Theorem 2.79 is a consequence of a general theorem on differential equations in $\mathbf{R}^n$.

2.84 Theorem. *Let* $F : \mathbf{R}^n \times \mathbf{R}^n \to \mathbf{R}^n$ *be a smooth map. We consider the differential system*

$$\frac{d^2x}{dt^2} = F(x, \frac{dx}{dt}), \qquad (*)$$

where x is a map $x : I \subset \mathbf{R} \to \mathbf{R}^n$.
Then for each point (x_0, v_0) in $\mathbf{R}^n \times \mathbf{R}^n$, there exist a neighborhood $U \times V$ of this point and $\epsilon > 0$ such that, for $(x, v) \in U \times V$, the equation $(*)$ has a unique solution $x_v :] - \epsilon, \epsilon[\to \mathbf{R}^n$ with initial conditions $x_v(0) = x$ and $x'_v(0) = v$.
Moreover, the map $X : U \times V \times] - \epsilon, \epsilon[\to \mathbf{R}^n$ defined by $X(x, v, t) = x_v(t)$ is smooth (dependance on initial conditions).
Proof. See for example [Sp], t.1. ∎

2.85 Corollary: *For m_0 in M, there is a neighborhood U of m_0 and $\epsilon > 0$ such that, for $m \in U$ and $v \in T_mM$ with $| v | < \epsilon$, there exists a unique geodesic $c_v :] - 1, 1[\to M$ with initial conditions $c_v(0) = m$ and $c'_v(0) = v$.*
Moreover, the map $C : TU \times] - 1, 1[\to M$ defined by $C(v, t) = c_v(t)$ is smooth.
Proof. Particular case of the previous theorem. Note that if $t \to c_v(t)$ is a solution of $(*)$, then $t \to c_v(kt)$ yields another solution. ∎

In the following we will denote by c_v the maximal geodesic with initial conditions $c_v(0) = m$ and $c'_v(0) = v$ (where $v \in T_mM$). The set $\Omega \subset TM$ of vectors v such that $c_{v(1)}$ is defined is an open subset of TM containing the null vectors $0_m \in T_mM$ of all the fibres.

2.86 Definition. The *exponential map* $\exp : \Omega \subset TM \to M$ is defined by $\exp(v) = c_v(1)$. We denote by $\exp_m$ its restriction to one tangent space T_mM. From this very definition, we see that $T_m \exp_m = \mathrm{Id}_{T_mM}$.
The use of the word "exponential" is explained by the following:

2.87 Example. We now turn to the Riemannian manifold $(SO(n), g)$, where g is a bi-invariant metric (that is invariant under left and right translations). The map $\exp_{Id} : T_{Id}SO(n) \to SO(n)$ coincides with the matrix exponential map defined by

$$\exp_A = Id + A + .. + \frac{A^p}{p!} + ...$$

under the identification of $T_{Id}SO(n)$ with the vector space of (n, n) skew symmetric matrices (see 2.90).

2.88 Proposition. i) *The map* $\exp : \Omega \to M$ *is smooth,*
ii) *for* $m_0 \in M$, *the map* $\Phi : \Omega \to M \times M$ *defined by*

$$\Phi(v) = \left(\pi(v), \exp_{\pi(v)}(v) \right)$$

is a local diffeomorphism from a neighborhood W *of* 0_{m_0} *in* TM *onto a neighborhood of* (m_0, m_0) *in* $M \times M$.

Part ii) states that any two points which are close enough to each other are joined by a unique short geodesic.

Proof. The first assertion is an immediate consequence of 2.85.

The second one is a consequence of the inverse function theorem. in a local chart (U, ϕ) around m_0, the map $\Phi : TU \approx U \times \mathbf{R}^n \to M \times M$ is given by $\Phi(v = (x, u)) = (x, \exp_x v)$. Let us compute the jacobian matrix of Φ at $(m_0, 0)$: for m_0 fixed and t small enough we have $\Phi(m_0, vt) = (m_0, c_v(t))$, hence

$$T_{0_{m_0}} \Phi \left(\frac{\partial}{\partial u^i} \right) = \left(0, \frac{\partial}{\partial u^i} \right).$$

If on the contrary we let x vary with $u \equiv 0$, we get $\Phi(m, 0) = (m, m)$, and hence

$$T_{0_{m_0}} \Phi \left(\frac{\partial}{\partial x^i} \right) = \left(\frac{\partial}{\partial x^i}, \frac{\partial}{\partial u^i} \right).$$

Finally the jacobian matrix of Φ is (in the previous basis):

$$\begin{pmatrix} Id & 0 \\ Id & Id \end{pmatrix}.$$

This matrix being invertible, we deduce that Φ is a local diffeomorphism around 0_{m_0}. ∎

As a consequence, we note that if U *is the open set* $U = \Omega \cap T_{m_0} M$, *the map* $\exp_{m_0|U} = \Phi_{|U}$ *is a local diffeomorphism around* 0_{m_0}. ∎

2.89 Corollary. *For* $m_0 \in M$, *there exist a neighborhood* U *of* m_0 *in* M *and* $\epsilon > 0$ *such that:*

i) *for* $x, y \in U$, *there exists a unique vector* $v \in T_x M$ *with* $| v | < \epsilon$ *and* $\exp_x v = y$. *We denote by* c_v *the corresponding geodesic.*

ii) *This geodesic depends smoothly on the parameters, that is the map defined by* $C(x, y, t) = c_v(t)$ *is smooth.*

iii) *For* $m \in U$, *the map* $\exp_m$ *is a diffeomorphism between the ball* $B(0_m, \epsilon) \subset T_m M$ *and its image in* M.

Remark. Neglecting the parametrization, part i) of 2.89 just says that any two points of U are joined by a unique geodesic of length less than ϵ.

Proof. i) After restricting the neighborhood W defined in 2.88, we can assume that $\pi(W)$ has compact closure. Using the continuity of g_m as a function of m, we know that, for ϵ small enough, there exists an open subset $m_0 \in V \subset M$ such that $W' = \bigcup_{m \in V} B(0_m, \epsilon) \subset W$. Now the map Φ is a diffeomorphism from W' onto its image, which is an open set around (m_0, m_0) and then contains a product $U \times U$, where U is an open set containing m_0. Hence, for $(x, y) \in U \times U$, there exists a unique $v \in W'$ such that $\Phi(v) = (x, y)$, that is a unique $v \in T_x M$ with $\exp_x v = y$.

ii) Just notice that

$$c_v(t) = \exp_x(t\Phi^{-1}(x, y)) = C(x, y, t).$$

iii) Since $B(0_m, \epsilon) \subset W'$, Φ is a diffeomorphism from this ball on its image. But, for m fixed, $\Phi(v) = (m, \exp_m v)$ and hence $\exp_m$ is a diffeomorphism from $B(0_m, \epsilon)$ on its image. ∎

Remark. Refining a little the proof of 2.89, one can show (cf. [dCa 2]) that for any $m \in M$, there exists $\epsilon > 0$ such that any geodesic ball with center m and radius $R < \epsilon$ is geodesically convex (that is any two points of this ball are joined by a unique geodesic of length less than R, and this geodesic is contained in the ball). The real ϵ is the *convexity radius* at m. If the manifold is compact, one can chose an ϵ valid for all the points of the manifold. For example, the convexity radius of the standard sphere is $\frac{\pi}{2}$.

2.89 bis Normal coordinates.

The map $\exp_m : B(0, \epsilon) \to M$ is a diffeomorphism onto its image: it yields a local chart for M around m. Let $(e_1, ..., e_n)$ be an orthonormal basis of $T_m M$, and $(x_1, ..., x_n)$ be the associated local coordinates. They are called *normal coordinates*. Let us prove that

$$g\left(\frac{\partial}{\partial x^i}, \frac{\partial}{\partial x^j}\right)_{|m} = \delta_{ij} \quad \text{and} \quad \left(D_{\frac{\partial}{\partial x^i}} \frac{\partial}{\partial x^j}\right)_{|m} = 0$$

(these equalities are only valid at m).

Identifying the tangent space to $T_m M$ at $v \in T_m M$ with $T_m M$ itself, we have by construction $(T_{0_m} \exp_m).e_i = e_i$ since, if c_i is the geodesic $c_i(t) = \exp_m t e_i$,

$$\frac{d}{dt}(c_i(t))_{|t=0} = e_i = \frac{d}{dt}(\exp_m t e_i)_{|t=0} = T_{0_m} \exp_m \cdot e_i :$$

this yields the first assertion.

On the other hand, the geodesic c_v associated to the vector $v = \sum_{i=1}^n v^i e_i$ is given in local coordinates by $c_v(t) = (x^i(t) = t v_i)_{(i=1, \cdots n)}$. The differential equation satisfied by the geodesics (2.78) yields for any i:

$$\sum_{j,k} \Gamma^i_{j,k}(c_v(t)) v^j v^k = \frac{d^2 x^i}{dt^2} + \sum_{j,k} \Gamma^i_{j,k}(c_v(t)) \frac{dx^j}{dt} \frac{dx^k}{dt} = 0.$$

Hence at $t = 0$, we get for any i: $\sum_{j,k} \Gamma^i_{j,k}(m) v^j v^k = 0$, and this for any vector $v \in T_m M$: the Christoffel symbols at m are forced to be zero. Of course, we have also $\partial_k g_{ij} = 0$. As we claimed in the introduction of this chapter, Riemannian metrics does not admit any local invariant of order one.

2.90 Exercise: the exponential map on Lie groups.

Let G be a Lie group. A *bi-invariant metric* on G is a Riemannian metric $\langle,\rangle$ on G for which left and right translations are isometries.

a) Show that it is equivalent to have a bi-invariant metric on G or an $Ad(G)$-invariant scalar product on $\underline{G} = T_eG$. Recall that any compact Lie group has a bi-invariant metric (2.47). We will see that on the contrary, $Sl_n\mathbf{R}$ and $O(n,1)$ have no bi-invariant metrics (2.108, 3.86).

In the following, G will be a Lie group equipped with a bi-invariant metric $\langle,\rangle$. We denote by i the diffeomorphism of G defined by $i(h) = h^{-1}$.

b) Let $u \in T_eG$, and c be the maximal geodesic of G satisfying the initial conditions $c(0) = e$ and $c'(0) = v$. Compute D_ei, and deduce that for $|\,t\,|$ small enough $c(-t) = c(t)^{-1}$. Conclude that c is defined on $\mathbf{R}$ and that $c : (\mathbf{R},+) \to (G,.)$ is a group homomorphism.

c) Prove then that the geodesics of G are the integral curves of all left invariant vector fields, and that the maps exp (on the group) and Exp_e (on the Riemannian manifold) coincide.

d) Let D be the Levi-Civita connection associated with $\langle,\rangle$. For two left invariant vector fields X and Y, show that

$$D_XY = \frac{1}{2}[X,Y].$$

Let us now turn to a more involved example.

2.90 bis Exercise. Let H be the Heisenberg group, that is the multiplicative subgroup of $(3,3)$-matrices

$$\begin{pmatrix} 1 & x & z \\ 0 & 1 & y \\ 0 & 0 & 1 \end{pmatrix}.$$

a) Identifying H and $\mathbf{R}^3$ as manifolds, show that the vector fields

$$A = \frac{\partial}{\partial x}, \quad B = \frac{\partial}{\partial y} + x\frac{\partial}{\partial z}, \quad \text{and} \quad C = \frac{\partial}{\partial z}$$

are left invariant under the H group action. Compute their brackets.

b) We equip H with the Riemannian metric g such that for $m \in M$ the triple of vectors A_m, B_m and C_m is an orthonormal basis for T_mH. This metric is left invariant. Is it bi-invariant?

c) Express the action of the Levi-Civita connection associated to g on the vector fields A, B and C, deduce the equation satisfied by the geodesics and compute them explicitly. Are they translated of one parameter subgroups of H?

d) Let Z be the subgroup of matrices

$$\begin{pmatrix} 1 & 0 & z \\ 0 & 1 & 0 \\ 0 & 0 & 1 \end{pmatrix}.$$

Show that Z is a normal subgroup of H and that H/Z is group isomorphic to $\mathbf{R}^2$. Deduce a Riemannian submersion $p : (H, g) \to (\mathbf{R}^2, \text{can})$, with totally geodesic fibers. What are the horizontal geodesics of (H, g)?

2.C.3 Riemannian manifolds as metric spaces.

2.91 Definition-proposition. *Let (M, g) be a connected Riemannian manifold. We define, for $x, y \in M$, $d(x, y)$ as the infimum of the lengths of all piecewise C^1 curves from x to y. Then d is a distance on M, which gives back the topology of M.*

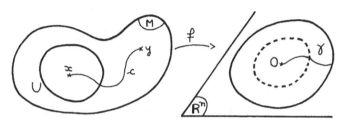

Fig. 2.11. Prop. 2.91: $d(x, y) \neq 0$ if $x \neq y$.

Proof. i) Let $x \in M$ and E_x be the set of the points of M joined to x by a piecewise C^1 curve. E_x is open in M (use charts!) and so is $E \backslash E_x = \bigcup_{(y \notin E_x)} E_y$. Since $x \in E_x$, E_x is non empty, hence $E_x = M$ and d is well defined.

ii) Since any curve can be parametrized backwards and forwards, we have $d(x, y) = d(y, x)$. The triangle inequality comes from the juxtaposition of piecewise C^1 curves from x to y, and from y to z, to get a piecewise C^1 curve from x to z.

iii) Plainly $d(x, x) = 0$. Let us prove the converse assertion. Let $x \neq y$. There exists a chart (U, f) around x with $y \notin U$ (M is Hausdorff), $f(x) = 0 \in \mathbf{R}^n$, and $f(U) = B(0, 1)$. Denote by h the metric induced by f on $B(0, 1)$ from g, and let $\| \cdot \|$ be the Euclidean norm in $\mathbf{R}^n$. Since $B(0, \frac{1}{2})$ is compact, there exist $\lambda, \mu > 0$ such that for $p \in B(0, \frac{1}{2})$ and $u \in T_pM$

$$\lambda \| u \|^2 \leq h_p(u, u) \leq \mu \| u \|^2 .$$

If c is a curve from x to y, using connectedness we see that c meets $f^{-1}(S(0, \frac{1}{2}))$ at a (first) point t. Denote by γ the curve of $\mathbf{R}^n$ defined by $\gamma = f \circ c_{|[0,t]}$. Then

$$L(c) = \int_0^t \left[g_{c(s)}(c'(s), c'(s)) \right]^{\frac{1}{2}} ds = \int_0^t \left[h_{\gamma(s)}(\gamma'(s), \gamma'(s)) \right]^{\frac{1}{2}} ds$$

$$= \int_0^t \sqrt{\lambda} \, \| \gamma'(s) \| \, ds \geq \sqrt{\lambda} \Big| \| \gamma(t) \| - \| \gamma(0) \| \Big|,$$

and hence $L(c) \geq \frac{\sqrt{\lambda}}{2}$ and $d(x, y) \geq \frac{\sqrt{\lambda}}{2} > 0$: d is actually a distance on M.

iv) We now prove that the topology of the manifold and the metric space M coincide. It is sufficient to show that on a family of domains of charts for M, the distance d and the distance induced by the chart from the Euclidean distance on $\mathbf{R}^n$ define the same topology: back with the notations of iii) we first show that, for $y \in M$ with $f(y) \in B(0, \frac{1}{2})$,

$$\sqrt{\lambda} \parallel f(y) \parallel \leq d(x, y) \leq \sqrt{\mu} \parallel f(y) \parallel$$

(this proves the bicontinuity at x of f). To this purpose, let us consider a curve c from x to y, parametrized on $[0, 1]$. If c is contained in $f^{-1}(B(0, \frac{1}{2}))$, we can consider $\gamma = f \circ c$, and then

$$L(c) = \int_0^1 [g_{c(s)}(c'(s), c'(s))]^{\frac{1}{2}} ds = \int_0^1 [h_\gamma(s)(\gamma'(s), \gamma'(s))]^{\frac{1}{2}} ds \geq \int_0^1 |\gamma'(s)| \, ds,$$

hence $L(c) \geq \sqrt{\lambda} \parallel \gamma(1) \parallel = \sqrt{\lambda} \parallel f(y) \parallel$.
If the curve c goes out $f^{-1}(B(0, \frac{1}{2}))$ at time t,

$$L(c) \geq \sqrt{\lambda} \parallel \gamma(t) \parallel = \frac{\sqrt{\lambda}}{2} \geq \sqrt{\lambda} \parallel f(y) \parallel .$$

On the other hand, if γ is the segment from $f(x)$ to $f(y)$, and if $c = f^{-1} \circ \gamma$, then

$$L(c) \leq \sqrt{\mu} \int_0^1 |\gamma'(s)| \, ds \leq \sqrt{\mu} \parallel f(y) \parallel . \blacksquare$$

The following theorem refines 2.85.

2.92 Theorem. Let $m_0 \in M$. There exist a neighborhood U of m_0 and $\epsilon > 0$ such that, for any $m, p \in U$, there is a unique geodesic c of length less than ϵ from m to p. More, $L(c) = d(m, p)$.

Proof. Consider the neighborhood U and the real ϵ defined in 2.89. The assertion on existence and uniqueness of c has already been proved. We have to check that for any other curve γ from m to p, $L(\gamma) \leq L(c)$, with equality if and only if c and γ coincide.
We will use the local chart around m defined by the diffeomorphism

$$\exp_m : B = B(0, \epsilon) \to B'.$$

Using polar coordinates in $T_m M$, we build the diffeomorphism

$$f :]0, \epsilon[\times S^{n-1} \to B' \setminus \{m\}$$

defined by $f(r, v) = \exp_m(rv)$. For $v \in S^{n-1}$ fixed, the curve $c_v : r \to f(r, v)$ is a geodesic.

2.93 Gauss lemma. *The curve c_v is normal to the hypersurfaces $f(\{r\} \times S^{n-1})$, that is g is given in "polar coordinates" by*

$$g = dr^2 + h_{(r,v)},$$

where $h_{(r,v)}$ is the metric induced by g at $f(r,v)$ on $f(\{r\} \times S^{n-1})$.
Be careful: g is not a product metric, since $h_{(r,v)}$ is a metric on S^{n-1} which depends on r.

Proof of the lemma. Let X be a vector field on S^{n-1}. Still denote by X the vector field defined on $B \setminus \{0\}$ by $X_{rv} = X_v$, and let $\tilde{X}$ be defined on B by $\tilde{X}_{rv} = r X_v$ ($v \in \mathbf{S}^{n-1}$). The vector field Y induced on $B' \setminus \{m\}$ from $\tilde{X}$ by f is

$$Y_{f(r,v)} = T_{(r,v)} f(0, X_v) = T_{r.v} \exp_m \cdot (r X_v).$$

To prove the lemma, it is sufficient to prove that the vector fields Y and $\frac{\partial}{\partial r}$ are everywhere orthogonal. We note that along c_v we have $\frac{\partial}{\partial r} = c'_v$ and hence, for v fixed,

$$\frac{d}{dr} g(Y, c'_v(r)) = g(D_{c'_v} Y, c'_v) + g(Y, D_{c'_v} c'_v)$$
$$= g(D_Y c'_v, c'_v) + g([c'_v, Y], c'_v) + g(Y, D_{c'_v} c'_v)$$
$$= 0$$

Indeed, we have $D_{c'_v} c'_v = 0$ and $g(D_Y c'_v, c'_v) = \frac{1}{2} Y.|c'_v|^2 = 0$, since c_v is a geodesic. Moreover,

$$D_{c'_v} Y - D_Y c'_v = [c'_v, Y] = f_*[\frac{\partial}{\partial r}, (O, X_v)] = 0$$

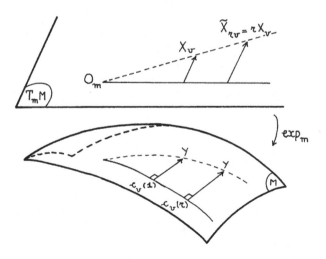

Fig. 2.12. Gauss lemma

Setting $a(r) = g(Y, c'_v(r))$, we have

$$a'(r) = \frac{1}{r}a(r), \quad \text{with} \quad a'(0) = g(f_*X, c'_v(0)) = 0,$$

for the exponential is isometric at the origin, and $\frac{d}{dr}$ is normal to X. Hence $a(r) \equiv 0$ and we are done. ∎

End of the proof of the theorem.

Recall that a change of parametrization does not change the length of a curve. Let $c : [0,1] \to U$ be the unique geodesic of length less than ϵ from m to p, and let $\gamma : [0,1] \to M$ be another curve from m to p.

i) If γ goes out the ball B' at a first point s, then $L(\gamma) > \epsilon$: let us indeed work in polar coordinates and set, for $0 \leq t \leq s$ $\gamma = f(r(t), v(t))$. Then

$$L(\gamma) > \int_0^s \left[g_{\gamma(t)}(\gamma'(t), \gamma'(t)) \right]^{\frac{1}{2}} dt \geq \int_0^s | r'(t) | \, dt \geq \epsilon.$$

ii) If γ is contained in B', we write in an analogous way:

$$L(\gamma) \geq \int_0^1 \left[r'(t)^2 + h_{(r(t),v(t))}(\gamma'(t), \gamma'(t)) \right]^{\frac{1}{2}} dt \geq r(1) - r(0) = L(c),$$

with equality if and only if the function r is monotonous, and if $\frac{dv}{dt} \equiv 0$. If we assume that γ is parametrized with constant velocity l and $L(\gamma) = L(c)$, we get $\gamma(t) = f(lt, v)$ and hence $\gamma \equiv c$. ∎

2.94 Corollary. *A curve $c : I \subset \mathbf{R} \to M$, parametrized proportionally to arclength, is a geodesic if and only if for any $t \in I$ there exists $\epsilon > 0$ such that $d\big(c(t), c(t + \epsilon)\big) = L(c_{|[t,t+\epsilon]})$.*

Proof. For $t \in I$, let us consider the neighborhood U of $c(t)$ defined in 2.93. Inside this neighborhood, the minimizing curves, parametrized with constant velocity, are exactly the geodesics. Hence if c is a geodesic, it locally minimizes the distance to $c(t)$. Conversely, if c is minimal between $c(t)$ and $c(t + \epsilon) \in U$, c is a geodesic between t and $c(t + \epsilon)$. This is valid for any t, hence c is a geodesic. ∎

2.94 bis Remark. The previous corollary ensures that geodesics are locally distance minimizing. This property is generally not global (for example, the geodesics of S^n are only minimizing before they come to the antipodal point). A globally distance minimizing geodesic called a *minimal geodesic*. The following alternative terminology is suggestive: a minimal geodesic is called a *segment* if it is defined on a compact interval, a *half-line* if it is defined on non-bounded interval but not on $\mathbf{R}$, a *line* if it is defined on $\mathbf{R}$.

Here is an example:

2.95 Proposition. *Let (M, g) be a complete non-compact Riemannian manifold, and $m \in M$. Then there exists a half-line from m.*

Proof. Let $p_n \in M$ be a sequence of points going to infinity. Since (M, g) is complete, there exists a minimal geodesic from m to p_n, let $\gamma_n : [0, t_n] \to M$, with $\gamma_n(t) = \exp_m(tv_n)$ (where $v_n \in S_m$ –the unit sphere in $T_m M$, and $t_n \to \infty$). Now S_m is compact, so that there exists a converging subsequence $v_n \to v \in S_m$. We claim that $\gamma : [0, \infty[\to M$ is a minimal geodesic: this is a straightforward consequence of the continuity of $\rho : S_m \to]0, \infty]$ (see 2.111, 2.112). ∎

2.96 Definition. The *energy* of a piecewise C^1 curve $c : [a, b] \to M$ is defined by

$$E(c) = \frac{1}{2} \int_a^b |c'(t)|^2 \, dt.$$

This quantity, unlike the length, depends on the parametrization.

2.97 Proposition. *Let $c : [a, b] \to M$ be a piecewise C^1 curve.*
i) *If c is minimal (that is $L(c) = d(c(a), c(b))$), and if c is parametrized proportionally to arclength, then c is a geodesic.*
ii) *If for any curve γ from $c(a)$ to $c(b)$ we have $E(\gamma) \geq E(c)$, then c is a geodesic (in particular, c is parametrized proportionally to arclength).*
Proof. i) Let $t_1, t_2 \in [a, b]$. If γ was a curve shorter than c from $c(t_1)$ to $c(t_2)$, then the curve α defined by

$$\alpha(t) = c(t) \quad \text{for} \quad t \in [a, t_1] \cup [t_2, b] \quad \text{and} \quad \alpha(t) = \gamma(t) \quad \text{otherwise},$$

would be shorter than c. Hence if c is distance minimizing between $c(a)$ and $c(b)$, it is also locally minimizing, and 2.96 yields the result.

ii) Let $\gamma : [k, l] \to M$ be a curve from $c(a)$ to $c(b)$. Cauchy-Schwarz inequality yields

$$2E(\gamma) = \int_k^l |\gamma'(t)|^2 \, dt \geq \left[\int_k^l |\gamma'(t)| \, dt \right]^2 \left(\int_k^l dt \right)^{-1},$$

hence $E(\gamma) \geq \frac{L(\gamma)^2}{2(k-l)}$, with equality if and only if $|\gamma'(t)|$ is constant.

Fig. 2.13. Proposition 2.97

Hence if the curve c is energy minimizing between $c(a)$ and $c(b)$, it is parametrized proportional to arclength. More, if $c_0 : [a, b] \to M$ is a curve from $c(a)$ to $c(b)$, parametrized proportionally to arclength, we have

$$L(c_0) = \left[2E(c_0)(b - a)\right]^{\frac{1}{2}} \geq \left[2E(c)(b - a)\right]^{\frac{1}{2}} = L(c),$$

then $L(c) = d\big(c(a), c(b)\big)$, and c is a geodesic. ■

2.C.4 An invitation to isosystolic inequalities

We are already in a position to have a feeling about some deep questions in Riemannian geometry.

2.98 Theorem. *On a compact Riemannian manifold, there exists, in any non trivial free homotopy class C, a smooth and closed geodesic c, whose length is minimal in C.*

Proof. Recall that a free homotopy is a continuous map $H : S^1 \times [0, 1] \to M$. Since M is compact, there exists $\epsilon > 0$ such that for any $m \in M$ the map $\exp_m$ is a diffeomorphism from the ball $B(0_m, \epsilon)$ onto its image (2.85).
Let l be the infimum of the lengths of all the curves $\gamma \in C$. Since M is compact, l is non zero. Indeed, equip C with the compact open topology. Then the function which to any curve assign its length (possibly infinite) is lower semi-continuous, and achieves its infimum. If this infimum were zero, C would be a trivial class. Let $(c_j)_{j \in \mathbf{N}}$ be a family of curves in the class C, parametrized on $[0, 1]$ with constant velocity, and such that $\lim L(c_j) = l$ ($j \to \infty$). Let $0 = t_0 < \ldots < t_p = 1$ be a subdivision of $[0, 1]$ with $t_{i+1} - t_i < \frac{\epsilon}{l}$. Using 2.89, we can assume that the c_j are geodesics segments on the intervals $[t_i, t_{i+1}]$. Since M and S^{n-1} ($n = \dim M$) are compact, we can assume (take a subsequence of (c_j) if necessary) that the sequences $(c_j(t_i))$ and $(c'_j(t_i))$ ($j \in \mathbf{N}$) are converging. Hence the curves (c_j) converge to a limit c, with $c \in C$ and $L(c) = l$. Since c is of minimal length in its class, we can not shorten it locally, and 2.97 yields that c is a smooth geodesic. ■

Remark. Do not mistake closed (or periodic) geodesics (these expressions are synonymous) for geodesic loop, which are geodesics $c : [a, b] \to M$ such that $c(a) = c(b)$, without any further assumption: that is, $c'(a) \neq c'(b)$ in general.

2.99 Definition. The *systole* $\mathrm{sys}(M, g)$ of a compact non simply connected (M, g), is the greatest lower bound of $\mathrm{length}(c)$, for all the closed geodesics which are not homotopic to a point.
The length of a closed geodesic is bigger the half of the injectivity radius of (M, g) (cf. 2. 116 for the definition) therefore $\mathrm{sys}(M, g) > 0$.
If you think of an embedded torus in the Euclidean space, it is intuitively clear that, if the systole is given, the area cannot be to small.

2.100 Theorem (Loewner). *For any Riemannian metric* g *on* T^2,

$$\frac{\text{Area}(g)}{\text{sys}^2(g)} \geq \frac{\sqrt{3}}{2},$$

and equality is achieved if and only if g *is isometric to the flat equilateral torus.*

The proof is a consequence of the following lemma.

Lemma. *Let* g_0 *a flat metric on* T^2, *and* $f^2 g_0$ *a conformal metric. Then*

$$\frac{\text{Area}(f^2 g_0)}{\text{sys}^2(f^2 g_0)} \geq \frac{\text{Area}(g_0)}{\text{sys}^2(g_0)},$$

and equality is achieved if and only if f *is constant.*

Proof. Let (e_1, e_2) be basis of the lattice defining g_0. Denote by a, b the lengths of the vectors, by α their angle, and by P the parallelogram $\{se_1 + te_2, 0 \leq s, t \leq 1\}$. Then

$$\text{Area}(f^2 g) = \int_P f^2 \sin \alpha \, ds dt = \int_0^a \sin \alpha \left(\int_0^b f^2 dt \right) ds.$$

Now, for any s the curve $t \mapsto se_1 + te_2$ gives in the quotient a closed curve which is not homotopic to a point, so that

$$\int_0^b f(s, t) dt \geq b \, \text{sys}(f^2 g_0).$$

Therefore, using Schwarz inequality, we infer that

$$\text{Area}(f^2 g_0) \geq \int_0^a \frac{\sin \alpha}{b} \left(\int_0^b f(s, t) dt \right)^2 ds \geq \frac{\text{Area}(g_0)}{b^2} \text{sys}^2(f^2 g_0).$$

The claimed inequality follows, since the above estimate works for any basis of the lattice. ∎

Proof of the theorem (sketchy).* Any Riemannian metric on T^2 is isometric to some $f^2 g_0$, where g_0 is flat. Indeed, on an oriented surface, the data of a conformal class of metric is equivalent to the data of a complex structure : both are given by the rotation of angle $+\pi/2$ on each tangent plane. The uniformization theorem (cf. [J2], ch.4) says then that any complex structure on T^2 is given by $\mathbf{C}/\Lambda$, where Λ is a lattice.* This result can also be proved by using real (non-linear) analysis (Melvin Berger, [Bg]).

We are now reduced the find the minimum of $\frac{\text{Area}(g_0)}{\text{sys}^2(g_0)}$ among flat metrics g_0. Using the classification of flat 2-tori, i.e. the metric classification of lattices of the Euclidean plane (cf. 2.24), we see that the minimum is achieved for the hexagonal lattice. ∎

This is just the point of departure of a very rich theory. The general question is: given a compact non simply connected manifold M^n, does there exist a positive constant $c(M)$ such that, for any metric g on M, $\mathrm{vol}(g) \geq c(M)\mathrm{sys}^n(g)$ ("isosystolic inequality"). We have just seen that for T^2 the answer is yes, and the optimal constant is known. There is a similar result for $P^2\mathbf{R}$: the "best metric" is the canonical one. The proof uses the very same idea of sweeping with geodesics. This argument also works for the Klein bottle (C. Bavard, [Bv]), but the situation is more involved: the infimum of $\mathrm{Area}(g)/\mathrm{sys}^2(g)$ is achieved for a singular metric.

There is also an isosystolic inequality for surfaces of higher genus, but the best constant is not known; extremal metrics are known to be singular (cf. the article [Ci] of E. Calabi for details).

In higher dimension, it is easy to find manifolds for which such an inequality cannot hold: just take $S^1 \times S^2$, equip it with a product metric, and make the factor S^2 very small. M. Gromov has singled out a class of manifolds, which satisfies an isosystolic inequality. This class includes tori and real projective spaces. His arguments are very involved, and the best constant is far from being known, even for T^3. For details, see [Gr4] and [Br4].

The manifolds for which the infimum of $\mathrm{vol}(g)/\mathrm{sys}^n(g)$ is zero have been characterized by I. Babenko (cf. also [Bb], and the appendix by M. Katz in [Gr1] for the story of higher dimensional systoles).

2.C.5 Complete Riemannian manifolds, Hopf-Rinow theorem

2.101 Definition. A Riemannian manifold (M, g) is *geodesically complete* if any geodesic of M can be extended to a geodesic defined on all $\mathbf{R}$.

2.102 Examples. a) A Riemannian manifold, with a point taken away, is not complete.

b) The half-plane $P = \{(x, y) \in \mathbf{R}^2/y > 0\}$, equipped with the metric induced by the Euclidean metric of $\mathbf{R}^2$ is not complete. It becomes complete when equipped with the metric $\frac{1}{y^2}(dx^2 + dy^2)$ (2.107 b). This kind of argument can be used to prove that any manifold carries a complete Riemannian metric.

2.103 Theorem (Hopf-Rinow). *Let (M, g) be a Riemannian manifold.*

i) *Let $m \in M$. If the map $\exp_m$ is defined on the whole $T_m M$, then any point of M can be joined to m by a minimal geodesic.*

ii) *If the manifold is geodesically complete, any two points of M can be joined by a minimal geodesic.*

Remark. The minimal geodesic in the theorem is of course not unique (take for example the geodesics joining two antipodal points of the sphere). On the other hand, the converse of this theorem is false: on an open canonical half-sphere, any two points are joined by a -unique- minimal geodesic, but this manifold is not geodesically complete.

Proof. The first assertion implies the second one. Denote by d the distance induced on M by the metric g. We first prove the following:

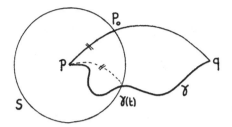

Fig. 2.14. Hopf-Rinow: across the distance-sphere

2.104 Lemma. *Let $p, q \in M$, and S be the sphere of radius δ and center p in (M, d). For δ small enough, there exists $p_0 \in S$ such that:*

$$d(p, p_0) + d(p_0, q) = d(p, q).$$

Proof. Theorem 2.92 ensures that if δ is small enough, we have $S = \exp_p(S(0_p, \delta))$, where $S(0_p, \delta)$ is the Euclidean sphere of radius δ in T_pM. The sphere S being compact, there exists a point $p_0 \in S$ such that $d(p_0, q) = d(S, q)$.

Now, let γ be a curve from p to q. If $\delta < d(p, q)$, the curve γ meets the sphere S at a point $\gamma(t)$, and

$$L(\gamma) \geq d(p, \gamma(t)) + d(\gamma(t), q) \geq d(p, p_0) + d(p_0, q),$$

and hence $d(p, q) \geq d(p, p_0) + d(p_0, q)$. The reversed inequality comes from the triangle inequality. ∎

Proof of the theorem. Let $m, q \in M$, and assume that $\exp_m$ is defined everywhere on T_mM. We are looking for a minimal geodesic from m to q. Use the previous lemma: for δ small enough, there exists $m_0 \in M$ such that

$$d(m, m_0) = \delta \quad \text{and} \quad d(m, m_0) + d(m_0, q) = d(m, q).$$

Let $v \in T_mM$ be such that $\exp_m y = m_0$, and let $c(t) = \exp_m(tv)$. The curve c is a geodesic defined on $\mathbf{R}$.

Let $I = \{t \in \mathbf{R} / d(q, c(t)) + t = d(m, q)\}$: we know that $\delta \in I$. Let $T = \sup(I \cap [0, d(m, q)])$: $T \in I$ since I is closed, and we want to prove that $T = d(m, q)$. Assume indeed that $T < d(m, q)$, and use the previous lemma for the points $c(T)$ and q: there exists $\epsilon > 0$ and $p_0 \in M$ such that $d(c(T), p_0) = \epsilon$ and

$$d(p_0, q) = d(c(T), q) - d(c(T), p_0) = d(m, q) - T - \epsilon,$$

and hence

$$d(m, p_0) \geq d(m, q) - d(q, p_0) \geq T - \epsilon.$$

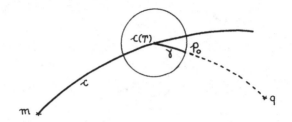

Fig. 2.15. Hopf-Rinow: minimizing the distance form p to q

Let γ be the minimal geodesic from $c(T)$ to p_0: we have

$$L(c)_{|[0,T]} + L(\gamma) = d(m, p_0).$$

Now we know by 2.97 that γ extends c, and that $p_0 = c(T + \epsilon)$. Hence $d(m,q) \in I$, that is $d(c(m,q)) = 0$. ∎

2.105 Corollary. *Let (M, g) be a connected Riemannian manifold. The following are equivalent:*
i) *(M, g) is geodesically complete,*
ii) *for any $m \in M$, the map $\exp_m$ is defined everywhere on $T_m M$,*
iii) *there exists $m \in M$ such that the map $\exp_m$ is defined everywhere on $T_m M$,*
iv) *any closed and bounded subset of (M, d) is compact (d is the metric induced by g on M),*
v) *the metric space (M, d) is complete.*
Proof. The implications i) $\Rightarrow$ ii) $\Rightarrow$ iii) are obvious.
Let us show that iii) $\Rightarrow$ iv): let $K \subset M$ be closed and bounded and assume that the map $\exp_m$ is defined on $T_m M$. Then 2.104 shows that $K \subset \exp_m (B(0_m, R))$ for $R > d(m, K)$. Since

$$K' = \exp_m^{-1}(K) \cap \overline{B}(0_m R)$$

is closed and bounded, hence compact in $T_m M$, and $K = \exp_m K'$, then K is also compact.
iv) $\Rightarrow$ v). A Cauchy sequence in (M, d) is bounded, hence contained in a compact ball of M: therefore it has exactly one cluster value and the sequence is convergent.
v) $\Rightarrow$ i). Let c be a geodesic of (M, g), parametrized by arclength. We want to prove that its domain of definition I is open and closed in $\mathbf{R}$. The local existence and uniqueness theorem ensures that if $c(T)$ is defined, then $c(T+t)$ is also defined for t small enough: hence I is open.
Now let (t_n) be a sequence of elements of I, converging to t: we want to prove that $t \in I$. Since

$$d(c(t_p), c(t_q)) \leq | t_p - t_q |,$$

the sequence $(c(t_n))$ is Cauchy and converges to a limit $m \in M$. Let U be the neighborhood of m defined in 2.89: there exists $\epsilon > 0$ such that any normal geodesic starting inside U is defined at least on $]-\epsilon, \epsilon[$. Conclude by chosing t_n with $|t_n - t| < \frac{\epsilon}{2}$ and $c(t_n) \in U$: the geodesic c is defined till time $(t + \frac{\epsilon}{2})$, and hence $t \in I$: I is open. ∎

2.105 bis Some remarks about non-complete Riemannian manifolds. Let (M, g) a non-complete Riemannian manifold. Then M, viewed as a metric space, admits a metric completion $\overline{M}$. There are of course obvious examples. On $\mathbf{R}^2 \setminus \{0\}$, the flat metric $dr^2 + r^2 d\theta^2$ is not complete. Of course, polar coordinates yield a singularity for $r = 0$, but this is a fake singularity: taking cartesian coordinate, we can extend the metric to the Euclidean metric on $\mathbf{R}^2$. Another example is

$$\left(\mathbf{R}^2, \frac{4(dx^2 + dy^2)}{(1 + x^2 + y^2)^2}\right).$$

Using a stereographic projection (compare with 2.11) it can be proved that this Riemannian manifold is isometric to the standard sphere with one point removed: the metric completion is nothing but the standard sphere.

In general, the metric completion will not be a manifold. To see that, consider any singular compact submanifold of the Euclidean space, and equip the regular part with the induced metric. A simple example is the half-cone of revolution $\cot^2 \alpha(x^2 + y^2) - z^2 = 0, z > 0$ in $\mathbf{R}^3$. Using the parametrization $(r, \theta) \mapsto (r \cos\theta, r \sin\theta, r \cot\alpha)$, the metric is given by $(1 + \cot^2 \alpha)dr^2 + r^2 d\theta^2$. Here the metric completion is certainly not Riemannian. Indeed, the length of small geodesic circle of radius ρ whose center is the vertex is $2\pi\rho \sin\alpha$: it is not equivalent to its Euclidean analogue, in contradiction with 2.89 bis.

It is worth-noting that singular Riemannian metrics have not been much studied.

The following result will be useful for the proof of Hadamard-Cartan theorem (3.87).

2.106 Proposition. *Let $p : (M, g) \to (N, h)$ be a local isometry. If (M, g) is geodesically complete, then p is a Riemannian covering map.*

Proof. Let $m \in N$ and $p^{-1}(m) = \{\overline{m}_i, i \in I\}$. Let r be such that $\exp_m : B(0_m, r) \to B(m, r) = U$ (ball of center m and radius r for the geodesic distance d on M), and $U_i = B(\overline{m}_i, r)$.

Clearly, $\bigcup_{i \in I} U_i \subset p^{-1}(U)$. Fix $i \in I$, take $w \in T_{\overline{m}_i}M$, $v = D_{\overline{m}_i}p.w \in T_m M$, and let γ and $\overline{\gamma}$ be the geodesics respectively issued from m and $\overline{m}_i$, satisfying the initial conditions $\gamma'(0) = v$ and $\overline{\gamma}'(0) = w$. Since p is a local isometry, $\gamma = p \circ \overline{\gamma}p$, so that we have a commutative diagramm

$$
\begin{array}{ccc}
B(0_{\overline{m}_i}, r) & \xrightarrow{\exp_{\overline{m}_i}} & B(\overline{m}_i, r) = U_i \\
{\scriptstyle D_{\overline{m}_i}p} \downarrow & & \downarrow {\scriptstyle p} \\
B(0_m, r) & \xrightarrow{\exp_m} & B(m, r) = U
\end{array}
$$

Now the map $\exp_m \circ D_{\overline{m}_i} p$ being a diffeomorphism between $B(0_{\overline{m}_i}, r)$ and U, the map

$$p : \exp_{\overline{m}_i}(B(0_{\overline{m}_i}, r)) \to U$$

is also a diffeomorphism, and we conclude by noticing that, with (2.103) and since (M, g) is geodesicaly complete,

$$\exp_{\overline{m}_i}(B(0_{\overline{m}_i}, r)) = B(\overline{m}_i, r) = U_i.$$

We still have to prove that $p^{-1}(U) \subset \bigcup_{i \in I} U_i$. Let $\overline{q} \in p^{-1}(U)$ and $q = p(\overline{q})$. Let $\gamma : [0, s] \to N$ be the unique minimal geodesic from q to m, $v = \gamma'(0) \in T_q M$ and $w \in T_{\overline{q}} M$ be the unique vector such that $D_{\overline{q}} p.w = v$. Let $\overline{\gamma}(t) = \exp_{\overline{q}}(tw)$: this geodesic is defined on $\mathbf{R}$ since (M, g) is complete. On the other hand, $p \circ \overline{\gamma} = \gamma$ and hence $p \circ \overline{\gamma}(s) = \gamma(s) = m$: then there exists $i \in I$ such that $\overline{\gamma}(s) = \overline{m}_i$, and $\overline{q} \in B(\overline{m}_i, r) = U_i$. ∎

Remark. The proof yields that (N, h) is also geodesically complete.

2.107 Counter-example. The canonical projection from S^n onto $P^n\mathbf{R}$, restricted to the sphere with a point taken away, is a local isometry (for the canonical metrics), but is not a covering map.

2.108 Exercises. a) Show that if (M, g) is complete and connected, and if (N, h) is connected and simply connected, any map $f : M \to N$ such that, for $m \in M$, the map $T_m f$ is an isometry from $(T_m M, g_m)$ onto $(T_{f(m)} N, h_{f(m)})$, is an isometry.

b) Show that an homogeneous Riemannian space is always geodesically complete.

c) Show that on any compact connected Lie group G equipped with its bi-invariant metric, the map $\exp : \underline{G} \to G$ is surjective.

d) Show that there is no bi-invariant metric on $Sl(2, \mathbf{R})$ (see also 3.86).

e) Show that a geodesic of $(P^n\mathbf{R}, \text{can})$ is minimal if and only if its length is less or equal to $\frac{\pi}{2}$.

f) If f and g are two isometries of a connected Riemannian manifold (M, g), such that, for some $a \in M$, $f(a) = g(a)$ and $T_a f = T_a g$, then $f = g$.

2.C.6 Geodesics and submersions, geodesics of $P^n\mathbf{C}$:

Let $p : (\tilde{M}, \tilde{g}) \to (M, g)$ be a Riemannian submersion. Recall that for any $\tilde{m} \in \tilde{M}$, the tangent space $T_{\tilde{m}}\tilde{M}$ is decomposed into the direct sum of the "vertical subspace" $V_{\tilde{m}} = \text{Ker} T_{\tilde{m}} p$ and of its orthogonal complement, the "horizontal subspace" $H_{\tilde{m}}$. Moreover, if $p(\tilde{m}) = m$, the map $T_{\tilde{m}} p : H_{\tilde{m}} \to T_m M$ is an isometry (2.28).

2.109 Proposition. (i) *Let $\tilde{c}$ be a geodesic of $(\tilde{M}, \tilde{g})$. If the vector $\tilde{c}'(0)$ is horizontal, then $\tilde{c}'(t)$ is horizontal for any t, and the curve $p \circ \tilde{c}$ is a geodesic of (M, g), of same length than $\tilde{c}$.*

ii) *Conversely, let $\tilde{m} \in \tilde{M}$ and c be a geodesic of (M, g) with $c(0) = p(\tilde{m})$. Then there exists a unique local horizontal lift $\tilde{\gamma}$ of c such that $\tilde{\gamma}(0) = \tilde{m}$, and $\tilde{\gamma}$ is also a geodesic of $(\tilde{M}, \tilde{g})$.*

iii) *If $(\tilde{M}, \tilde{g})$ is complete, so is (M, g).*

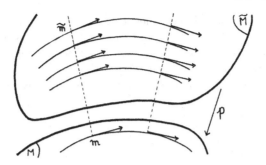

Fig. 2.16. Horizontal geodesics

Proof. We first prove that a Riemannian submersion shortens the distances, that is for $\tilde{x}, \tilde{y} \in \tilde{M}$,

$$d\big(p(\tilde{x}), p(\tilde{y})\big) \leq d(\tilde{x}, \tilde{y}).$$

It is sufficient to prove that, if $\tilde{\gamma}$ is a curve of $\tilde{M}$ then $L(\tilde{\gamma}) \geq L(p \circ \tilde{\gamma})$, which is the case, since the map $T_{\tilde{m}}p$ lessens the norms: indeed, if $v = v^V + v^H$ (decomposition with respect to $H_{\tilde{m}}$ and $V_{\tilde{m}}$), we have:

$$\mid v \mid^2 = \mid v^V \mid^2 + \mid v^H \mid^2 \geq \mid v^H \mid^2 = \mid T_{\tilde{m}}p \cdot v \mid^2 .$$

Let us turn to ii). Be given the geodesic c, we build $\tilde{\gamma}$ as follows: for ϵ small enough, the segment of geodesic $V = c(]t_0 - \epsilon, t_0 + \epsilon[)$ is a one dimensional submanifold of M (image under the map $\exp_{c(t_0)}$ of a segment of straight line). From 1.91, we know that $\tilde{V} = p^{-1}(V)$ is a submanifold of $\tilde{M}$. We define then an horizontal vector field X on $\tilde{V}$ by

$$X(\tilde{x}) = \big(T_{\tilde{x}}p\big)^{-1} \cdot \big(c'(p(\tilde{x})\big),$$

where $T_{\tilde{x}}p$ denotes the isomorphism between $H_{\tilde{x}}$ and $T_{p(\tilde{x})}M$. For any $\tilde{m} \in \tilde{V}$, there exists a unique integral curve $\tilde{\gamma}$ of X through $\tilde{m}$. Beginning at $\tilde{m}$ above $m = c(t_0)$, we defined $\tilde{\gamma}$ on a neighborhood of t_0. The curve $\tilde{\gamma}$ is a geodesic since, first $\mid \tilde{\gamma}'(t) \mid = \mid c'(t) \mid =$ constant, and second:

$$L(\tilde{\gamma})_{|[t,t+s]} = L(c)_{|[t,t+s]} = d\big(c(t), c(t+s)\big) \leq d\big(\tilde{\gamma}(t), \tilde{\gamma}(t+s)\big)$$

for s small enough (the curve $\tilde{\gamma}$ is, like c, locally minimal).

It is hopeless to lift the whole geodesic c: the manifold (M, g) may be complete, and $(\tilde{M}, \tilde{g})$ not (take for example $p : \mathbf{R}^2 \setminus \{0\} \to \mathbf{R}^2$ with the canonical metrics).

We now prove i). Let $v = T_{\tilde{c}(0)}p \cdot \tilde{c}'(0)$, c be the geodesic of (M, g) with initial conditions $c'(0) = v$, and $\tilde{\gamma}$ be an horizontal lift of c starting at $\tilde{c}(0)$. We have just seen that $\tilde{\gamma}$ is a geodesic. But, by construction, $\tilde{\gamma}'(0) = \tilde{c}'(0)$ and hence the geodesics $\tilde{c}$ and $\tilde{\gamma}$ coincide on their common interval of definition. Hence the set of parameters where the geodesic $\tilde{c}$ is horizontal, and where it is a lift of c is an open set containing 0. These two conditions being also closed, they are satisfied on the maximal interval of definition of $\tilde{c}$.

Part iii) is an immediate consequence of what we have just seen. ∎

2.109 bis Exercise. Let c be a geodesic of (M, g). Then (c, c') is a geodesic of TM for the metric of 2.B.6.

2.110 Example: the geodesics of $P^n\mathbf{C}$.

The geodesics c of $P^n\mathbf{C}$ with $c(0) = p(\tilde{x}) = x$ are the curves $t \to p((\cos t)\tilde{x} + (\sin t)\tilde{v})$, where p is the Riemannian canonical submersion

$$(S^{2n+1}, \text{can}) \to (P^n\mathbf{C}, \text{can}),$$

and $\tilde{v}$ is a vector orthogonal to $\tilde{x}$ and $i\tilde{x}$. The great circle $\tilde{c}(t) = (\cos t)\tilde{x} + (\sin t)\tilde{v}$ is then horizontal (since orthogonal to all the $e^{i\theta}\tilde{x}$). Note that $c'(0) = v = T_{\tilde{x}}p.\tilde{v}$, and that the geodesic c is periodic of period π, since

$$c(t + \pi) = p(-\tilde{c}(t)) = p(\tilde{c}(t)) = c(t)\cdot$$

Each $T_x(P^n\mathbf{C})$ is equipped with the complex structure coming from the complex structure of the complex subspace $H_{\tilde{x}} = \{x, ix\}^\perp$ of $\mathbf{C}^{n+1}$ (2.30).

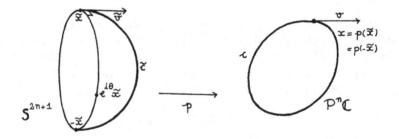

Fig. 2.17. From S^{2n+1} to $P^n\mathbf{C}$

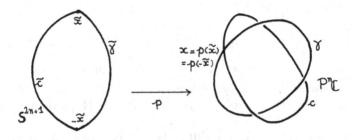

Fig. 2.18. Case a): the geodesics meet at time π

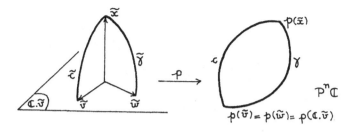

Fig. 2.19. Case b): the geodesics meet at time $\frac{\pi}{2}$

Let c and γ be two geodesics of $P^n\mathbf{C}$ starting at $x = p(\tilde{x})$, with $c'(0) = v$ and $\gamma'(0) = w$ (assume v and w have length 1), and let $\tilde{c}$ and $\tilde{\gamma}$ be their respective horizontal lifts through $\tilde{x}$: these curves are two geodesic of S^{2n+1}.

a) If v and w do not belong to the same complex line, the geodesics c and γ meet at the first time π: indeed, $\tilde{c}$ and $\tilde{\gamma}$ meet at the point $-\tilde{x}$, and for $t \in]0, \pi[$, $p(\tilde{c}(t)) \neq p(\tilde{\gamma}(t))$.

b) On the contrary, if there exists $\lambda \in \mathbf{C}$ with $\lambda w = v$, the geodesics meet for the first time at time $\frac{\pi}{2}$ since

$$\gamma\left(\frac{\pi}{2}\right) = p\left(\tilde{\gamma}\left(\frac{\pi}{2}\right)\right) = p(\tilde{v}) = p(\tilde{w}) = p\left(\tilde{c}\left(\frac{\pi}{2}\right)\right) = c\left(\frac{\pi}{2}\right).$$

A normal geodesic c is hence minimal on the interval $[t_0, t_0 + t]$ if and only if $t \le \frac{\pi}{2}$. If indeed $t \le \frac{\pi}{2}$, there exists no shorter geodesic from $c(t_0)$ to $c(t_0 + t)$. But from 3.103, there exists a minimal geodesic between these two points: hence c is minimal. On the contrary for $t > \frac{\pi}{2}$, the geodesic $s \to c(t_0 - s)$ joins $c(t_0)$ to $c(t_0 + t)$ and is of length $\frac{\pi}{2} - t$, hence is shorter.

2.C.7 Cut-locus

As a consequence of Hopf-Rinow's theorem, we have the following.

2.111 Corollary. *Let (M, g) be a complete Riemannian manifold, and c be a geodesic of (M, g):*

i) *if there exists no geodesic shorter than c from $c(a)$ to $c(b)$, then c is minimal on $[a, b]$;*

ii) *if there exists a geodesic of of the same length as c from $c(a)$ to $c(b)$, then c is no more minimal on any bigger interval $[a, b + \epsilon]$;*

iii) *if c is minimal on an interval I, it is also minimal on any subinterval $J \subset I$.*

Proof. (i) Hopf-Rinow theorem says that there exists a minimal geodesic γ from $c(a)$ to $c(b)$: if c is the shortest geodesic between these two points, c is minimal.

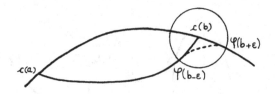

Fig. 2.20. proposition 2.111 ii)

(ii) If γ and c are two geodesics of same length from $c(a)$ to $c(b)$, the curve ϕ defined by

$$\phi(t) = \gamma(t) \quad \text{if} \quad t \in [a, b] \,, \text{ and } \quad \phi(t) = c(t) \quad \text{if} \quad t \geq b$$

joins $c(a)$ to $c(b + \epsilon)$.

Chose $0 < \epsilon' < \epsilon$: 2.103 says that there exists a minimal geodesic s from $\phi(b - \epsilon')$ to $\phi(b + \epsilon')$, whose length is stricly less than the length of ϕ between these two points (ϕ is indeed not a geodesic since it is not smooth at $\phi(b)$).

Then there exists a curve shorter than ϕ from $c(a)$ to $c(b + \epsilon)$, and since the lengths of c and γ between $c(a)$ and $c(b)$ are equal, c is no more minimal after $c(b)$.

(iii) Direct consequence of the triangle inequality. ∎

Let (M, g) be a *complete* Riemannian manifold. For $v \in T_m M$, denote by c_v the unique geodesic satisfying $c_v'(0) = v$ ($c_v(t) = \exp_m tv$). Let

$$I_v = \{t \in \mathbf{R}/ c_v \text{ is minimal on } [0, t]\}.$$

The interval I_v is closed (use the fact that any geodesic is locally minimal, and iii) of the previous corollary): let $I_v = [0, \rho(v)]$ (with $\rho(v)$ possibly infinite). For M compact, $\rho(v)$ is bounded by the diameter of (M, g) when v is of length one.

Note that, if $w = \lambda v$, then $\rho(v) = \lambda \rho(w)$ (the fact that a geodesic is minimal or not does not depend on its velocity). Hence we can restrict the study of the map ρ to the unit bundle of (M, g).

One can prove, for m fixed in M, that the map $v \to \rho(v)$ is continuous on S_m. When v goes through S_m, the $\rho(v)$ are bounded below by a strictly positive real number ([K-N] t.2, p.98). Let

$$U_m = \left\{v \in T_m M/ \mid v \mid < \rho\left(\frac{v}{\mid v \mid}\right)\right\} = \{v \in T_m M/\rho(v) > 1\}.$$

Then U_m is an open neighborhood of 0_m in $T_m M$, with boundary ∂U_m.

2.112 Definition. Let $m \in M$. The *cut-locus* of m is defined as

$$\mathrm{Cut}(m) = \exp_m(\partial U_m) = \{c_v(\rho(v).v), v \in S_m\}.$$

2.113 Proposition. *For any $m \in M$, we have:*

$$M = \exp_m(U_m) \cup \mathrm{Cut}(m),$$

where the union is disjoint.

Proof. From the Hopf-Rinow theorem, we know that for any $x \in M$, there exists a minimal geodesic c_v from m to x. Assume that c_v is parametrized on $[0, 1]$. In particular, $\rho(v) \geq 1$, and hence $v \in \overline{U}_m$. This proves the first assertion.

Let us now prove that the union is disjoint. Let $x \in \exp_m(U_m) \cap \mathrm{Cut}(m)$: since $x \in \exp_m(U_m)$, there exists a geodesic c with $c(0) = m$, $c(a) = x$, and which is minimal on $[0, a + \epsilon]$. On the other hand, since $x \in \mathrm{Cut}(m)$, there exists a geodesic γ with $\gamma(0) = m$, $\gamma(b) = x$, and which is not minimal after b: both of these geodesics being minimal from m to x, they have the same length between m and x. They are also distinct: conclude by contradiction with (2.111 ii)) (c cannot be minimal after the point x). $\blacksquare$

Remarks. We will see in 3.77 that the map $\exp_m$ is a diffeomorphism from U_m onto its image. This will prove that $\exp_m(U_m)$ is diffeomorphic to a ball in $\mathbf{R}^n$: the manifold M is hence obtained by gluing together an n-dimensional ball on the cut-locus of a point ($n = \dim M$): $\mathrm{Cut}(m)$ is a deformation retract of $M \setminus \{m\}$.

2.114 Examples. a) cut-locus of S^n.
All the geodesics are minimizing before distance π. For $m \in S^n$, we have $U_m = B(0_m, \pi)$, $\exp_m(U_m) = S^n \setminus \{-m\}$, and $\mathrm{Cut}(m) = \{-m\}$: the cut-locus is reduced to a single point.

b) cut-locus of $P^n\mathbf{R}$.
All the geodesics minimize before distance $\frac{\pi}{2}$, hence $U_m = B(0_m, \frac{\pi}{2})$. We see that $\exp_m(U_m)$ is the image under the canonical projection p of S^n onto $P^n\mathbf{R}$ of the upper hemisphere in S^n. We also see that $\mathrm{Cut}(m)$ is the image under p of the equator of S^n, and hence is a submanifold isometric to $(P^{n-1}\mathbf{R}, \mathrm{can})$ (2.21). In $P^2\mathbf{R}$, $\mathrm{Cut}(m)$ is the "line at infinity".

c) cut-locus of $P^n\mathbf{C}$.
Let p be the canonical Riemannian submersion from (S^{2n+1}, can) onto $(P^n\mathbf{C}, \mathrm{can})$. We have seen in 2.110 that all the geodesics of $P^n\mathbf{C}$ are minimizing before length $\frac{\pi}{2}$: hence if $x = p(\tilde{x}) \in P^n\mathbf{C}$, we have $U_x = B_x(0_x, \frac{\pi}{2})$. Denote by $H_{\tilde{x}}$ the complex subspace of complex dimension n, normal to $\mathbf{C} \cdot \tilde{x}$ in $\mathbf{C}^{n+1}$: then $H_{\tilde{x}} \cap S^{2n+1} = S_{\tilde{x}}$ (sphere in $H_{\tilde{x}}$, which is a submanifold of S^{2n+1}, isometric to S^{2n-1}) and hence $\mathrm{Cut}(x) = p(S_{\tilde{x}})$ is isometric to $P^{n-1}\mathbf{C}$ (if indeed c is a normal geodesic from $\tilde{x}$, then $c(t) = p(\cos t\tilde{x} + \sin t\tilde{v})$ with $\tilde{v} \in S_{\tilde{x}}$: the cut-locus of $\tilde{x}$ is the set of all points $c(\frac{\pi}{2})$, that is the set of points $p(\tilde{v})$ for $v \in S_{\tilde{x}}$).

d) cut-locus of the hyperbolic space H^n.
Let γ be a geodesic of $H^n \subset \mathbf{R}^{n+1}$ through m: γ is the intersection of H^n with a two plane of $\mathbf{R}^{n+1}$ containing the origin 0, and m (2.10 and 2.80). For $x \in H^n$, there is a unique geodesic from m to x: it is the intersection of H^n

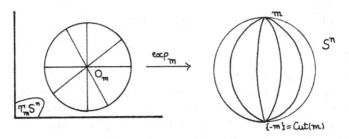

Fig. 2.21. Cut-locus of the sphere

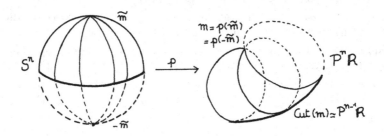

Fig. 2.22. Cut-locus of the real projective space

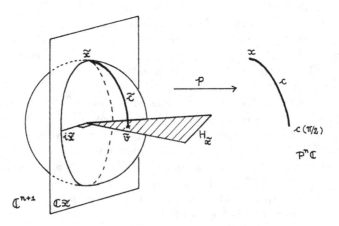

Fig. 2.23. Cut-locus of the complex projective space

with the 2 dimensional vector space of $\mathbf{R}^{n+1}$ generated by x and m. Then Hopf-Rinow's theorem (2.103) yields that this geodesic is minimal. The whole geodesics are then minimal, and the cut-locus of any point is empty.

e) cut-locus of 2 dimensional flat tori.

Rectangle tori: Let (T^2, g) be the Riemannian quotient of $(\mathbf{R}^2, \mathrm{can})$ by a lattice Γ generated by two orthogonal vectors a and b: this torus is the Riemannian product of two circles.

Consider a rectangle cell R of the lattice, and denote by $\tilde{m}$ the center of this cell. For any $m \in T^2$, there exists a Riemannian covering map $p : (\mathbf{R}^2, \mathrm{can}) \to (T^2, g)$ with $p(\tilde{m}) = m$. Identifying $\mathbf{R}^2$ with $T_m T^2$, the map from $\mathbf{R}^2$ to T^2 defined by $t \to p(\tilde{m} + tv)$ coincides with the map $\exp_m$ since the geodesics from m are parametrized by $t \to p(\tilde{m} + tv)$.

Let $x \in T^2$ and $\tilde{x} \in R$ with $p(\tilde{x}) = x$. There are infinitely many geodesics from m to x: they are the images under p of the line segments in $\mathbf{R}^2$ joining $\tilde{x}$ to all the points $\tilde{m} + \gamma$ with $\gamma \in \Gamma$. Among these geodesics, the shortest is the one which is the projection of the line segment $[\tilde{m}, \tilde{x}]$: indeed, the point of $\tilde{m} + \Gamma$ the nearest to $\tilde{x}$ is $\tilde{m}$, since the mediatrices of the segments $[\tilde{m}_1, m]$ ($\tilde{m}_1 \in \tilde{m} + \Gamma$) divides the plane $\mathbf{R}^2$ into two half-planes, one of them containing both $\tilde{m}$ and $\tilde{x}$.

Now let c be a geodesic with $c(0) = m$, and $\tilde{c}$ be the corresponding geodesic with $\tilde{c}(0) = \tilde{m}$ ($p \circ \tilde{c} = c$). The geodesic c is minimal before $\tilde{c}$ goes out of R, and is no more minimal afterwards.

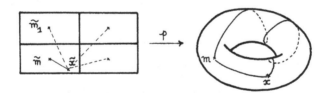

Fig. 2.24. A rectangular torus: before the cut-locus

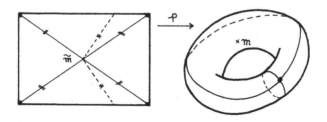

Fig. 2.25. The cut-locus of a flat rectangular torus

Hence $U_m \sim \overset{\circ}{R}$, $\partial U_m \sim \partial R$, $\mathrm{Cut}(m) = p(\partial R) = S^1 \times \{1\} \cup \{1\} \times S^1$, and

$$\exp_m(\partial U_m) = p(\overset{\circ}{R}) = \{(e^{i\theta}, e^{i\phi}) \in S^1 \times S^1 / \theta \neq 0, \phi \neq 0\}.$$

There is a unique minimal geodesic from m to a point of $\exp_m(U_m)$, four minimal geodesics from m to $p(0,0)$, and two minimal geodesics from m to the other points of $\mathrm{Cut}(m)$ (see figure 2.25).

General flat tori: Let Γ be a lattice in $\mathbf{R}^2$, with generators a_1 and a_2 satisfying the conditions of 2.24 ($a_1 = (1,0)$ and $a_2 \in \mathcal{M}$), and consider the canonical Riemannian covering map from $(\mathbf{R}^2, \mathrm{can})$ to (T^2, can).

Let $m \in T^2$ and $\tilde{m} \in \mathbf{R}^2$ with $p(\tilde{m}) = m$. Once again, identifying $T_m T^2$ with $\mathbf{R}^2$, we have $\exp_m(v) = p(\tilde{m} + v)$. Let c be the geodesic satisfying $c'(0) = v \in T_m T^2$: c is the image under p of the geodesic $\tilde{c}(t) = \tilde{m} + tv$. The geodesic c will be minimal between m and $c(t)$ if and only if, for any $\gamma \in \Gamma$, $d(\tilde{m}, \tilde{c}(t)) \leq d(\tilde{m} + \gamma, \tilde{c}(t))$. Hence

$$U_m = \{v \in \mathbf{R}^2 / \|v\| < \|v + \gamma\|, \gamma \in \Gamma\}.$$

Divide each parallelogram (fundamental domains) of the lattice into two triangles whose sides are equal and parallel to the sides of the triangle $(0, a_1, a_2)$. Consider the six triangles which admit 0 as a common vertex: they yield an hexagon around 0, and the mediatrix of the six segments $[0, a_1]$, $[0, a_2]$, $[0, a_2 - a_1]$, $[0, -a_1]$, $[0, -a_2]$ and $[0, a_1 - a_2]$ also yield an hexagon H around 0. The points v lying inside this hexagon are the points nearer to 0 than to the points a_1, a_2, $a_2 - a_1$, $-a_1$, $-a_2$ and $a_1 - a_2$. The interior of H is then equal to U_m: one can indeed show that for any $k \in \Gamma \setminus 0$, the mediatrix of the

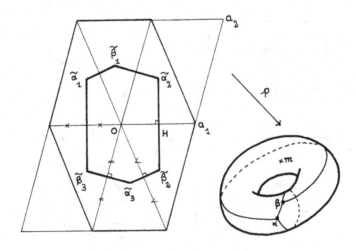

Fig. 2.26. The cut-locus of a generic flat torus

segment $[0, k]$ lies outside H or is tangent to it, hence for v in the interior of H, $\| v \| < \| v - k \|$. Finally, $\mathrm{Cut}(m) = p(H)$ (see figure 2.26).

The opposite sides of H are identified by p, the vertex are identified three by three, which gives two points on $\mathrm{Cut}(m)$: these two points are joined to m by three minimal geodesics, the other points of the cut locus being joined to m by two minimal geodesics.

Remarks. a) The fact that the intersection of the half-planes bounded by the mediatrices of the segments $[0, k]$, and containing zero is an hexagon is non trivial (although using only elementary geometrical properties of the triangle).

b) All the examples above are homogeneous spaces: hence the cut loci of two different points are the same. This fact the following important but straightforward consequence.

2.115 Proposition.*Any homogeneous Riemannian metric is complete.*
Proof. The function $m \mapsto d\big(m, \mathrm{Cut}(m)\big)$ is constant.∎

2.116 Definition. The infimum of $d\big((m, \mathrm{Cut}(m)\big)$, when m runs through M, is the *injectivity radius* of (M, g).
It is denoted by $i(M, g)$.

If M is compact, a continuity argument (cf. [C-E], p.94) shows that $i(M, g)$ is always strictly positive. From 3.77, we know that the metric balls $B(m, r)$, with $r < i(M, g)$, are diffeomorphic to $\mathbf{R}^n$. The injectivity radius of H^n is clearly infinite; if we take the quotient of the Poincaré half-plane by the translations $(x, y) \mapsto (x + n, y), n \in \mathbf{Z}$, we get a complete Riemannian manifold whose injectivity radius is zero. Finding lower bounds for $i(M, g)$ involves curvature estimates (cf. 3.80). A rather complete discussion can be found in [C-E], chapter 5.

2.117 Definition. *The* *diameter* *of a Riemannian manifold* (M, g) *is the number*

$$\mathrm{diam}(M, g) = \sup_{x, y \in M} d(x, y).$$

Example. For the sphere, the real and complex projective spaces equipped with their canonical Riemannian metric, the injectivity radius is equal to the diameter.*The same property is true for the quaternionic projective space and for the Cayley plane, and it is conjectured that these are the only examples (cf. [B1] for details).*

2.118 Exercise. For a compact (M, g), let p, q be two points such that $d(p, q) = \mathrm{diam}(M, g)$. Show that there are at least 2 geodesic segments from p to q. Check what happens for the series of examples in 2.114.

2.C.8 The geodesic flow

It is well known that a second order differential equation can be considered as a first order one, by considering the first derivatives as unknown functions. To do that on a manifold, for the equation of geodesics for instance, there is a prize to pay. We must work with the tangent bundle. If morever one does not want to use local coordinates, a rather heavy formalism is needed. For that reason, our treatment will be down to earth. The reader ma consult the third chapter of [A-L], and also the first chapter of [B1] for more details and a coordinate-free treatment. Roughly speaking, the equation of geodesics is the Lagrange equation for the energy. However, since the cotangent bundle T^*M carries a canonical symplectic structure, using the isomorphism between T^*M and TM provided by the metric, the Hamitonian point of view turns out to be very convenient. In this paragraph, basic notions of symplectic geometry are freely used. They can be found for instance in [Ar] or [A-L], ch. I.

If (M, g) is a complete Riemannian manifold, we can assign to any point (m, u_m) of TM the position and velocity at time t of the geodesic c such that $c(0) = m$ and $c'(0) = u_m$. Set

$$\Phi_t(m, u_m) = \left(\exp_m tu_m, \frac{d}{dt} \exp_m tu_m \right).$$

Clearly, Φ_t is a one parameter group of diffeomorphisms of TM, and also of the unit bundle UM.

2.119 Definition. The one parameter group Φ_t is the *geodesic flow* of (M, g); its infinitesimal generator is the *geodesic spray*.

If $(q^i)_{1 \leq i \leq n}$ denote local coordinates for M, and $(q^i, \dot{q}^i)_{1 \leq i \leq n}$ the corresponding coordinates for TM (here we use the standard notations of variation calculus), the corresponding expression for the spray Ξ is

$$\Xi(q, \dot{q}) = \sum_{i=1}^{n} \dot{q}^i \frac{\partial}{\partial q^i} - \sum_{i=1}^{n} \Gamma_{jk}^i(q) \dot{q}^j \dot{q}^k \frac{\partial}{\partial \dot{q}^i}.$$

This is just another way of writing the differential equation of geodesics.

2.120 The geodesic flow of the sphere.
For the standard sphere S^2, the unit tangent bundle US^2 can be identified with $SO(3)$ (since an orthonormal frame of the Euclidean 3-space is determined by the first two vectors. and the geodesic flow Φ_t is given by the right multiplication by the matrix

$$\begin{pmatrix} \cos t & -\sin t & 0 \\ \sin t & \cos t & 0 \\ 0 & 0 & 1 \end{pmatrix}.$$

If we lift this action to the universal cover, we get the *Hopf action*

$$\tilde{\Phi}_t(z, z') = (e^{it} z, e^{it} z') \quad \text{on} \quad S_r^3 = \{(z, z') \in \mathbf{C}^2, |z|^2 + |z'|^2 = r^2\}$$

The search for the "good value" of r (when S^2 has radius 1) is left to the reader.

Moreover, the metric on US^2 we have seen in 2.B.6 can be identified with "the" bi-invariant metric on $SO(3)$, which is also the canonical metric of

$P^3\mathbf{R}$, so that the geodesic flow acts by isometries.* The reader can check that this property characterizes manifolds with constant *positive* curvature.*

2.121 The geodesic flow of the hyperbolic plane.
In the beginning, if we use the hyperboloid model, the discussion is very similar. The unit tangent bundle can now be identified with $SO_o(1,2)$, and the geodesic flow with the right multiplication by

$$\begin{pmatrix} \cosh t & \sinh t & 0 \\ \sinh t & \cosh t & 0 \\ 0 & 0 & 1 \end{pmatrix} = \exp tL,$$

where

$$L = \begin{pmatrix} 0 & 1 & 0 \\ 1 & 0 & 0 \\ 0 & 0 & 0 \end{pmatrix}$$

We are not interested in the explicit form of the natural metric on UH^2. For our purpose, it is enough to know that isometries of H^2 lift to isometries of UH^2. Moreover, if an isometry of H^2 is defined by some $g \in SO_o(1,2)$, its lift to $UH^2 \simeq SO_o(1,2)$ is the left translation by g. So our metric is some left-invariant metric on $SO_o(1,2)$.
To understand the metric behavior of Φ_t, we can look at the differential

$$T_{m,u}\Phi_t : T_{m,u}UH^2 \to T_{\Phi_t(m,u)}UH^2.$$

Identify (m,u) with the unit of $SO_o(1,2)$. Using the isometry given by the left multiplication by $\exp tL$, one is reduced to the study of

$$\operatorname{Ad} \exp(-tL) : so_o(1,2) \to so_o(1,2).$$

Take the basis of $so_o(1,2)$ made with the matrices L, X^+, X^-, where

$$X^+ = \begin{pmatrix} 0 & 0 & 1 \\ 0 & 0 & 1 \\ 1 & -1 & 0 \end{pmatrix}, \quad X^- = \begin{pmatrix} 0 & 0 & 1 \\ 0 & 0 & -1 \\ 1 & 1 & 0 \end{pmatrix}$$

We have $[L, X^+] = X^+$ and $[L, X^-] = X^-$, therefore

$$\operatorname{Ad} \exp(-tL) \cdot X^+ = e^{-t}X^+$$
$$\operatorname{Ad} \exp(-tL) \cdot X^- = e^{t}X^-$$

(By the way, we also have $[X^+, X^-] = 2L$, so that, setting $H = 2L$, we get a down to earth proof that the Lie algebras $so_o(1,2)$ and $sl(2,\mathbf{R})$ are isomorphic. Sooner or later, the reader will get aware of the pervading role of $Sl(2,\mathbf{R})$ in mathematics.)
Of course, the flow leaves its own trajectories invariant. The above relations say that, in $T_{m,u}UH^2$, the tangent line to the flow admits a transversal plane

which is the direct sum of two lines, one of them being exponentially stable, the other exponentially instable.

This remarkable property has been singled out by D. Anosov, and flows which share it are called Anosov flows. Isometric quotients of H^2 have of course the same property. Moreover, on a compact space, it is preserved by a small deformation of the metric. * It can be proved (cf. [K-H], ch. 17) that the geodesic flow of Riemannian manifolds with negative curvature is Anosov, and this is crucial for proving the ergodicity of the geodesic flow. *

We now come to the symplectic interpretation of Ξ.

2.122 Definition. Let M be a smooth manifold, T^*M its cotangent bundle, and π the canonical projection. The *Liouville form* λ of T^*M is the one-form on T^*M definded by

$$\lambda_\xi(V) = \xi\big(T_\xi\pi(V)\big) \qquad \text{where } \xi \in T^*M \text{ and } V \in T_\xi T^*M.$$

In other words, λ can be characterized by the following property: for any $\alpha \in \Omega^1(M)$, $\alpha^*\lambda = \alpha$ (recall that α can be viewed as a section of T^*M). Denoting by (q^i, p_i) the local coordinates on T^*M associated with the (q^i), we have

$$\lambda = \sum_{i=1}^n p_i dq^i$$

This formula has the following straightforward consequence.

2.123 Proposition. *The differential of the Liouville form is a symplectic form on T^*M.* ∎

Now, a (pseudo)-Riemaniann metric on M provides an ismorphism between T and T^*M, and therefore a symplectic form Ω on TM. Moreover:

2.124 Theorem. *Let $H : TM \rightarrow \mathbf{R}$ be the energy function, defined by $H(v_m) = \frac{1}{2}g_m(v_m, v_m)$. Then the geodesic spray is the symplectic gradient (i.e the Hamiltonian) of H. Namely, $i_\Xi\Omega = -dH$. In particular, the geodesic flow preserves Ω.*

Proof. We must prove that the differential system of geodesics is equivalent to the Hamilton equations on the cotangent space. In local coordinates on T^*M, we have

$$H = \frac{1}{2}\sum_{i,j} g^{ij}p_i p_j$$

and the differential system given by the Hamiltonian of H is

$$\frac{dq^i}{dt} = \frac{\partial H}{\partial p_i} \qquad \text{and} \qquad \frac{dp_i}{dt} = -\frac{\partial H}{\partial q^i} \quad (1 \leq i \leq n)$$

The first n equations just say that

$$\frac{dq^i}{dt} = \sum_{j=1}^n g^{ij}p_i,$$

i.e. that $\frac{dq}{dt} = p^{\sharp}$. We must check that the next n ones yield the familiar equations of geodesics.
We have

$$
\begin{aligned}
\frac{dp_i}{dt} &= -\frac{1}{2} \sum_{k,l=1}^{n} \partial_i g^{kl} p_k p_l \\
&= \frac{1}{2} \sum_{k,l,u,v=1}^{n} g^{ku} \partial_i g_{uv} g^{vl} p_k p_l \\
&= \frac{1}{2} \sum_{u,v=1}^{n} \partial_i g_{uv} \frac{dq^u}{dt} \frac{dq^v}{dt}
\end{aligned}
$$

Here, we have just used the formula for the derivative of the inverse of a matrix. On the other hand,

$$
\frac{dp_i}{dt} = \frac{d}{dt} \left(\sum_{k=1}^{n} g_{ik} \frac{dq^k}{dt} \right) = \sum_{k,l=1}^{n} \partial_l g_{ik} \frac{dq^k}{dt} \frac{dq^l}{dt} + \sum_{r=1}^{n} g_{ir} \frac{d^2 q^r}{dt^2}.
$$

From these two formula, we infer that

$$
\begin{aligned}
\sum_{r=1}^{n} g_{ir} \frac{d^2 q^r}{dt^2} &= \frac{1}{2} \sum_{k,l=1}^{n} \partial_i g_{kl} \frac{dq^k}{dt} \frac{dq^l}{dt} - \sum_{k,l=1}^{n} \partial_l g_{ik} \frac{dq^k}{dt} \frac{dq^l}{dt} \\
&= \frac{1}{2} \sum_{k,l=1}^{n} (\partial_i g_{kl} - \partial_k g_{il} - \partial_l g_{ik}) \frac{dq^k}{dt} \frac{dq^l}{dt},
\end{aligned}
$$

which is exactly what we want! ∎
Remark. It is of course possible to give a coordinate-free proof, cf.[B1] ch.1 for instance. But it requires a rather heavy formalism about the second tangent bundle.

There is a similar result for the restriction of the flow to the unit (co)tangent bundle.

2.125 Theorem. *The $(2n - 1)$-form $\lambda \wedge d\lambda^n$ is a volume form on the unit (co)tangent bundle, which is preserved by the geodesic flow.*
Proof. Up to sign,

$$
\lambda \wedge d\lambda^n = \left(\sum_{i=1}^{n} (-1)^i p_i dp_1 \wedge \cdots \widehat{dp_i} \wedge \cdots dp_n \right) \wedge \left(dq^1 \wedge \cdots \wedge dq^n \right),
$$

which proves the first part. Now,

$$
L_{\Xi} \lambda = d i_{\Xi} \lambda + i_{\Xi} d\lambda.
$$

But $i_{\Xi} \lambda = \sum_{i=1}^{n} p_i \dot{q}^i = 2H$, so that $L_{\Xi} \lambda = dH$ vanishes on the unit (co)tangent bundle. ∎

Therefore, Symplectic Geometry is relevant for studying the dynamics of geodesic flows. It is also useful when explicit computations are needed. One avoids the tedious computation of the Γ^i_{jk}. For example, let us revisit the geodesics of a revolution surface (cf. 2.82). With our present notations, the energy, as a function on the cotangent space, is

$$H(u, \theta, p_u, p_\theta) = \frac{1}{2}\left[p_u^2 + \frac{1}{a^2(u)}p_\theta^2\right].$$

The second group of equations gives here

$$\frac{dp_u}{dt} = \frac{a'(u)}{a^3(u)}p_\theta^2, \quad \frac{dp_\theta}{dt} = 0.$$

In particular, we recover that $a^2(u)\frac{d\theta}{dt}^2 = p_\theta$ is constant along any geodesic.

2.126 Definition. A *first integral* of the geodesic flow is a smooth function $f : TM \to \mathbf{R}$ which is left invariant by the flow.

The metric, viewed as a function on TM is of course a first integral. Of course, using the isomorphism between TM and T^*M given by the metric, we can work with the cotangent bundle as well. For instance, a function $f : T^*M \to \mathbf{R}$ which is fiberwise linear is just a vector field on M.

2.127 Exercise. A vector field on M defines a first integral of the geodesic flow of (M, g) if and only if it is a Killing field.

∗ In the presence of non trivial first integral, the geodesic flow on a 2-manifold defines a so-called *integrable system* (cf.[Ar]): the levels of (f, H) give a (possibly singular) foliation of TM by Lagrangian tori. Even though the description of the Lagrangian tori for revolution surfaces is a nice exercise, this case is elementary. It turns out that the geodesic flow of an ellipsoid $(x/a)^2 + (y/b)^2 + (z/c)^2 = 1$, with a, b, c *distinct* is still integrable, and that situation is already fairly involved (cf. [Kl], 3.5 for instance). Another important notion concerning geodesic flows (as measure preserving dynamical systems) is *ergodicity*. It means that any invariant subset has either zero or full measure (here the measure has been defined in 2.125). This property is in some way the opposite of integrability. A classical example of Riemannian metrics with ergodic geodesic flow is given by compact quotients of the hyperbolic plane (E. Hopf, cf. [Hf]). More generally, the geodesic flow of compact manifolds with negative curvature is ergodic. Surprising examples of ergodic metrics on T^2 or S^2 were given more recently.

In fact, the flows which are the most difficult to understand are those which are neither integrable, nor ergodic. For instance, the celebrated KAM-theory (after A. Kolmogoroff, V.I. Arnold and J. Moser), cf. [Ar], predicts the persistence, after a small perturbation of the flow, of "many" invariant tori. See also [A-F-L] for information about the disappearance of tori after a big enough perturbation.∗

2.D A glance at pseudo-Riemannian manifolds

It is a good question, when studying a result in Riemannian geometry, to ask whether positive-definiteness is really necessary. However, pseudo-Riemannian geometry is not only a free intellectual exercise. It presents many unexpected, counter-intuitive and fascinating features. We will focus on the Lorentzian (signature $(1, n)$) case. It is of course very important because of General Relativity. Moreover, we are not aware of significant results in the higher signature case.

2.D.1 What remains true?

Firstly, if (M, g) is an oriented pseudo-Riemannian manifold, a volume form v_g is assigned to g, by prescribing that $(v_g)_m(e_1, \cdots, e_n) = 1$ for any direct pseudo-orthonormal basis of $T_m M$, exactly like in the Riemannian case (cf. 2.7). If (x_i) is a local coordinate system compatible with the orientation,

$$v_g = \sqrt{|\det(g_{ij}|}dx_1 \wedge \cdots \wedge dx_n.$$

Like in the Riemannian case, orientability is not necessary, and g defines a canonical measure (cf. 3.H).

However, we have to be careful. In the Riemannian case, any submanifold N of M inherits such a measure, since the restriction of g to N is Riemannian. But if g is not positive definite, its restriction to a submanifold may be singular.

Secondly, we have pointed out in B. that the Levi-Civita connection does exist, provided g is everywhere non degenerate. Geodesics are defined in the same way, by the condition $D_{c'}c' = 0$. Therefore, we still have a local existence theorem for geodesics, and an exponential map. For any $m \in M$, it is indeed true that there exists a neighborhood U of $0 \in T_m M$ such that $\exp_m$ is a diffeomorphism on its image. However, *no metric description of U is available*, and this can be rather misleading, see for instance 2.141 below.

A consequence of the existence of the Levi-Civita connection is that pseudo-Riemannian metrics are "rigid" or "finite type" structures (we won't define these terms, whose meaning is fairly intuitive, and refer to [D'A-G] for details). Here are two illustrations of this property.

2.127 **Proposition.** *Let (M, g) a connected pseudo-Riemannian manifold and $f : M \to M$ an isometry. Suppose that for some $m \in M$, $f(m) = m$ and $T_m f = Id$. Then f is the identity map.*

Proof. The subset of the points $p \in M$ such that $f(p) = p$ and $T_p f = Id$ is clearly closed. The local exsitence theorem for geodesics implies that it is open. It is non empty by assumption, therefore $f(p) = p$ everywhere.∎

2.128 **Corollary.** *The isometry group of an n-dimensional pseudo-Riemannian manifold (M, g) is a Lie group of dimension at most $n(n + 1)/2$.*

The proof, for which we already refered to [Hn], is the same as in the Riemannian case. * Moreover, if the dimension in actually $n(n+1)/2$, it is easily checked that the curvature is constant. In the Riemannian case, (M, g) is then isometric to $(\mathbf{R}^n, \text{can}), (S^n, \text{can}), (P^n\mathbf{R}, \text{can})$, since the other quotients give too small isometry groups. The same kind of result holds in the pseudo-Riemannian case. The manifolds for which the maximal dimension is achieved are pseudo-Riemannian versions of spheres, Euclidean and hyperbolic space. We shall meet them soon.*

2.D.2 Space, time and light-like curves

It is time to introduce some vocabulary, which comes from Relativity.

2.129 Definition. A tangent vector v to (M, g) is *space-like* if $g(v, v) > 0$, *time-like* if $g(v, v) < 0$, *light-like*[2] if $g(v, v) = 0$ but $v \neq 0$. A curve on (M, g) is space, time or light-like if its tangent vectors are.

We can assign to a space or time-like curve $c : [a, b] \to M$ its *pseudo-length*

$$\int_a^b \sqrt{\pm g(c'(t), c'(t))}\, dt,$$

which clearly does not depend on the parametrization.

NB. In General Relativity, for an observer, that is a time-like curve, the differential element

$$\sqrt{-g(c'(t), c'(t))}\, dt$$

is the so-called *proper time*).

But the main new feature of pseudo-Riemannian geometry is the existence of light-like curves. They are conformally invariant, since (M, g) and (M, fg), where $f \in C^\infty(M)$ does not vanish, have the same light-like curves. * In dimension 2, the light-like curves define two transverse foliations.* A further property occurs.

2.130 Proposition. *The light-like curves of a 2-dimensional Lorentzian manifold are unparametrized geodesics.*

Proof. This is a direct consequence of the uniqueness theorem for geodesics.∎

2.131 Proposition *In any dimension and signature, the light-like unparametrized geodesics are conformally invariant.*

Proof. Here we use the symplectic interpretation of the geodesic flow we have seen in 2.C.8. The geodesics, when viewed as curves on T^*M, are trajectories of the Hamiltonian of the energy H. The restriction of the canonical symplectic form on T^*M to a level set $H = h$ has a 1-dimensional kernel, and defines a line field on this level set. The trajectories of this line-field are just the geodesics whose energy is h. Now, the level $H = 0$ is clearly conformally invariant. ∎

Remark. The reader who is unfamiliar with elementary symplectic geometry can prove this property by a direct computation (cf. 3.129 for the Levi-Civita of a conformal (pseudo)-metric).

---

[2] Many authors use the word null for light-like. We find it rather confusing

**2.D.3 Lorentzian analogs of Euclidean spaces,
spheres and hyperbolic spaces**

The pseudo-Riemannian of Euclidean space is of course $(\mathbf{R}^n, \sum_{i,j} a_{ij} dx^i dx^j)$, where the quadratic form

$$\sum_{i,j} a_{ij} dx^i dx^j$$

is non-degenerate. Sylvester theorem tells us that there are $[n/2] + 1$ such spaces (geometrically speaking, there is no difference between g and $-g$). The Levi-Civita connection is the same for all these spaces, the geodesics are straight lines and the isometry group is the semi-direct product of $\mathbf{R}^n$ with $O(p,q)$ (here (p,q) denotes the signature). If the signature is $(1, n-1)$, this space is called *Minkowski space*, after Hermann Minkowski, who was a professor of Albert Einstein and was involved in the outcome of Special Relativity. Let us now define the analogs of spheres. We begin with dimension 2.

2.132 **Definition.** *The 2-dimensional de Sitter space Σ is the submanifold of the 3-dimensional Minkowski space defined by $-x_0^2 + x_1^2 + x_2^2 = 1$, provided with the induced pseudometric.*

It is diffeomorphic to $\mathbf{R} \times S^1$. The same algebraic argument as for the hyperbolic plane shows that the induced metric has signature $(1,1)$. The group $O(1,2)$ acts isometrically, and is in fact (check it) the full isometry group.

For finding the geodesics, the same method as in 2.80 c) and d) works. Let P be a 2-dimensional vector space. If the restriction to P of $\langle x, x \rangle$ is non-degenerate, the orthogonal symmetry with respect to P is well defined. Two cases are to be considered

- If this restriction is positive definite, $P \cap \Sigma$ is an ellipse, which defines a space-like geodesic.
- If this restriction has signature $(1,1)$, $P \cap \Sigma$ is an hyperbola. Each of its two branches defines a time-like geodesic.

We are left with the case where this restriction is degenerate. Then $P \cap \Sigma$ consists in 2 parallel straight lines. It is easy to check that they are light-like. Therefore, using 2.130, one infers that each of them defines a light-like geodesic. It is also possible to conclude by using the following elementary property.

2.133 **Exercise.** Let (M, g) be a pseudo-Riemannian manifold, and N a pseudo-Riemannian submanifold. If a curve is a geodesic for (M, g) and is contained in N, it is also a geodesic for $(N, g_{|N})$.

In the above description, we used our Euclidean eyes. For the Euclidean geometer, Σ is a one-sheeted hyperboloid (cf. figure 1.1), and all that we have said can be interpreted from this point of view (compare with [Br2]) However, for the inhabitants of Σ, all the space-like geodesics are the same: they are closed and have the same pseudo-length. It is also worth-noting that, in spite of its friendly aspect, Σ *is not geodesically connected*. To see that, just take two points p and q which sit in a Lorentz plane P, but belong to different branches of the hyperbola $P \cap \Sigma$. There is no geodesic arc from p to q, since such an arc would be contained in P.

2.134 Exercise. Write down the universal Lorentzian covering of Σ, and study its geodesics.

2.135 Flat pseudo-Riemannian tori.
Defining them is easy. Equip $\mathbf{R}^n$ with a non-degenerate symmetric bilinear form g of signature (p, q) and take a lattice Γ. Denote by $p : \mathbf{R}^n \to \mathbf{R}^n/\Gamma$ the covering map of the torus $\mathbf{R}^n/\Gamma$. The argument of 2.22 proves that there is a unique pseudo-Riemannian metric g_Γ on $\mathbf{R}^n/\Gamma$ such that $p^*g_\Gamma = g$. This simple formalism hides a very intricate situation, even in dimension 2.

To get all the possible cases, one can us stick to the standard lattice $\mathbf{Z}^2 \subset \mathbf{R}^2$ and vary the quadratic form g. We shall give a significant example. Let $A \in Sl(2, \mathbf{Z})$ be a matrix with real distinct eigen-values, for example $A = \begin{pmatrix} 2 & 1 \\ 1 & 1 \end{pmatrix}$

Let $\lambda, \frac{1}{\lambda}$ be these eigenvalues, and e, f corresponding eigenvectors. Define a symmetric bilinear form b on $\mathbf{R}^2$ by prescribing

$$b(e, e) = b(f, f) = 0, \ b(e, f) = 1,$$

so that A is an isometry for the $(1, 1)$ quadratic form associated with b. Now, b and A go to the quotient. We get a flat Lorentzian torus, equipped with an isometry $\overline{A}$ which generates a non-compact group (in $\mathrm{Diff}(T^2)$, for the compact open topology). See [K-H], 3.2.e, for a detailed study of the dynamical properties of this map. See also the papers of W. Goldman for amazing higher dimensional flat examples.

2.136 Exercise. Infer from the above construction that there are infinitely many non-diffeomorphic compact 3-manifolds which carry a flat Lorentzian metric (here again we have a strong contrast with the positive definite case: Bieberbach theorem (cf. [Wol]) says that are in any dimension finitely may compact flat Riemannian manifolds. This comes from the fact that $Gl(n, \mathbf{Z}) \cap O(n)$, in contrast with $Gl(n, \mathbf{Z}) \cap O(1, n - 1)$, is finite).

Remark. The example of 2.135 illustrates a puzzling question. There is no reason why the isometry group of a compact pseudo-Riemannian (but non Riemannian) manifold should be compact. But finding examples of non-compact groups is not so easy. By the way, it can be proved that the neutral component of the isometry group of a Lorentzian 2-torus is compact.

Now we turn to higher dimensonal examples. Let us equip $\mathbf{R}^{n+1}$ with the quadratic forms

$$\langle X, X \rangle_1 = -x_0^2 + \sum_{i=1}^n x_i^2$$

$$\langle X, X \rangle_2 = -x_0^2 - x_1^2 + \sum_{i=2}^n x_i^2.$$

2.137 Definitions. *The de Sitter (resp. Anti de Sitter)*[3] *space of dimension* n *is the submanifold of* $\mathbf{R}^{n+1}$ *defined by the equation* $\langle X, X \rangle_1 = 1$ (resp. $\langle X, X \rangle_2 = -1$), equipped with the pseudo-Riemannian (Lorentzian) metric induced by $\langle \, , \, \rangle_1$ (resp. $\langle \, , \, \rangle_2$).

These spaces will be respectively denoted by dS^n and AdS^n. The intersection with a space-like hyperplane, equipped with the induced metric, is a $(n-1)$-sphere for de Sitter, and a $(n-1)$-hyperbolic space for Anti de Sitter.

Like the sphere and the hyperbolic space, these spaces possess an isometry group which has the maximal dimension $n(n+1)/2$: these groups are respectively $O(1,n)$ and $O(2, n-1)$. Their geodesics can be obtained in the same way as Σ. It is also clear that the 2-dimensional de Sitter and Anti de Sitter spaces are the same. Therefore, the first Anti de Sitter space to study is the 3-dimensional one. Like Σ, it has the charm of being non-simply connected (since it is diffeomorphic to $S^1 \times \mathbf{R}^2$).

It turns out that AdS^3 admits a very pleasant description. Namely, on the Lie algebra of $Sl(2, \mathbf{R})$, the bilinear form $(X, Y) \mapsto \operatorname{tr}(XY)$ is non degenerate, Ad-invariant, and has signature $(1, 2)$ (∗ this form is in fact proportional to the Killing form, but we shall not need this result ∗). Therefore (compare with 2.90), $Sl(2, \mathbf{R})$ carries a bi-invariant Lorentzian metric. It is not completely trivial (but not difficult either) to prove that it is isometric to AdS^3. Anyhow, we won't bother about the proof, and will study $Sl(2, \mathbf{R})$ with this metric for itself.

Since the left and right translations are isometries, the isometry group contains $Sl(2, \mathbf{R}) \times Sl(2, \mathbf{R})$, and has the maximal dimension 6. The method of 2.90, which does not use positive-definiteness, proves that the geodesics issued from the unit are one-parameter subgroups. Now, we shall see in III.L that $Sl(2, \mathbf{R})$ admits discrete subgroups Γ such that the quotient manifold $Sl(2, \mathbf{R})/\Gamma$ is compact. In fact, $PSl(2, \mathbf{R})/\Gamma$ is the unit bundle of P/Γ (where P denotes, as usually, the Poincaré half-plane). Now, when we quotient by the right action of Γ, the left translations go to the quotient.

Therefore, the compact 3-dimensional Lorentzian manifold $SL(2, \mathbf{R})/\Gamma$ *has a non-compact isometry group*. This time, the neutral component is already non-compact.

In contrast with these examples, we have the following beautiful result.

Theorem. (G. d'Ambra, [D'A]). *The isometry group of a compact* simply connected *analytic Lorentzian manifold is compact.*

Analyticity is probably a technical assumption, but it is worth-noting that this result is false in higher signature (cf. [D'A] again). For a further study of isometry groups of Lorentzian manifolds, see [Ad].

To finish this discussion, let us mention the "Calabi-Markus phenomena".

2.138 Exercise. The de Sitter space admits no compact quotients by isometries.

---

[3] We are not responsible for this terrible terminology!

2.D.4 (In)completeness

Hopf Rinow theorem basically said that, for a Riemannian manifold, the distance completeness and the completeness of the geodesic spray on TM are equivalent. In the present situation, we have no distance and this question does not make sense.

2.139 Definition. A pseudo-Riemannian manifold (M, g) is *complete* if the geodesic spray is a complete vector field on TM, that is if any geodesic can be extended to a geodesic defined on all $\mathbf{R}$.

The flat, de Sitter and Anti de Sitter spaces are complete, and so are their quotients. It is easy to see, because the geodesics are explicitly known. In the Riemannian case, it can be proved that a compact manifold is complete by restricting the geodesic spray to UM, and invoking the fact that a vector field on a compact manifold is complete. This argument breaks down in the pseudo-Riemannian case. However, in the presence of sufficiently many first integrals, it may happen that some compact submanifold of TM is invariant under the geodesic flow. Thanks to this phenomena, we have the following result, whose proof is left as an exercise.

2.140 Theorem (J. Marsden). *A compact homogeneous pseudo-Riemannian manifold is complete*

Hint: first prove that, if a Lie group acts transitively on a connected manifold M, so does its neutral component.

Strangely enough at first sight, homogeneity alone does not guarantee completeness.

2.141 Example. Consider the half-plane $\{(x, y) \in \mathbf{R}^2, y > 0\}$, provided with the Lorentz metric $dxdy$. It is homogeneous, since the transforms $(x, y) \mapsto (x+a, y)$ and $(x, y) \mapsto (e^t x, e^{-t} y)$ generate a transitive group of isometries. But it is not complete: just take a line which goes across the x axis! This example is only superficially surprising: if you revisit the proof that a homogeneous Riemannian manifold is complete (cf. 2.115), you will notice a crucial metric argument.

Compactness does not guarantee completeness either. A classical example is the *Clifton-Pohl* torus. On the Lorentzian manifold $(\mathbf{R}^2 \setminus \{0\}, \frac{2dxdy}{x^2+y^2})$, the homotheties $h_\lambda(x, y) = (\lambda x, \lambda y)$ act as isometries. The subgroup generated by h_2 defines a proper and free action of $\mathbf{Z}$. The quotient $\mathbf{R}^2 \setminus \{0\}/\mathbf{Z}$ inherits the deck Lorentz metric. To see it is not complete, we use the following elementary, but useful property.

2.142 Lemma. *Let $p : (\tilde{M}, \tilde{g}) \to (M, g)$ a pseudo-Riemannian covering, that is a covering map such that $p^*g = \tilde{g}$. Then g is complete if and only if $\tilde{g}$ is.* ∎

Now, the parallel to the axis are unparemeterized geodesics (of course, for the axis themselves, we must take half-lines). The Hamiltonian is

$$H(x, y, p_x, p_y) = (x^2 + y^2)p_x p_y.$$

Along the line $y = a$, $p_x = 0$, Hamilton equation gives

$$\frac{dp_y}{dt} = 2yp_xp_y = 0,$$

so that, along this line, $p_y = \left(\frac{x'}{x^2+a^2}\right)$ is a non zero constant. Therefore the function $x(t)$ goes to infinity in finite time.

The half-lines through the origin deserve a particular attention: their images in the torus are circles: the Clifton-Pohl torus admits four light-like closed curves, which does not prevent the corresponding geodesics from being incomplete! This phenomena is particular to light-like geodesics: if a space or time-like closed C^1 curve is an unparametrized geodesic, it is easily checked, using the pseudo-arc-length parametrization, that we have a periodic complete geodesic.

2.143 **Exercise.** Prove that the Lorentzian metric $2dxdy + f(x)dy^2$ on the 2-torus $\mathbf{R}^2/\mathbf{Z}^2$ is not complete as soon as the (smooth 1-periodic) function f vanishes somewhere.

Much information about Lorentz surfaces can be found in [We'].

It can be said that for pseudo-Riemannian metrics, completeness is rather rare and difficult to obtain.

2.144 **Theorem** (Y. Carrière, [Ca]). *A compact flat Lorentzian manifold is complete*

This result has been extended by Klingler to the constant curvature case. The proof resorts to dynamics more than geometry. It uses signature $(1, n-1)$ in a crucial way, and the corresponding question for higher signatures is completely open.

2.145 **Remark.** This theorem is difficult, although the assumptions look so strong. But the discrete subgroups of the isometry group of the Minkowski space display rich and strange properties. For example, Margulis has found (cf. [G-M]), in the 3-dimensional case, the amazing example of a subgroup which acts properly and is isomorphic to the free subgroup with two generators.

2.D.5 The Schwarzschild model

In General Relativity, the universe is modelled by a Lorentzian manifold. In the absence of matter the Ricci curvature (cf. next chapter) vanishes.

The first explicit example, besides the trivial example of the Minkowski space, was the Schwarzschild solution, given by

$$g = -\left(1 - \frac{2M}{r}\right)dt^2 + \left(1 - \frac{2M}{r}\right)^{-1}dr^2 + r^2d\sigma^2$$

Here $d\sigma^2 = d\theta^2 + \sin^2\theta d\phi^2$ is the standard (Riemannian) metric on S^2. Even if we do not know (or pretend we do not know) its relativistic origin, we can look at the strange and beautiful properties of this metric. We refer to [B2] for more mathematical details (in particular for curvature computations), and to [M-S-T] for a thorough relativistic discussion.

It is defined on $\mathbf{R} \times \left(\mathbf{R}^3 \setminus (\{0\} \cup S^2(2M))\right)$. Clearly, the isometry group contains $SO(3)$, and the orbits are the 2-spheres $r = cte, t = cte$. The induced metric on such a 2-sphere is (proportional to) the standard metric, and the area is $4\pi r^2$, which gives an intrinsic interpretation of the r coordinate.

2.146 Newtonian approximation

The Schwarzschild metric can model the gravitational field of a star. We shall see that M can be interpreted as the mass. This explains the $2M$ in the formula. The motion of a planet is described by a timelike geodesic.

Let c be such a geodesic. If for some value of the parameter τ the S^2-component of $c'(\tau)$ does not vanish, then the usual uniqueness argument shows that this geodesic is contained in the totally geodesic submanifold $\mathbf{R}\times]2M,+\infty[\times S^1$, where S^1 is the great circle in S^2 tangent to this component. For this submanifold, the metric is

$$g = -\left(1 - \frac{2M}{r}\right)dt^2 + \left(1 - \frac{2M}{r}\right)^{-1}dr^2 + r^2 d\phi^2$$

From 2.127, since $\partial/\partial t$ and $\partial/\partial \phi$ are Killing fields, we have the first integrals

$$\left(1 - \frac{2M}{r}\right)\frac{dt}{d\tau} = c_1$$

$$r^2 \frac{d\phi}{d\tau} = c_2$$

The former says that, if the trajectory stays in a region where r is large, t, which can be viewed as the Newtonian universal time, is proportional to the proper time τ of the planet.

The latter is the same as the first integral which gives Kepler's area law in the Newtonian situation.

The energy integral is

$$-\left(1 - \frac{2M}{r}\right)\left(\frac{dt}{d\tau}\right)^2 + \left(1 - \frac{2M}{r}\right)^{-1}\left(\frac{dr}{d\tau}\right)^2 + r^2\left(\frac{d\phi}{d\tau}\right)^2 = -1$$

Eliminating $\frac{dt}{d\tau}$, we get

$$\left(\frac{dr}{d\tau}\right)^2 + \left(1 - \frac{2M}{r}\right)r^2\left(\frac{d\phi}{d\tau}\right)^2 = \frac{2M}{r} - 1 + c_1^2,$$

which can also be written as

$$\left(\frac{dr}{d\tau}\right)^2 + r^2\left(\frac{d\phi}{d\tau}\right)^2 = \frac{2M}{r} - 1 + c_1^2 + \frac{2Mc_2^2}{r^3}$$

Recall that in the Newtonian situation the energy first integral is just

$$\left(\frac{dr}{d\tau}\right)^2 + r^2\left(\frac{d\phi}{d\tau}\right)^2 = \frac{2M}{r} + cte$$

In our situation, we have an extra term which can be neglected at large distances.

This computation can be pushed forward, to obtain the famous perihelion procession, which was observed for Mercury in the 19th century (cf. [B2], ch. 3 and [M-S-T]).

Here we shall content ourselves with a very rough description. A *circular orbit* is defined as a time-like geodesic such that r is constant. Then $\frac{dt}{d\tau}$ and $\frac{d\phi}{d\tau}$ are of course constant. Using Hamilton equations (cf. 2.125), we see that $\frac{\partial H}{\partial r} = 0$, which gives

$$-\frac{2M}{r^2}\left(\frac{dt}{d\tau}\right)^2 + 2r\left(\frac{d\phi}{d\tau}\right)^2 = 0.$$

Moreover, if τ is the proper time of the orbit,

$$-\left(1 - \frac{2M}{r}\right)\left(\frac{dt}{d\tau}\right)^2 + r^2\left(\frac{d\phi}{d\tau}\right)^2 = -1.$$

Eventually, we get

$$\left(\frac{dt}{d\tau}\right)^2 = \left(1 - \frac{3M}{r}\right)^{-1}$$
$$\left(\frac{d\phi}{d\tau}\right)^2 = \frac{M}{r^3}\left(1 - \frac{3M}{r}\right)^{-1}$$

Such orbits exist if and only if $r > 3M$. They are periodic. If $T = 2\pi/\frac{d\phi}{d\tau}$ is the period, we have

$$T^2 = \frac{4\pi^2}{M}r^3\left(1 - \frac{3M}{r}\right)^{-1}$$
$$\simeq \frac{4\pi^2}{M}r^3 \quad \text{for big enough } r.$$

This is just Kepler's third law. In fact, the period is longer than the Newtonian one. This phenomena is analogous, for a circular orbit, to the perihelion procession.

Remark. In this discussion, it might look strange that we compare a mass with a distance. If one wants a numerical comparison with the Newtonian situation, one must replace t by ct' and M by Gm/c^2, where G is the gravitation constant and c the speed of light.

2.147 The Kruskal-Szekeres extension

Let us give some evidence that the sphere $S^2(2M)$ of radius $2M$ in $\mathbf{R}^3$ is a "fake singularity". Set $r = (1 + m/2R)^2 R$. The metric becomes

$$-\frac{\left(1 - \frac{M}{2R}\right)^2}{\left(1 + \frac{M}{2R}\right)^2}dt^2 + \left(1 + \frac{M}{2R}\right)^4 g_{\mathbf{R}^3}$$

For $R = M/2$, which corresponds to $r = 2M$, it is smooth (but singular).

For any given $m \in S^2$, the map $(t, x) \mapsto (t, R_m x)$, where R_m is some Euclidean rotation in $\mathbf{R}^3$ of axis m, is clearly an isometry for g. Its fixed point set is a totally geodesic (piece of) plane. Geodesics which are contained in such planes are called *radial geodesics*. To find them amounts to find the geodesics of the Lorentzian plane (t, r) for the metric

$$-\left(1 - \frac{2M}{r}\right)dt^2 + \left(1 - \frac{2M}{r}\right)^{-1}dr^2$$

We have the two first integrals

$$\left(1 - \frac{2M}{r}\right)\frac{dt}{d\tau} = c_1$$

$$-\left(1 - \frac{2M}{r}\right)\left(\frac{dt}{d\tau}\right)^2 + \left(1 - \frac{2M}{r}\right)^{-1}\left(\frac{dr}{d\tau}\right)^2 = c_2$$

If the geodesic is time or light-like, clearly $c_1 \neq 0$. In the light-like case, after multiplying the first equation by $1 - \frac{2M}{r}$, we see that $\frac{dr}{d\tau} = \pm c_2$: starting from an initial position (t_0, r_0) with $r_0 > 2M$ in a direction with decreasing r, a position with $r = 2M$ is reached in a finite time τ. However, t becomes infinite, since

$$t(r) = \int_{r_0}^r \pm\left(1 - \frac{2M}{\rho}\right)^{-1}d\rho = r + 2M\log(\frac{r}{2M} - 1) + cte$$

2.148 Exercise. Show that the same property occurs for time-like geodesics (try to minimize computations).

∗ A further evidence comes from curvature computations. It turns out that the formulas for the curvature tensor are perfectly well-behaved for $r = 2M$ (see [M-S-T], ch. 31, pp. 820 ∗

Set $r^* = r + 2M\log\left(\frac{r}{2M} - 1\right)$. We have just seen that the radial light-like geodesics are given by $r^* \pm t = cte$. So it is tempting to take $r^* \pm t$ as new coordinates instead of r and t. Setting $\tilde{u} = t - r^*$, $\tilde{v} = t + r^*$ we get

$$d\tilde{u}d\tilde{v} = dt^2 - \frac{dr^2}{\left(1 - \frac{2M}{r}\right)^2}$$

and we can write the Scharzschild metric as

$$-\left(1 - \frac{2M}{r}\right)d\tilde{u}d\tilde{v} + r^2 d\sigma^2.$$

It is still singular for $r = 2M$. However, noticing that

$$\exp\frac{\tilde{v} - \tilde{u}}{4M} = \exp\frac{r^*}{2M} = \left(\frac{r}{2M} - 1\right)\exp\frac{r^*}{2M},$$

and setting

$$v = \exp\frac{\tilde{v}}{4M} = \exp\frac{t + r^*}{4M} \qquad (2.3)$$

$$u = \exp-\frac{\tilde{u}}{4M} = \exp\frac{r^* - t}{4M} \qquad (2.4)$$

we get

$$g = \frac{32M^3}{r}\exp\left(-\frac{r}{2M}\right)du\,dv + r^2 d\sigma^2,$$

which is now regular for $r = 2M$.

Lastly, if we set $U = (v + u)/2$ and $V = (v - u)/2$, we have

$$g = \frac{32M^3}{r}\exp\left(-\frac{r}{2M}\right)(dU^2 - dV^2) + r^2 d\sigma^2.$$

Using 2.3 and 2.4, we see that

$$uv = U^2 - V^2 = \exp\frac{r^*}{2M} = \left(\frac{r}{2M} - 1\right)\exp\frac{r}{2M}.$$

This equation determines r uniquely and smoothly, and shows that we must have $V^2 - U^2 < 1$.

2.149 **Definition** The *Kruskal-Szekeres extension* of the Scharzchild metric is the Lorentzian metric on $\mathcal{U} \times S^2$, where $\mathcal{U} = \{(U, V) \in \mathbf{R}^2, V^2 - U^2 < 1,$ given by

$$g = \frac{32M^3}{r}\exp\left(-\frac{r}{2M}\right)(dU^2 - dV^2) + r^2 d\sigma^2,$$

with

$$V^2 - U^2 = \left(1 - \frac{r}{2M}\right)\exp\frac{r}{2M}.$$

Our computations show that

$$U = \left(\frac{r}{2M} - 1\right)^{1/2}\exp\frac{r}{4M}\cosh\frac{t}{4M}$$

$$V = \left(\frac{r}{2M} - 1\right)^{1/2}\exp\frac{r}{4M}\sinh\frac{t}{4M}$$

is an isometric embedding of the region $r > 2M$ of the Scharzshild space. By the way, we get another copy of this region by taking $(-U, -V)$.

The region $r < 2M$ can be embedded in a similar way. Performing the same computation, we see that

$$\exp\frac{\tilde{v} - \tilde{u}}{4M} = \left(1 - \frac{r}{2M}\right)\exp\frac{r^*}{2M}.$$

Exchanging U and V, we have an embedding given by

$$U = \left(1 - \frac{r}{2M}\right)^{1/2}\exp\frac{r}{4M}\sinh\frac{t}{4M}$$

$$V = \left(1 - \frac{r}{2M}\right)^{1/2}\exp\frac{r}{4M}\cosh\frac{t}{4M}$$

and of course the twin embedding given by $(-U, -V)$.

2.150 Exercise. What is the extension of the Killing field $\frac{\partial}{\partial t}$?

This metric becomes singular for $V^2 - U^2 = 1$. Morevoer, lightlike geodesics are carried by the lines $U \pm V = cte$, which are parallel to the asymptotes of the hyperbola $V^2 - U^2 = 1$. The computations of 2.147 show that any lightlike radial geodesic either reaches the upper branch in a finite (proper) time, or comes from the lower branch a in finite time. Moreover, any time-like curve which enters the region $V > 0, V^2 - U^2 < 1$ (which corresponds to the region $0 < r < 2M$ in the Schwarzschild model) stays there and eventually reaches the singular surface $V^2 - U^2 = 1$. This is the simplest example of a "black hole".

It can also be checked that it is impossible to connect the two copies of the "exterior" Schwarzschild model by a time or light like geodesic.

Moreover, the metric admits no C^2-extension: indeed, a direct computation (cf. [B2] or [M-S-T] again) shows that the squared norm of the curvature tensor goes to ininity as r goes to zero (or as $V^2 - U^2$ goes to 1). The (non)-existence of a continuous or C^1 extension is an open problem.

2.D.6 Hyperbolicity versus ellipticity

We conclude this section about Lorentz geometry by some heuristic (and sloppy) remarks. For the basic properties of partial differentials equations which are alluded to, see [J1], [Ta] or the chapter 10 of [Sp].

Riemannian geometry belongs to the elliptic world. The two main manifestations of ellipticity are finite dimension and regularity properties. For instance, the space of conformal structures on T^2 has dimension 2; in any dimension, Bieberbach theorems says in particular that, for any n, the number of topological types of flat (with zero curvature) compact n-dimensional Riemannian manifolds is finite.

The Laplace operator $\Delta = -\mathrm{div}(\mathrm{grad})$, which is naturally associated with the Euclidean structure, and can be defined for any Riemannian manifold (cf. 4.7 below), is *elliptic*. Any local solution of $\Delta f = 0$ is smooth (and even analytic), and any solution on a compact manifold is constant. The adapted problem is the Dirichlet problem, that is solving $\Delta f = 0$ in a compact domain, the value of f on the boundary being given.

In constrast with that situation, 2.143 provides an infinite dimensional set of conformal Lorentz structures on T^2, and 2.136 an infinity of topological types for compact Lorentz flat 3-manifold. Indeed, Lorentz geometry belongs to the *hyperbolic* world. Isometries of a Lorentz manifold, even compact, may exhibit an hyperbolic dynamic (cf. 2.135 again).

For the Minkowski space, the analog of the Laplace operator is the *wave operator*

$$\Box = -\frac{\partial^2}{\partial t^2} + \frac{\partial^2}{\partial x^2} + \frac{\partial^2}{\partial y^2} + \frac{\partial^2}{\partial z^2}$$

This operator, whose definition extends to Lorentz manifolds, in the same way as the Laplace operator extends to Riemannian manifold, is the prototype of an hyperbolic operator. There is no regularity theorem, and the space of solutions is infinite dimensional.

The adapted problem is the Cauchy problem, that is solving $\Box f = 0$, the values of f and $\partial_t f$ on a hypersurface being given.

Not surprisingly, one of the main problems of mathematical relativity is also a (non linear!) Cauchy problem. The data are a Riemannian manifold (S, g), together with a symmetric 2-form h, and the question is to find a Lorentz manifold (M, G) (the manifold M itself is unknown!) such that S is a space-like hypersurface of M, g being the induced metric and h the second form (cf. 5.2) of the embedding.

3

Curvature

3.1 A parallel vector field in $\mathbf{R}^2$ is just a constant field. Now, on a surface, there are generally no (even local) parallel vector fields. How much the parallel transport of a field along a small closed curve differ from the identity is measured in terms of the *curvature* of the surface, a function $k : M \to \mathbf{R}$.

Now, on an n-dimensional manifold, the effect of the parallel transport along small closed curves lying in different "2-planes" depends on these very planes, and actually involves a $(1,3)$-tensor, the *curvature tensor* of the Riemannian manifold.

In this chapter, we first introduce the curvature tensor as the defect of symmetry of the second covariant derivative. Indeed, this defect of symmetry corresponds to the defect of parallelism for infinitesimal rectangles.

Then we turn back to the study of geodesics. They were introduced in Chapter II as auto-parallel curves, and their local minimizing property was proved afterwards. Here we show that the critical points for the length (or energy) functional are exactly the geodesics. The second order behavior of these functionals near a geodesic involves the curvature tensor, and leads to the notion of *Jacobi fields*, which are infinitesimal variations of geodesics along a given one. These are used to give typical computations of curvature.

Finally, we present various results concerning the relations between the curvature and the topology of a Riemannian manifold, and close the chapter with a discussion of surfaces with constant negative curvature. We give in particular three proofs (two of them in a geometric spirit, and the last one analytic) of the existence of a metric of constant curvature -1 on a surface with negative Euler-characteristic.

For an historical account on the discovery of this notion of curvature by Gauss, the reader is invited to read the remarkable [Dom], and the second volume of [Sp]. Very convincing motivations are also given in [M-S-T].

Last, not least, the notion of negative (or positive) curvature can be developed from a purely metric point of view. It is one of the beautiful realizations of the Russian school. The idea is to take a suitable property which occurs in the Riemannian case as a definition (cf. 3.142 for the negative case). There are marvelous books on that subject, for instance the textbook [BBI], and the more advanced [B-H] and [BGS].

3.A The curvature tensor

3.A.1 Second covariant derivative

Let Z be a vector field on a Riemannian manifold (M, g). From 2.58 the covariant derivative of the $(1, 1)$-tensor DZ is the $(1, 2)$ tensor defined by

$$(D(DZ))(x, y) = (D^2_{x,y})Z_m = D_x(D_Y Z)_m - (D_{D_x Y})Z_m$$

where Y is a vector field such that $Y_m = y$. We already met in 2.64 the second covariant derivative of a function, which is a symmetric 2-tensor. This symmetry property is no more true for the second derivative of a tensor. However, $(D^2_{x,y} Z - D^2_{y,x} Z)_m$ only depends on Z_m.

3.2 Proposition. *The map*

$$(X, Y, Z) \to D^2_{X,Y} Z - D^2_{Y,X} Z$$

from $\Gamma(TM) \times \Gamma(TM) \times \Gamma(TM)$ into $\Gamma(TM)$ is $C^\infty(M)$-linear.
Proof. For the first two arguments, this property is just a consequence of the definition of the covariant derivative. For the third one, write

$$D^2_{X,Y} Z - D^2_{Y,X} Z = D_X D_Y Z - D_{D_X Y} Z - D_Y D_X Z + D_{D_Y X} Z$$
$$= D_X D_Y Z - D_Y D_X Z - D_{[X,Y]} Z$$

Using the property 2.49 i) of connections, we see that

$$D^2_{X,Y} fZ - D^2_{Y,X} fZ = f(D^2_{X,Y} - D^2_{Y,X} Z) + aZ,$$

where the function a equals $X.(Y.f) - Y.(X.f) - [X, Y].f$. Therefore $a = 0$ from the very definition of the bracket. ∎

3.3 Definition. The *curvature tensor* of a Riemannian manifold (M, g) is the (1,3)-tensor defined by

$$R_m(x, y)z = \left(D^2_{Y,X} Z - D^2_{X,Y} Z\right)_m$$
$$= \left(D_Y(D_X Z) - D_X(D_Y Z) + D_{[X,Y]} Z\right)_m$$

where $X, Y, Z \in \Gamma(TM)$ are such that $X_m = x$, $Y_m = y$, $Z_m = z$.

3.4 Remarks. i) This means that $D_{[X,Y]} Z - [D_X, D_Y]Z = R(X, Y)Z$. Compare with the Lie derivative (cf. 1.112), for which $[L_X, L_Y] = L_{[X,Y]}$.
ii) The opposite sign convention for R is used quite often, in [K-N] or [Sp] for instance. Our convention may look unnatural for vector fields, but it gives the

"good" sign for covariant tensors. Indeed, in $\alpha \in \Omega^1(M)$, it is easy to check that

$$((D^2_{X,Y} - D^2_{Y,X})\alpha)(Z) = \alpha(R(X,Y)Z).$$

Moreover, this kind of commutation formula is used more often in the covariant case (cf. 4.13). Further reasons for our choice will be given later on (3.7, 3.82).

3.A.2 Algebraic properties of the curvature tensor

It will be useful to introduce the $(0,4)$-tensor associated with R. This tensor will be also denoted by R. It is defined by $R_m(x,y,z,t) = g_m(R_m(x,y)z,t)$. The point m will be omitted if there is no ambiguity.

3.5 Proposition. *For any $x,y,z,t \in T_m M$, we have*
i) $R(x,y,z,t) = -R(y,x,z,t) = -R(x,y,t,z)$;
ii) $R(x,y,z,t) + R(y,z,x,t) + R(z,x,y,t) = 0$;
iii) $R(x,y,z,t) = R(z,t,x,y)$.
Proof. The first equality in i) is clear. For the second one, take vector fields Z and T such that $Z_m = z$ and $T_m = t$, and compute the Hessian of the function $g(Z,T)$. We find

$$g(D^2_{x,y}Z,T) + g(D_x Z, D_y T) + g(D_y Z, D_x T) + g(Z, D^2_{x,y}T).$$

The antisymmetric part of this expression is just $R(x,y,z,t) + R(x,y,t,z)$, and it must be zero.
Relation ii), known as the first Bianchi identity, follows from the Jacobi identity. A direct computation on vector fields shows that

$$R(X,Y)Z + R(Y,Z)X + R(Z,X)Y = -([[X,Y],Z] + [[Y,Z],X] + [[Z,X],Y]).$$

Relation iii) comes from i) and ii) by using (clever) algebraic manipulations which are left to the reader (see also [Mi 2], p.54 or [Sp], t.2). ∎

We can summarize properties i) and iii) by saying that R_m defines a symmetric bilinear form ρ_m on $\Lambda^2 T_m M$. This form is obtained by setting

$$\rho_m(x \wedge y, z \wedge t) = R_m(x,y,z,t). \tag{3.6}$$

Now, equip $\Lambda^2 T_m M$ with the scalar product associated with g_m. This scalar product, which will be also denoted by g_m, is given by

$$g_m(x \wedge y, z \wedge t) = \det \begin{pmatrix} g_m(x,z) & g_m(x,t) \\ g_m(y,z) & g_m(y,t) \end{pmatrix}.$$

Then ρ_m can be identified with a self-adjoint endomorphism of $\Lambda^2 T_m M$, which shall be called the *curvature operator*.

Let $G_m^2 M$ be the 2-Grassmannian of $T_m M$, and

$$G^2 M = \bigcup_{m \in M} G_m^2 M$$

(of course $G^2 M$ is a fiber bundle over M, but this point of view is not necessary). If P is a 2-plane of $T_m M$ with a basis (x, y), then

$$\frac{R_m(x, y, x, y)}{g_m(x \wedge y, x \wedge y)} = \frac{R_m(x, y, x, y)}{g_m(x, x)g_m(y, y) - (g_m(x, y))^2}$$

does not depend on the basis.

3.7 **Definition.** The *sectional curvature* of a Riemannian manifold (M, g) is the function $P \to K(P)$ on $G^2 M$ we obtained just above. The sectional curvature of the plane $\mathbf{R}x + \mathbf{R}y$ will be denoted by $K(x, y)$. If the basis $\{x, y\}$ is orthonormal, $K(x, y) = R_m(x, y, x, y)$.

Remark. This is the reason for our convention for the curvature tensors. Even though Riemannian geometers may disagree for the definition of R (for instance, the convention of [K-N] and [Sp] is opposite to ours), all of them say that the standard sphere has positive sectional curvature (we shall see that very soon). Then, if you follow the opposite convention for curvature, you must define the sectional curvature as $R_m(x, y, y, x)$, which we find less natural.

3.8 **Theorem.** *The sectional curvature determines the curvature tensor.*

Proof. A direct computation, using the Bianchi identity (3.5 ii) in a crucial way, shows that for $\alpha = \beta = 0$ the second derivative

$$\frac{\partial^2}{\partial \alpha \partial \beta} \left(R(x + \alpha z, y + \beta t, x + \alpha z, y + \beta t) - R(x + \alpha t, y + \beta z, x + \alpha t, y + \beta z) \right)$$

is equal to $6R(x, y, z, t)$. ∎

For example in dimension 2, we have $G^2 M = M$ and the curvature tensor is just

$$R_m(x, y, z, t) = K(m) \left(g_m(x, z)g_m(y, t) - g_m(x, t)g_m(y, z) \right). \qquad (3.9)$$

This formula can be checked directly, remarking that $\Lambda^2 T_m M$ is 1-dimensional. The curvature tensor is completely determined by the function K, which is called *the Gaussian curvature.*

Let us consider now the general case, and suppose that $K(P)$ is constant on each Grassmannian $G_m^2 M$. As the tensor

$$R_m^o(x, y, z, t) = g_m(x, z)g_m(y, t) - g_m(x, t)g_m(y, z)$$

clearly satisfies the properties of Proposition 3.5, Theorem 3.8 shows that $R = fR^o$, where $f \in C^\infty(M)$.

3.10 Exercise. Using the second Bianchi identity (cf. 3.134), show that f is *constant* as soon as $\dim M \geq 3$.

3.11 Definition. A Riemannian manifold is said to be *with constant curvature* (resp. *with positive, negative curvature*), if the sectional curvature is constant (resp. positive, negative). A Riemannian manifold manifold whose curvature tensor (or sectional curvature, which amounts to the same by 3.8) vanishes is said to be *flat*.

3.11 bis Remark. In pseudo-Riemannian geometry, the sectional curvature cannot be defined for 2-plane which are singular for g. Anyhow, constant curvature makes sense, and is defined by saying that $R = fR^0$. The proof that f is constant if $\dim M \geq 3$ works with any signature.

However, the notions of positive or negative curvature don't make sense. It can be proved that the sectional curvature at a point takes both signs as soon as R is not proportional to R^0.

3.A.3 Computation of curvature: some examples

3.12 The *Euclidean space* $\mathbf{R}^n$ is flat. Indeed,

$$D_X D_Y Z = X.(Y.Z) \quad \text{and} \quad X.(Y.Z) - Y.(X.Z) = [X,Y].Z = 0.$$

We shall see later on (see the proof of 3.82) that every flat Riemannian manifold is locally isometric to $\mathbf{R}^n$.

3.13 If $p : (\tilde{M}, \tilde{g}) \to (M, g)$ is a Riemannian covering map, and if the vectors $(\tilde{x}_i)_{(1 \leq i \leq 4)}$ of $T_{\tilde{m}} \tilde{M}$ are projected on the vectors $(x_i)_{(1 \leq i \leq 4)}$ of $T_m M$, then

$$\tilde{R}_{\tilde{m}}(\tilde{x}_1, \tilde{x}_2, \tilde{x}_3, \tilde{x}_4) = R_m(x_1, x_2, x_3, x_4).$$

Indeed, p is a local isometry of a neighborhood of $\tilde{m}$ onto a neighborhood of m. This explains why the Riemannian manifolds we studied in 2.22 are called flat tori.

3.14 Proposition. *The sectional curvatures of (S^n, can) and (H^n, can) are constant.*

Proof. The natural action of the group $SO(n+1)$ (resp. $SO_o(n,1)$) on S^n (resp. H^n) is isometric and transitive. The isotropy group of $m = (1, 0, ..., 0)$ is the group of matrices of type

$$\begin{pmatrix} 1 & 0 \\ 0 & A \end{pmatrix},$$

where 0 abusively denotes $(1, n)$ and $(n, 1)$-matrices, and $A \in SO(n)$. As the tangent space at m is orthogonal to Om for the Euclidean (resp. Lorentzian) metric of $\mathbf{R}^{n+1}$, the action of the isotropy group on $T_m M$ is given by $(A, x) \to Ax$. We simply get the natural representation of $SO(n)$ on $\mathbf{R}^n$.

In particular, this representation is transitive on 2-planes, therefore the sectional curvature is constant on $G^2(T_m S^n)$ (resp. $G^2(T_m(H^n))$). Since we are working with Riemannian homogeneous spaces, this constant is the same at every point. ∎

Remark. We shall see in 3.47 and 3.48 that these constants are $+1$ and -1.

3.15 Curvature of a product.
Let $(M, g) = (M_1, g_1) \times (M_2, g_2)$ be a Riemannian product. If X_1 and X_2 are vector fields on M_1 and M_2 respectively, define a vector field X on $M_1 \times M_2$ by setting

$$X_{m_1, m_2} = (X_1, 0)_{m_1, m_2} + (0, X_2)_{m_1, m_2}.$$

We shall abusively write $X = X_1 + X_2$. If Y is another vector field of type $Y_1 + Y_2$, where $Y_i \in \Gamma(TM_i)$, then 2.52 shows that

$$D_X Y = D^1_{X_1} Y_1 + D^2_{X_2} Y_2.$$

Here D, D^1 and D^2 denote the canonical connections of (M, g), (M_1, g_1) and (M_2, g_2) respectively. Using similar notations, the curvature tensor will be given by

$$R_m(x, y, z, t) = R^1_{m_1}(x_1, y_1, z_1, t_1) + R^2_{m_2}(x_2, y_2, z_2, t_2) :$$

just compute R_m using decomposed vector fields whose values at m are x, y, z, t.

In particular, the sectional curvature of a *mixed plane*, i.e. of a plane generated by $(x_1, 0)_m$ and $(0, x_2)_m$, will always *vanish*. For example, the sectional curvature of $(S^2, \text{can}) \times (S^2, \text{can})$ is clearly non-negative, but not strictly positive. It was conjectured by H. Hopf in the fifties, that $(S^2, \text{can}) \times (S^2, \text{can})$ carries no metric whose sectional curvature is everywhere (strictly) positive.

3.16 Local coordinates.
If a Riemannian metric is given in local coordinates by $g = \sum_{i,j} g_{i,j}(x) dx^i dx^j$, the curvature tensor is given by

$$R^l_{ijk} = \partial_i \Gamma^l_{jk} - \partial_j \Gamma^l_{ik} + \sum_r (\Gamma^r_{jk} \Gamma^l_{ir} - \Gamma^r_{ik} \Gamma^l_{jr}).$$

This formula is very unpleasant and not so useful. However, it gives an interpretation of the curvature as an *obstruction to integrability for the horizontal distribution*. Namely, equip TM with the Riemannian metric of g_T of 2.B.6, and recall that $p : (TM, g_T) \to (M, g)$ is a Riemannian submersion. Consider two vector fields X and Y on M, and denote by $\tilde{X}$ and $\tilde{Y}$ their horizontal lifts. Then

$$\left[\tilde{X}, \tilde{Y} \right]^{\text{Vert}}_{v_m} = -R(X_m, Y_m) v_m$$

This is just a consequence of our formula for R^l_{ijk}: just take $X = \frac{\partial}{\partial x^i}, Y = \frac{\partial}{\partial x^j}$, and compute.

There is of course a more conceptual proof, but it requires rather long developments (compare with [Sp], t.2), and the first chapter of [B1]).

3.17 Proposition. *If G is a Lie group equipped with a bi-invariant metric g, and if X, Y, Z, T are left invariant vector fields, then*

$$R(X, Y, Z, T) = \frac{1}{4}g([X, Y], [Z, T]).$$

In particular, G has non-negative sectional curvature.
Proof. We have seen in 2.90 that $D_X Y = \frac{1}{2}[X, Y]$. Therefore

$$R(X, Y)Z = \frac{1}{4}[Y, [X, Z]] - \frac{1}{4}[X, [Y, Z]] + \frac{1}{2}[[X, Y], Z] = \frac{1}{4}[[X, Y], Z].$$

On the other hand, since g is biinvariant,

$$g_e(\mathrm{Ad}\phi_t X(e), \mathrm{Ad}\phi_t Y(e)) \equiv g_e(X, Y)$$

holds, where ϕ_t is the local group associated to Z. Differentiating yields

$$g([Z, X], Y) + g(X, [Z, Y]) = 0$$

so that

$$R(X, Y, Z, T) = \frac{1}{4}g([[X, Y], Z], T) = \frac{1}{4}g([X, Y], [Z, T]),$$

and $R(X, Y, X, Y) = \frac{1}{4}g([X, Y], [X, Y])$ is non-negative. $\blacksquare$
See 3.86 for a group-theoretical consequence of this property.

3.A.4 Ricci curvature, scalar curvature

From a naive view-point, since the curvature tensor is rather complicated, we want to define more simple tensors. More conceptual motivations will be given in III.K.

3.18 Definition. The *Ricci curvature* of a Riemannian manifold (M, g) is the trace of the endomorphism of $T_m M$ given by $v \to R_m(x, v)y$.
It follows from proposition 3.5 that we get a symmetric $(0, 2)$-tensor. We shall denote the Ricci curvature by Ric. We shall see in 3.H.4 that it measures the difference between the volume elements of (M, g) and of the Euclidean space.

3.19 Definition. The *scalar curvature* of a Riemannian manifold is the trace of the Ricci curvature. We get a function on the manifold (M, g), which shall be denoted by $Scal$ or s.
If $(e_i)_{1 \le i \le n}$ is an orthonormal basis of $T_m M$ then

$$Ric_m(x, y) = \sum_{i=1}^{n} R_m(x, e_i, y, e_i) \quad \text{and}$$

$$Scal(m) = \sum_{i,j=1}^{n} R(e_i, e_j, e_i, e_j) = \sum_{i,j=1}^{n} K(e_i, e_j).$$

3.20 Examples: i) In dimension 2, the curvature tensor is entirely given by the scalar curvature, and $Scal(m) = 2K(m)$.

ii) In dimension 3, the curvature tensor is entirely given by the Ricci curvature (cf. III.K).

iii) For a manifold with constant sectional curvature K, we have $Ric = (n-1)Kg$ and $Scal = n(n-1)K$.

iv) If $g_1 = tg$, where t is a positive constant, the corresponding (0,4)-curvature tensor R_1 is just tR (since $D_1 = D$). Therefore

$$K_1(P) = t^{-1}K(P), \quad Ric_1 = Ric, \quad Scal_1 = t^{-1}Scal.$$

3.21 Proposition. *Let G/H be an isotropy irreducible homogeneous space, equipped with "its" G-invariant Riemannian metric. Then there exists a constant λ such that $Ric = \lambda g$.*

Proof. The argument is the same as in 2.43, when we proved the "uniqueness" of the G-invariant metric. Diagonalizing $Ric_{[e]}$ with respect to $g_{[e]}$, we first see that $Ric_{[e]} = \lambda g_{[e]}$. Then $Ric = \lambda g$ because of homogeneity. ∎

Riemannian manifolds whose Ricci curvature is proportional to the metric have been studied extensively (see [B2]). They are called *Einstein manifolds*. In the Lorentzian case, they are important in General Relativity. Indeed, the Einstein equation in vacuum is given by $Ric = 0$. An important example is the Schwarzschild metric we have met in 2.D.5.

3.22 Exercise. A Riemannian manifold (M, g) is said to satisfy condition (F) if for any m and any orthonormal basis $(e_i)_{1 \leq i \leq n}$ of $T_m M$, one has

$$\text{for any } i: \quad \sum_{j,k \neq i} K(j,k) \geq 0$$

(we suppose that $n \geq 3$). Show that condition (F) holds if and only if $Ric \leq \frac{1}{2}Scal_g$. Is $Scal_g$ positive ? What happens if $n = 3$?

3.B First and second variation of arc-length and energy

To write down the equation of geodesics, we had to take the covariant derivative of vector fields along a curve; we are going to revisit this point of view, and see that geodesics are critical points for the length and the energy functionals, defined on a suitable space of curves. It is then natural to take a further derivative. To do that, we need to define covariant derivatives of vector fields along a family of curve, that is a parametrized surface. The technical machinery is a bit heavy, but in the end the results are just what is expected: just compare the results of 3.B.1 below with the definitions of the Levi-Civita connection and of the curvature tensor.

3.B.1 Technical preliminaries: vector fields along parametrized submanifolds

3.23 Definition. A *parametrized submanifold* of M is a smooth map H of a manifold N into M.

A basic example is given by smooth variations of a curve c.

3.24 Definition. A *variation* of a smooth curve $c : [a, b] \to M$ is a smooth map $H : [a, b] \times] - \epsilon, -\epsilon[\to M$ such that $H(s, 0) = c(s)$. The curve $s \to H(s, t)$ will be denoted by c_t.

3.25 Remark. Here, N is a manifold with boundary (see 4.1), but there will be no trouble. The picture shows that $H(N)$ is not usually a sub-manifold.

3.26 Definition. A *vector field along a parametrized submanifold* H is a smooth map $Y : N \to TM$ such that $Y(p) \in T_{H(p)}M$ for any $p \in M$. Let $\Gamma(H, TM)$ be the vector space of such vector fields.

3.27 Example: For a variation H of the curve c, such vector fields are given by

$$\frac{\overline{\partial}}{\partial s} = T_{(s,t)} H \cdot \frac{\partial}{\partial s} \quad \text{and} \quad \frac{\overline{\partial}}{\partial t} = T_{(s,t)} H \cdot \frac{\partial}{\partial t}.$$

These vector fields are respectively tangent to the curves $t = t_o$ and $s = s_o$ of the variation.

The following theorem is a generalization of 2.68. The proof is the same.

3.28 Theorem. *Let (M, g) be a Riemannian manifold, and $H : N \to M$ be a parametrized submanifold. For any pair (X, Y), where X is a vector field on N and Y a vector field along H, there is a unique vector field along H, denoted by $\overline{D}_X Y$, such that*
i) the map $(X, Y) \to \overline{D}_X Y$ is bilinear;
ii) for any smooth function f on N, we have

$$\overline{D}_{fX} Y = f \overline{D}_X Y \quad \text{and} \quad \overline{D}_X (fY) = (X.f)Y + f \overline{D}_X Y;$$

iii) if in a neighborhood of $H(x_o)$ in M there exists a vector field $\tilde{Y}$ such that, near x_o, we have $\tilde{Y}_{H(x)} = Y_x$, then

$$(\overline{D}_X Y)_{x_o} = \left(D_{T_{x_o} H \cdot X_{x_o}} \tilde{Y} \right)_{H(x_o)}.$$

For any vector field X on N, we get a vector field $\overline{X}$ along H by setting $\overline{X}_x = T_x H \cdot X_x$.

3.29 Proposition. *Let X and Y be vector fields on N, and U, V vector fields along H. Then*

$$\overline{D}_X \overline{Y} - \overline{D}_Y \overline{X} = \overline{[X, Y]},$$
$$X.g(U, V) = g(\overline{D}_X U, Y) + g(U, \overline{D}_X V)$$
$$R(\overline{X}, \overline{Y})U = \overline{D}_Y(\overline{D}_X U) - \overline{D}_X(\overline{D}_Y U) + \overline{D}_{[X,Y]} U$$

where R is the curvature tensor of (M, g).

Sketch of the Proof. First check that the three expressions

$$\overline{T}(X,Y) = \overline{D}_X \overline{Y} - \overline{D}_Y \overline{X} - \overline{[X,Y]}$$
$$\overline{D}g(X,U,V) = X.g(U,V) - g(\overline{D}_X U, V) - g(U, \overline{D}_X V)$$
$$\overline{R}(X,Y)U = \overline{D}_Y \overline{D}_X U - \overline{D}_X \overline{D}_Y U + \overline{D}_{[X,Y]}U$$

are $C^\infty(N)$-multilinear. Then it is enough to check the claimed formulas when X, Y are coordinate vector fields on N, and U, V restrictions to H of coordinates vector fields on M. ∎

3.B.2 First variation formula

If p and q are two points of M, denote by Ω_{pq} the set of smooth curves $c : [0,1] \to M$ such that $c(0) = p$ and $c(1) = q$. It is possible to equip Ω_{pq} with a smooth (infinite dimensional) structure. Then the critical points of the function $c \to E(c)$ provide information about the topology of M. That is the celebrated Morse theory, for which we refer to the marvelous book [Mi2]. We shall content ourselves with a variational study of the functions L and E in the neighborhood of a given curve c. The following lemma shows that the vector spaces of vectors fields along c which vanish at 0 and 1 can be viewed as $T_c\Omega_{pq}$.

3.30 Lemma. *Let $c : [0,1] \to M$ be a smooth curve. For any variation H of c, the formula*

$$Y(s) = T_{(s,0)}H \cdot \frac{\partial}{\partial t}$$

defines a vector field along c. If c has fixed ends, this vector field vanishes at the ends. Conversely, if Y is a vector field along c, there exists a variation H such that

$$Y_s = T_{(0,s)}H \cdot \frac{\partial}{\partial t}.$$

Moreover, if $Y(0) = Y(1) = 0$, we can take take H with fixed ends.

Proof. We have seen the first part in 3.27. Conversely, setting

$$H(s,t) = \exp_{c(s)} tY(s),$$

it is clear that $Y(s) = T_{(s,0)}H.\frac{\partial}{\partial t}$. A priori, H is is only defined in a neighborhood of $[0,1] \times 0$ in $[0,1] \times \mathbf{R}$. Using a compactness argument, we see that such a neighborhood contains some $[0,1] \times] - \epsilon, \epsilon[$. ∎

The *energy*, that was defined in 2.96, turns out to be much more convenient than the length.

3.31 Theorem (first variation formula).
i) *For any variation $(t,s) \to H(s,t) = c_t(s)$ of the curve c, one has*

$$\frac{d}{dt}E(c_t) = \left[g(Y(s), c'(s))\right]_0^1 - \int\limits_0^1 g\big(Y(s), D_{c'}c'(s)\big)ds$$

where $Y(s) = T_{(s,0)}H \cdot \frac{\partial}{\partial t}$.

ii) *The critical points of $E = \Omega_{pq} \to \mathbf{R}$, namely the curves such that*

$$\frac{d}{dt}E(c_t)\,|_{t=0}= 0$$

for any variation with fixed ends, are geodesics.
iii) *If c is parametrized with arc-length, for any variation c_t*

$$\frac{d}{dt}E(c_t)\,|_{t=0}= \frac{d}{dt}L(c_t)\,|_{t=0}\ .$$

iv) *A curve $c \in \Omega_{pq}$ is critical for L if and only if there exists a change of parameter α such that the curve $\gamma = c(\alpha)$ is a geodesic.*
Proof. i) Using 3.29 together with the notations of 3.27, we get

$$\frac{1}{2}\frac{d}{dt}g(c_t'(s), c_t'(s)) = g(\overline{D}_{\frac{\partial}{\partial t}}c_t', c_t')$$

$$= g(\overline{D}_{\frac{\partial}{\partial s}}\frac{\partial}{\partial t}, \frac{\partial}{\partial s}) = \frac{\partial}{\partial s}g(\frac{\partial}{\partial t}, \frac{\partial}{\partial s}) - g(\frac{\partial}{\partial t}, \overline{D}_{\frac{\partial}{\partial s}}\frac{\partial}{\partial s}).$$

The claimed formula follows by integration, since $Y(s) = \frac{\partial}{\partial t}(0, s)$.
ii) If c is a geodesic, then $D_{c'}c' = 0$ and c is critical for E, since the integrated terms vanish for variations with fixed ends. Conversely, if c is critical for E, take variations H such that $Y = fD_{c'}c'$, where f is a smooth function with $f(0) = f(1) = 0$. For any such function, it follows from i) that

$$\int_0^1 f\,g(D_{c'}c', D_{c'}c')ds = 0,$$

therefore $D_{c'}c' = 0$ and c is a geodesic.
iii) is straightforward.
iv) Reparametrizing c by arc-length, we get a curve γ which is still critical for L. Moreover, using iii) and ii), γ is a geodesic. ∎

3.32 Exercise. Let (M, g) be a complete Riemannian manifold. Let V be a closed submanifold of M, and $m \in M$. Show that there exists $p \in V$ such that $d(m, V) = d(m, p)$, and a geodesic c from m to p of length $d(m, V)$. Show that c meets V orthogonally at p.

3.33 Exercise. Replace Ω_{pq} by the space Ω'_{pq} of *piecewise smooth* curves from p to q. Give a formula analogous to 3.31 i), and show that the critical points of E are still (smooth) geodesics.

3.B.3 Second variation formula

3.34 Theorem. *Let $c : [0, L] \to M$ be a length-parametrized geodesic. Let H be a variation of c, and $Y(s, t)$ be the vector field along H given by $T_{(s,t)}H \cdot \frac{\partial}{\partial t}$.*

Then the second derivative of the energy $E(c_t)$ and of the length $L(c_t)$ are respectively

$$\left[g\big(\overline{D}_{\frac{\partial}{\partial t}}Y(s,0),c'(s)\big)\right]_0^L + \int_0^L \left(|\,Y'\,|^2 - R(Y,c',Y,c')\right)ds,$$

and

$$\left[g\big(\overline{D}_{\frac{\partial}{\partial t}}Y(s,0),c'(s)\big)\right]_0^L + \int_0^L \left(|\,Y'\,|^2 - R(Y,c',Y,c') - g(c',Y')^2\right)ds,$$

where $Y'(s) = \overline{D}_{\frac{\partial}{\partial s}}Y(s,0)$. Moreover, denoting by $\tilde{Y} = Y - g(c',Y)c'$ the normal component of Y, we have

$$\frac{d^2}{dt^2}L(c_t)_{|t=0} = \left[g\big(\overline{D}_{\frac{\partial}{\partial t}}Y(s,0),c'(s)\big)\right]_0^L + \int_0^L \left(|\,\tilde{Y}'\,|^2 - R(\tilde{Y},c',\tilde{Y},c')\right)ds.$$

Remark. The integrated part of those formulas measures the failure of the curves $t \to H(0,t)$ and $t \to H(L,t)$ to be geodesics.

Proof. Let X be the vector field $c_t'(s)$ along H. Using 3.29 ii) we have

$$\frac{\partial}{\partial t}g(X,X) = 2g(\overline{D}_{\frac{\partial}{\partial s}}Y,X),$$

therefore

$$\frac{\partial^2}{\partial t^2}g(X,X) = 2g(\overline{D}_{\frac{\partial}{\partial t}}\overline{D}_{\frac{\partial}{\partial s}}Y,X) + 2g(\overline{D}_{\frac{\partial}{\partial s}}Y,\overline{D}_{\frac{\partial}{\partial t}}X).$$

Since $[\frac{\partial}{\partial t},\frac{\partial}{\partial s}] = 0$ we have $\overline{D}_{\frac{\partial}{\partial t}}X = \overline{D}_{\frac{\partial}{\partial s}}Y$ in view of 3.29 i). Using 3.29 iii), we have also

$$\overline{D}_{\frac{\partial}{\partial t}}\overline{D}_{\frac{\partial}{\partial s}}Y = \overline{D}_{\frac{\partial}{\partial s}}\overline{D}_{\frac{\partial}{\partial t}}Y + R(\frac{\overline{\partial}}{\partial t},\frac{\overline{\partial}}{\partial s})Y.$$

Therefore

$$\frac{1}{2}\frac{\partial^2}{\partial t^2}g(X,X)_{|t=0} = g(\overline{D}_{\frac{\partial}{\partial s}}\overline{D}_{\frac{\partial}{\partial t}}Y,X) - R(Y,X,Y,X) + |\,Y'\,|^2,$$

which is also equal to

$$\frac{\partial}{\partial s}g(\overline{D}_{\frac{\partial}{\partial t}}Y,X) - R(X,Y,X,Y) + |\,Y'\,|^2.$$

Formula i) follows by integration from 0 to L. For ii), just write

$$\frac{d}{dt}L(c_t) = \int\limits_0^L \sqrt{g(X,X)}\,\frac{\partial}{\partial t}g(X,X)ds,$$

hence $\quad \dfrac{d^2}{dt^2}L(c_t)_{|t=0} = \int\limits_0^L \left(\dfrac{1}{2}\dfrac{\partial^2}{\partial t^2}g(X,X) - \dfrac{1}{4}(\dfrac{\partial}{\partial t}g(X,X))^2\right)ds.$

Indeed, for the points $H(s,0)$ we have $g(X,X) = 0$. The claimed formula for the second derivative of L follows from a direct computation. If $\tilde{Y} = Y - g(X,Y)X$, then $\tilde{Y}' = Y' - g(X,Y')X$, since $D_{c'}X = 0$. This gives our last claim. ∎

3.35 Synge theorem. *Any compact even dimensional orientable Riemannian manifold with strictly positive sectional curvature is simply connected.*

Remark. Looking at $(P^2\mathbf{R}, \mathrm{can})$, $(P^3\mathbf{R}, \mathrm{can})$ and flat tori, we see that we have an optimal result. On the other hand, it can be proved that any *non-compact* complete Riemannian manifold with strictly positive sectional curvature is contractible (cf. [C-E] ch.8).

Proof. Suppose that M is not simply connected. Then (cf. 2.98) there exists in any free homotopy class a closed geodesic $c : [0, L] \to M$ for which the minimum of length in that class is achieved. The parallel transport P along c from $c(0)$ to $c(L) = c(0)$ is orientation preserving and leaves the orthogonal E to $c'(0)$ invariant. Since E is odd-dimensional, there exists $y \in E$ such that $Py = y$.

Take the parallel vector field Y along c such that $Y(0) = y$, and a variation c_t of c associated with Y. Then $\frac{d}{dt}L(c_t)_{|t=0} = 0$ and

$$\frac{d^2}{dt^2}L(c_t)_{|t=0} = -\int_O^L R(c',Y,c',Y)ds$$

is strictly negative. Hence $L(c_t) < L(c)$ for t small enough, a contradiction. ∎

3.36 Remark. $*$ When removing the assumption about orientability, the theorem can be applied to the orientable covering of M. Therefore, the fundamental group of an even-dimensional Riemannian manifold with strictly positive sectional curvature has order 2 at most. For example, the manifold $P^2\mathbf{R} \times P^2\mathbf{R}$ admits no metric with strictly positive sectional curvature. Compare with 3.15. $*$

3.C Jacobi vector fields

3.C.1 Basic topics about second derivatives

3.37 If M is a smooth manifold and $m \in M$ is a critical point of a smooth function f on M, we can define the second derivative of f at m as follows.

Let u and v be tangent vectors at m, and take vector fields X and Y such that $X_m = x$ and $Y_m = y$. From the very definition of the bracket, since m is critical,

$$(X.Y.f)(m) - (Y.X.f)(m) = df_m([X,Y]_m) = 0.$$

Since $(X.Y.f)(m) = u.(Y.f)(m)$ and $(Y.X.f)(m) = v.(X.f)(m)$, it follows that $(X.Y.f)(m)$ only depends on u and v and defines a symmetric bilinear form on $T_m M$. This form shall be denoted $d^2 f_m$.

Remarks: i)*This definition works for infinite dimensional manifolds.*
ii) Do not confuse $d^2 f_m$ and the Riemannian Hessian Ddf we defined in 2.64. (Some authors call $d^2 f_m$ Hessian, which might be misleading). The second differential only depends on the differential structure, but is only defined at critical points. The Hessian is defined everywhere, but depends on the Riemannian structure. However, $Ddf_m = d^2 f_m$ at any critical point m.

To study f in the neighborhood of a critical point, the basic tool is Morse lemma (cf. [Mi2], p.6).

3.38 Lemma. *Let $m \in M$ be a critical point of a smooth function f. Suppose that the bilinear form $d^2 f_m$ is non degenerate. Then there exists local coordinates $(x^i)_{1 \leq i \leq n}$ in a neighborhood of m such that*

$$f(x^1, \cdots, x^n) = f(m) - (x^1)^2 - \cdots - (x^p)^2 + (x^{p+1})^2 + \cdots + (x^n)^2.$$

Notice that the integer p, which is just the index of the bilinear symmetric form $d^2 f_m$, does not depend on the coordinate system.

3.39 Definition. The point m is said to be *a non degenerate critical point of index p*. For instance, a non degenerate critical point of index zero is a local minimum.

3.C.2 Index form

All that was said above can be applied to the case where $M = \Omega_{pq}$ and f is the energy. Of course, Ω_{pq} is infinite dimensional. In [Mi2], J. Milnor introduces "finite dimensional approximations" of Ω_{pq} which are made of broken geodesics. J. In [Kl], W. Klingenberg, introduces a Hilbert manifold modeled on $H^1(I, \mathbf{R}^n)$. Four our purpose, it will be enough to mimick the finite dimensional situation. The tangent space at a curve c will be the space of vector fields along c. The critical points of the energy are the geodesics. We have seen in 2.92 that geodesics which are "short enough" provide minima for length and energy. But too long a geodesic (think of a great circle of S^n) will no longer be a local minimum. However, it turns out that the second derivative is a quadratic form with finite index. Namely, the maximal dimension of a space of vector fields along c such that $\frac{d^2}{dt^2}E < 0$ is finite. More explicitly, remark that if $m \in M$ is a critical point of a smooth function f and v a vector of T_mM, the second derivative is given by

$$d^2 f_m(v,v) = \frac{d^2}{dt^2} f(\gamma(t))_{|t=0},$$

where γ is any curve such that $\gamma(0) = m$ and $\gamma'(0) = v$. Therefore, we shall define the second differential of energy (or *index form I*) as

$$I(X,X) = \frac{d^2}{dt^2} E(c_t)_{|t=0},$$

where $dc_t/dt_{|t=0} = X$. It is a symmetric bilinear form on the vector space of vector fields along c.

3.40 Proposition. *If X and Y are vector fields along a length parametrized geodesic c which vanish at the ends of c, the second derivative of the energy is given by*

$$I(X,Y) = -\int_0^L g(X, Y'' + R(c',Y)c')\, ds$$

and the second derivative of the length by

$$I'(X,Y) = I(\tilde{X}, \tilde{Y}) = -\int_0^L g(\tilde{X}, \tilde{Y}'' + R(c', \tilde{Y})c')\, ds,$$

where $\tilde{X}$ and $\tilde{Y}$ are the normal components of X and Y respectively.
Proof. Use the equality $g(X',Y') = (g(X,Y'))' - g(X,Y'')$ and 3.34. Notice that $\frac{D}{ds}$ preserves normal and tangential components of a vector field along a geodesic. ∎

3.41 Remark. When $Y(s) = f(s)c'(s)$, we can take for the associated variation $H(s,t) = c(s + tf(s))$. It amounts to change the parameter of c. This explains why the variation formulas for length do not involve tangential components.
It is clear that I (resp. I'), as a bilinear form on $\Gamma(c, TM)$, is degenerate if and only if there exists a vector field along c (resp. a vector field along c which is normal to c) which vanishes at the ends and satisfies

$$Y'' + R(c',Y)c' = 0.$$

3.42 Definition. *A Jacobi field* along a geodesic c is a vector field along c which satisfies the above differential equation.

3.43 Theorem. *Let $c : [0, L] \to M$ be a geodesic in a Riemannian manifold (M,g). Then*
i) *for any $u, v \in T_{c(0)}M$, there exists one and only one Jacobi field along c such that $Y(0) = u$ and $Y'(0) = v$. If $Y(0) = 0$ and $Y'(0) = kc'(0)$, then*

$Y(s) = ksc'(s)$ *for any s. If* $Y(0)$ *and* $Y'(0)$ *are orthogonal to* $c'(0)$, *then* $Y(s)$ *is orthogonal to* $c'(s)$ *for any s. In particular, the vector space of Jacobi fields has dimension 2n, and the subspace of Jacobi fields which are normal to c has dimension* $2(n-1)$;

ii) *for given initial data u and v,* $Y(s)$ *only depends on* $c(s)$. *Namely, if* $\gamma(s) = c(\lambda s)$, *the Jacobi field along* γ *with the same initial data as* Y *is given by* $\tilde{Y}(\gamma(s)) = Y(c(\lambda s))$.

Proof. Take an orthonormal basis $(e_1, ..., e_n)$ of $T_{c(0)}M$ such that $e_1 = kc'(0)$. Using 2.74, we see that the parallel transport along c of the vectors e_i gives a field of orthonormal frames

$$s \to (X_1(s), ..., X_n(s))$$

along c, with $X_1(s) = kc'(s)$. Every Jacobi vector field Y along c is a linear combination of the X_i, say $Y = \sum_{i=1}^{n} y_i X_i$, whose coefficients y_i satisfy the differential system

$$y_i'' + \sum_{j=2}^{n} R(c', X_j, c', X_i) y_j = 0.$$

For given initial data $Y(0) = u$ and $Y'(0) = v$, the existence and uniqueness of Y come from standard results about linear differential systems.

If $Y(0) = 0$ and $Y'(0) = kc'(0)$, then $Y(s) = ksc'(s)$ since $(sc'(s))'' = 0$. The condition for $Y(0)$ and $Y'(0)$ to be orthogonal to c means that $y_1(0) = 0$ and

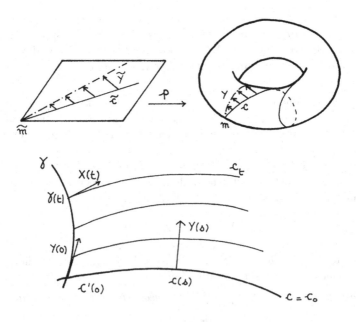

Fig. 3.1. Jacobi fields on a flat torus

$y_1'(0) = 0$. In that case $y_1(s) = 0$ for any s, since $y_1'' = 0$. The last claim is quite as elementary. ∎

3.44 Example. Let us give the Jacobi fields along a geodesic $\tilde{c}$ of $\mathbf{R}^n$ and along its projection c in a flat torus T^n. On $\mathbf{R}^n$, since $R = 0$, the Jacobi fields satisfy $Y'' = 0$. They are given by $Y(t) = u + tv$, where u and v are constant vector fields.

Now, since the projection p of $\mathbf{R}^n$ onto T^n is a local isometry, a Jacobi field along c is just the projection of a Jacobi field along $\tilde{c}$. In fact, if u and v are initial data as in theorem 3.43 above, Y is given by

$$Y(t) = tY_1(t) + Y_0(t),$$

where Y_0 and Y_1 are parallel vector fields along c such that $Y_0(0) = u$ and $Y_1(0) = v$.

3.C.3 Jacobi fields and exponential map

3.45 Proposition. *Let $c : [a, b] \rightarrow M$ be a geodesic, and H be a variation of c such that all the curves c_t are geodesics. Then*

$$Y(s) = \frac{\overline{\partial}}{\partial t} H(s, 0)$$

is a Jacobi field along c. Conversely, every Jacobi field can be obtained in that way.

Proof. We have

$$Y''(s) = \overline{D}_{\frac{\partial}{\partial s}} \overline{D}_{\frac{\partial}{\partial s}} \frac{\overline{\partial}}{\partial t}.$$

Performing two exchanges of s and t, we get (see 3.29i) and 3.29iii))

$$Y''(s) = \overline{D}_{\frac{\partial}{\partial t}} \overline{D}_{\frac{\partial}{\partial s}} \frac{\overline{\partial}}{\partial s} - R(\frac{\overline{\partial}}{\partial t}, \frac{\overline{\partial}}{\partial s}) \cdot \frac{\overline{\partial}}{\partial s}.$$

Since the curves c_t are geodesics, the first term vanishes and we get

$$Y''(s) = -R\big(c'(s), Y(s)\big)c'(s).$$

Conversely, take a Jacobi field Y along c, and the geodesic γ from $c(0)$ such that $\gamma'(0) = Y(0)$. Take parallel vector fields X_0 and X_1 along γ such that $X_0(0) = c'(0)$ and $X_1(0) = Y'(0)$. Set

$$X(t) = X_0(t) + tX_1(t) \quad \text{and} \quad H(s, t) = \exp_{\gamma(t)}\big(sX(t)\big).$$

It is clear that H is a variation of c such that the curves c_t are geodesics. Using the first part of the proof, we see that the derivative $Z(s)$ of H with respect to t at $(s, 0)$ defines a Jacobi field. It is easy to check that Z and Y satisfy the same initial conditions, so $Y = Z$ and we are done. ∎

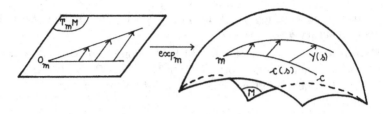

Fig. 3.2. Differentiation of the exponential map

3.46 Corollary. *Let m be a point of (M, g), and u, v be two tangent vectors at m. Let c be the geodesic $r \to \exp_m rv$, and Y be the Jacobi field along c such that $Y(0) = 0$ and $Y'(0) = u$. Then*

$$Y(r) = T_{rv} \exp_m \cdot ru.$$

(In other words, the tangent map to the exponential is described by Jacobi fields along radial geodesics).

Proof. In $T_m M$ equipped with its Euclidean metric g_m, the Jacobi field along the geodesic $s \to sv$ such that $X(0) = 0$ and $X'(0) = u$ is just $X(s) = su$. This Jacobi field is generated by the variation $H(s, t) = s(v + tu)$. Since all the curves in this variation are radial geodesics, the variation $\exp_m H(s, t)$ of c in M is also a geodesic variation, which is generated by the Jacobi field $Y(s) = T_{sv} \exp_m \cdot X(s)$. Clearly, $Y(0) = 0$ and $Y'(0) = u$, so that Y answers our claim. ∎

3.C.4 Applications

We are going to compute normal Jacobi vector fields such that $Y(0) = 0$ for a geodesic c of (S^n, can), (H^n, can) and $(P^n \mathbf{R}, \text{can})$. That will give us the curvature of these manifolds (we already derived from their structures of homogeneous spaces that they have constant curvature (see 3.14).

3.47 Curvature of the sphere S^n.

The geodesic from x with initial speed v is just $c(s) = \cos s.x + \sin s.v$. Take u in $T_x S^n$ orthogonal to v and with norm 1. The variation

$$H(s, t) = \cos s.x + \sin s(\cos t.v + \sin t.u)$$

is geodesic. Since the vector field along c defined by $U(s) = u$ is parallel (cf. 2.56), we see that the Jacobi field which is associated with H is given by $Y(s) = \sin s.U(s)$. Since Y satisfies the differential equation

$$R_{c(s)}(c', Y)c' = -Y'',$$

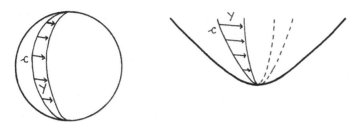

Fig. 3.3. Jacobi fields for S^n and H^n

we see, after noticing that $Y + Y'' = 0$, that the sectional curvature is constant and equal to 1, so that

$$R(u, v, w, z) = g(u, w)g(v, z) - g(u, z)g(v, w).$$

Namely, R is the tensor R^o of 3.9.

3.48 Curvature of the hyperbolic space H^n.

This time, the geodesic c is given by $c(s) = \cosh s.x + \sinh s.v$. Recall that v is orthogonal to x in $\mathbf{R}^{n+1}$ for the Lorentzian metric we have used to define H^n (cf. 2.10). Take a tangent vector u at x as above, and the geodesic variation

$$H(s, t) = \cosh s.x + \sinh s(\cos t.v + \sin t.u).$$

Using the same notations as above, we get the Jacobi field $Y(s) = \sinh s.U(s)$. This time $Y = Y''$, so that the sectional curvature is constant and equal to -1.

3.49 Curvature of the projective space $P^n\mathbf{R}$.

Using the Riemannian covering map p by the standard sphere and 3.13, we see that $P^n\mathbf{R}$ has constant (equal to 1) sectional curvature. The normal Jacobi fields are the projections by Tp of the normal Jacobi fields of S^n. They are given by $Y(s) = \sin s.U(s)$, where U is a normal parallel field along c. If we compare with the sphere, we have the following important difference:

on S^n, if $Y(0) = 0$, the first point for which Y vanishes again is the antipodal point of $c(0)$;

on $P^n\mathbf{R}$, this point is just $c(0)$, which is obtained when the parameter is equal to π.

3.50 Another application: the curvature of 2-dimensional Riemannian manifolds.

Let (M, g) be such a manifold, u be a tangent unit vector at m and c be the geodesic $r \rightarrow \exp_m ru$. In $T_m M$, take polar coordinates (r, θ) associated with an orthonormal basis (u, v). Using Gauss lemma (2.93), and restricting g to the complement of $\mathrm{Cut}_m M$, we get

$$\exp_m^* g = dr^2 + f^2(r, \theta)d\theta^2,$$

with $f(0, \theta) = 0$ and $f'_r(0, \theta) = 1$ (because of 3.89 bis). The parallel transport of v along c gives a vector field V. Clearly,

$$T_{(r,0)} \exp \cdot \frac{\partial}{\partial r} = c'(r) \quad \text{and} \quad T_{(r,0)} \exp \cdot \frac{\partial}{\partial \theta} = f(r,0)V(r).$$

This proves that $Y(r) = f(r,0)V(r)$ is the normal Jacobi field such that $Y(0) = 0$ and $Y'(0) = v$. Therefore, the curvature at $c(r)$ is

$$-\frac{\partial_r^2 f(r,0)}{f(r,0)}.$$

3.51 Exercise. Compute the curvatures of (S^2, can) and (H^2, can) in that way. Deduce another proof that (S^n, can) and (H^n, can) have constant sectional curvature.

3.D Riemannian submersions and curvature

3.D.1 Riemannian submersions and connections

3.52 Definitions. i) Let $p : (\tilde{M}, \tilde{g}) \to (M, g)$ be a Riemannian submersion. A vector field $\tilde{X}$ on $\tilde{M}$ is said to be *horizontal* if $\tilde{X}_{\tilde{m}} \in H_{\tilde{m}}$ for any $\tilde{m}$ in $\tilde{M}$ (recall that $H_{\tilde{m}}$ is the horizontal space introduced in 2.26).
ii) If Y is a vector field on M, the unique horizontal vector field $\tilde{Y}$ on $\tilde{M}$ such that $Y_{p(\tilde{m})} = T_{\tilde{m}}\tilde{p} \cdot \tilde{Y}_{\tilde{m}}$ for any $\tilde{m}$ is the *horizontal lift* of Y.
To compute the Levi-Civita connection of (M, g), we shall use formula 2.52 together with the following lemmas.

3.53 Lemma. *A vector field U on $\tilde{M}$ is vertical if and only if $U \cdot (f \circ p) = 0$ for any function f on M.*
Proof. Elementary. ∎

3.54 Lemma. *If $\tilde{X}$ and $\tilde{Y}$ are the horizontal lifts of vector fields X and Y on M, the vector field $[\tilde{X}, \tilde{Y}] - \widetilde{[X,Y]}$ is vertical; if U is vertical, $[\tilde{X}, U]$ is vertical.*
Proof. Set $\tilde{f} = f \circ p$. Then $([X, Y] \cdot f) \circ p = [\tilde{X}, \tilde{Y}] \cdot \tilde{f}$. Our first claim follows using 3.53. To prove the second claim, remark that $U \cdot \tilde{f} = 0$. Therefore,

$$[\tilde{X}, U] \cdot \tilde{f} = -U \cdot \tilde{X} \cdot \tilde{f}.$$

Now, if $\tilde{X}$ is the horizontal lift of X, the function $\tilde{X} \cdot \tilde{f}$ is equal to $(X \cdot f) \circ p$. It is constant on the fibers, so that $U \cdot \tilde{X} \cdot \tilde{f} = 0$. We conclude by using lemma 3.53 again. ∎

3.55 Proposition. *Let X and Y be vector fields on M, with horizontal lifts $\tilde{X}$ and $\tilde{Y}$. Then*

$$\tilde{D}_{\tilde{X}}\tilde{Y} = \widetilde{D_X Y} + \frac{1}{2}[\tilde{X}, \tilde{Y}]^v.$$

(Here, the superscript v denotes the vertical component). In particular, for any $\tilde{m}$ in $\tilde{M}$

$$(D_X Y)_{p(\tilde{m})} = T_{\tilde{m}} p(\tilde{D}_{\tilde{X}} \tilde{Y})_{\tilde{m}}.$$

Proof. If $\tilde{Z}$ is the horizontal lift of Z, using formula 2.52 and lemma 3.54, we get

$$\tilde{g}(\tilde{D}_{\tilde{X}} \tilde{Y}, \tilde{Z})_{\tilde{m}} = g(D_X Y, Z)_m.$$

If $U \in \Gamma(TM)$ is vertical, for the same reasons

$$\tilde{g}(\tilde{D}_{\tilde{X}} \tilde{Y}, U) = \frac{1}{2} \tilde{g}([\tilde{X}, \tilde{Y}], U).$$

Our claims are consequences of those two relations. ∎

3.56 **Remark:** There is no reason why $\tilde{D}_{\tilde{X}} \tilde{Y}$ should be horizontal. The vector field $[\tilde{X}, \tilde{Y}]$ is not generally horizontal either. However, using 1.114, it can be checked that

$$[\tilde{X}, \tilde{Y}]^v = [\tilde{X}, \tilde{Y}] - \widetilde{[X, Y]}$$

is a $(1, 2)$ tensor.*This tensor measures the non-integrability of the horizontal distribution. It is crucial when studying total spaces of Riemannian submersions, see e.g. [B2], and compare with 3.16.*

Before going on with the general study of Riemannian submersions, let us look further at the submersion

$$p : (S^{2n+1}, \text{can}) \rightarrow (P^n \mathbf{C}, \text{can}).$$

3.57 **Exercise.** Let $J \in \Gamma(\text{End}(P^n \mathbf{C}))$ be the tensor field of endomorphisms we introduced in 2.30. Show that J and the 2-form ω defined by

$$\omega_m(x, y) = g_m(J_m x, y)$$

are *parallel*.

3.D.2 Jacobi fields of $P^n \mathbf{C}$

3.58 Here $\tilde{M}$ is (S^{2n+1}, can) and M is $(P_n \mathbf{C}, \text{can})$. Denote by p the canonical Riemannian submersion (cf. 2.28) of $\tilde{M}$ on M, and by c a geodesic in M such that $c(0) = m$ and $c'(0) = v$. Let u be a unit vector in $T_m M$ which is orthogonal to v. Take $\tilde{m}$ in $\tilde{M}$ such that $p(\tilde{m}) = m$, and denote by $\tilde{u}$ and $\tilde{v}$ the horizontal vectors of the tangent space at $\tilde{m}$ which project onto u and v respectively. Recall that $\tilde{M}$ is viewed as the unit sphere S^{2n+1} in $\mathbf{C}^{n+1}$. Therefore

$$H_m = \{x \in \mathbf{R}^{2n+2}, \langle m, x \rangle = \langle im, x \rangle = 0\}.$$

The variation

$$\tilde{H}(s, t) = \cos s . \tilde{m} + \sin s (\cos t . \tilde{v} + \sin t . \tilde{u})$$

of the geodesic $\tilde{c}(s) = \cos s.\tilde{m} + \sin s.\tilde{v}$ of S^{2n+1} consists in horizontal geodesics. Therefore, its projection $H(s,t)$ is a geodesic variation of c. The Jacobi field $\tilde{Y}$ which is associated with $\tilde{H}$ is just $\sin s.\tilde{U}(s)$, where $\tilde{U}$ is the parallel vector field along $\tilde{c}$ such that $\tilde{U}(0) = u$.

The Jacobi field which is associated with H is given by

$$Y(s) = T_{\tilde{c}(s)}p \cdot \tilde{Y}(s) = \sin s \, T_{\tilde{c}(s)}p \cdot \tilde{U}(s).$$

Set $U(s) = T_{\tilde{c}(s)}p \cdot \tilde{U}(s)$.

First case: $\tilde{u}$ is orthogonal to $i\tilde{v}$. By 2.30, it amounts to say that u is orthogonal to Jv. Then $\tilde{U}(s)$ is orthogonal to $\tilde{c}'(s)$ and $i\tilde{c}'(s)$: it means that $\tilde{U}(s)$ is *horizontal*. Since $\tilde{D}_{\tilde{c}'}\tilde{U} = 0$, by 3.55 we have $D_{c'}U = 0$, hence

$$Y''(s) = -\sin s.U(s), \quad \text{thus} \quad Y(s) = \sin sU(s).$$

Now, the equation of Jacobi fields gives $R(c',Y)c' = Y$. This means that

$$R(v, u)v = u.$$

Second case: $\tilde{u} = i\tilde{v}$, which is the same as $u = Jv$. The vector field $\tilde{U}$ is no more horizontal. Its horizontal component at the point $\tilde{c}(s)$ is

$$i\tilde{v} - \langle i\tilde{v}, i\tilde{c}(s)\rangle i\tilde{c}(s) = \cos s(\cos s.i\tilde{v} - \sin s.i\tilde{m}) = \cos s.i\tilde{c}'(s).$$

Therefore $Y(s) = \sin s \cos s.J_{c(s)}c'(s)$.

3.59 Lemma. *We have $D_{c'}(Jc)' = 0$.*

Proof. Using 3.55, we see that $(D_{c'}(Jc'))_{c(s)} = T_{\tilde{c}(s)}p \cdot (\tilde{D}_{\tilde{c}'(s)}i\tilde{c}'(s))$. But

$$D_{\tilde{c}'(s)}i\tilde{c}'(s) = -\cos s.i\tilde{m} - \sin s.i\tilde{v} = -i\tilde{c}(s),$$

which means that $D_{\tilde{c}'}i\tilde{c}'$ is vertical. ∎

Another proof. $D_{c'}(Jc)' = (D_{c'}J)c' + JD_{c'}c' = 0$, for we have seen in 3.57 that J is parallel. ∎

These computations show that

$$Y''(s) = (\sin s \cos s)''J_{c(s)}c'(s) = -4Y(s).$$

Then $R(c', Jc')c' = 4Jc'$, or

$$R(v, Jv)v = 4Jv.$$

General case: Any unit vector of $T_m P^n \mathbf{C}$ which is orthogonal to v can be written as

$$u = \cos\alpha.u_0 + \sin\alpha.Jv,$$

where u_0 is orthogonal to iv. Then

$$R(v, u)v = \cos\alpha.u_0 + 4\sin\alpha.Jv,$$

and the sectional curvature of the plane generated by u and v is just

$$K(u,v) = 1 + 3\sin^2\alpha.$$

We have proved that K is strictly positive, and vary between 1 (for $K(u_0,v)$) and 4 (for $K(v,Jv)$). Now, we have seen in 2.42 that the isometry group of $P_n\mathbf{C}$ acts transitively on the unit tangent bundle. Therefore, the sectional curvature and the curvature tensor are completely determined.

3.60 **Exercise.** i) Recover those results by imitating the method of 3.51 and using suitable totally geodesics submanifolds of $P^n\mathbf{C}$.
ii) Using an orthonormal basis of $T_mP^n\mathbf{C}$ of type

$$\{e_1, Je_1, \cdots, e_n, Je_n\}$$

(where $g(e_i,e_j) = \delta_{ij}$), diagonalize the curvature operator.

3.D.3 O'Neill's formula

The formula which gives the curvature of $P^n\mathbf{C}$ is just a particular case of the following result, due to O'Neill.

3.61 **Theorem.** *Let $p : (\tilde{M},\tilde{g}) \to (M,g)$ be a Riemannian submersion, and X, Y be orthonormal vector fields on M with horizontal lifts $\tilde{X}$ and $\tilde{Y}$. Then*

$$K(X,Y) = K(\tilde{X},\tilde{Y}) + \frac{3}{4}\,|\,[\tilde{X},\tilde{Y}]^v\,|^2\,.$$

Proof. First remark that if U is vertical, lemma 3.54 shows that

$$\tilde{g}(\tilde{D}_U\tilde{X},\tilde{Y}) = \tilde{g}(\tilde{D}_{\tilde{X}}U,\tilde{Y}) + \tilde{g}([U,\tilde{X}],\tilde{Y}) = \tilde{g}(\tilde{D}_{\tilde{X}}U,\tilde{Y}).$$

Now, using 3.55, we get

$$\tilde{g}(\tilde{D}_{\tilde{X}}U,\tilde{Y}) = -\tilde{g}(U,\tilde{D}_{\tilde{X}}\tilde{Y}) = -\frac{1}{2}\tilde{g}([\tilde{X},\tilde{Y}]^v,U).$$

Eventually,

$$\tilde{g}(\tilde{D}_U\tilde{X},\tilde{Y}) = -\frac{1}{2}g([\tilde{X},\tilde{Y}]^v,U).$$

The claimed formula follows from a direct computation, using 3.55 and this last formula. We leave the details to the reader. See also [C-E] ch.3. ∎

3.62 **Remark.** From a "qualitative" view-point, O'Neill's formula just says that the basis of a Riemannian submersion carries "more curvature" than the total space. We shall see in 3.69 a geometric proof of this property.

Let G be a compact Lie group equipped with a bi-invariant Riemannian metric, and let H be a closed sub-group of G. Right-translations are isometries of G, and from 2.28 we see that, for a given bi-invariant metric, there exists a unique Riemannian metric on G/H that makes the canonical projection a Riemannian submersion. The left translations go to the quotient and give *isometries* of G/H, which is a Riemannian homogeneous space.

3.63 Definition. A *normal homogeneous* metric on G/H is a quotient metric of a bi-invariant metric on G by the right action of H.

3.64 Proposition. *If G/H is an isotropy irreducible homogeneous space, with G compact, "the" canonical metric of G/H is normal homogeneous.*

Proof. Equip G with a bi-invariant metric. The preceding discussion shows that such a metric goes to the quotient and gives a G-invariant metric on G/H. But we know from 2.43 that such a metric is unique up to a scalar. ∎

For a normal homogeneous space G/H, denote by $\underline{P}$ the orthogonal of $\underline{H}$ in $\underline{G}$. By definition, $\underline{P}$ is the horizontal space at e of the Riemannian submersion p from G to G/H.

3.65 Theorem. *Let G/H be a normal homogeneous space. Then*
i) the geodesics from $p(e)$ in G/H are projections of one parameter subgroups of type $\exp tX$, where $X \in \underline{P}$;
ii) the sectional curvature of G/H is given by

$$K(X,Y) = \frac{1}{4} \mid [X,Y]^{\underline{P}} \mid^2 + \mid [X,Y]^{\underline{H}} \mid^2$$

where $X,Y \in \underline{P}$ and $T_{p(e)}G/H$ is identified with $\underline{P}$.

Proof. Assertion i) is just a particular case of theorem 2.109, and ii) is a direct consequence of O'Neill's formula and formula 3.17 for the curvature of bi-invariant metrics on Lie groups. ∎

3.66 Exercise. Recover the sectional curvature of $P^n\mathbf{C}$ from theorem 3.65.

3.D.4 Curvature and length of small circles. Application to Riemannian submersions

Take a 2-plane P in the tangent space at m of a Riemannian manifold (M, g), and an orthonormal basis $\{u, v\}$ of P. Set

$$H(r, \theta) = \exp_m r(\cos\theta.u + \sin\theta.v).$$

We shall call *small circles* the curves $C_r : \theta \to H(r, \theta)$. For a given P, we shall give an asymptotic expansion of the function $L(C_r)$ near $r = 0$.

3.67 Lemma. *Let Y be a smooth vector field along a curve $r \to c(r)$ in M. Let $Y_0,...,Y_k$ be the parallel vector fields along c such that $Y_k(0) = D_{c'}^{(k)}Y(0)$. For any k,*

$$Y(r) = \sum_{i=0}^{k} \frac{r^i}{i!} Y_i(r) + o(r^k).$$

Proof. Just apply Taylor-Young formula to the components of Y in a parallel frame along c. ∎

3.68 Theorem. *The function $L(C_r)$ admits the asymptotic expansion*

$$L(C_r) = 2\pi r\left(1 - \frac{K(P)}{6}r^2 + o(r^2)\right).$$

In particular, if $L(C_r^o)$ denotes the length of the corresponding Euclidean circle, $K(P)$ is the limit when r goes to 0 of

$$\frac{6}{r^2}\left(1 - \frac{L(C_r)}{L(C_r^o)}\right).$$

Proof. Denote by Y_θ the Jacobi field along the geodesic $r \to H(r, \theta)$ which is given by the variation H. From 3.45, one has

$$L(C_r) = \int_0^{2\pi} | Y_\theta(r) | \, d\theta.$$

Set $Y_0 = Y$. Clearly, $Y(0) = 0$ and $Y'(0) = v$. From the equation of Jacobi fields, we also get $Y''(0) = 0$. Now, taking the derivative of Jacobi equation, we get

$$Y'''(0) = -\left(\frac{D}{dt}R(c'(t), Y(t))c'(t)\right)_{|t=0} = -R(u, v)u,$$

since $Y(0) = 0$ (which rules out the term involving the derivative of R), and $D_{c'}c' = 0$. We have similar results for Y_θ, replacing u and v by $\cos\theta u + \sin\theta v$ and $-\sin\theta u + \cos\theta v$ respectively. From the preceding lemma, we get

$$| Y_\theta(r) | = r - \frac{K(P)}{6}r^3 + o(r^3),$$

so that

$$L(C_r) = 2\pi r\left(1 - \frac{K(P)}{6}r^2 + o(r^2)\right)$$

as claimed. ■

Using this formula, we can deduce the sectional curvature of (S^n, can) and (H^n, can) from a direct computation of $L(C_r)$. We also get a geometric proof of a weakened O'Neill's theorem.

3.69 Theorem. *Let $p : (\tilde{M}, \tilde{g}) \to (M, g)$ be a Riemannian submersion. Let $\tilde{p}(m) = m$ and let u and v be tangent vectors at m whose horizontal lifts at $\tilde{m}$ are $\tilde{u}$ and $\tilde{v}$. Then*

$$K(u, v) \geq K(\tilde{u}, \tilde{v}).$$

Proof. We can suppose that u and v, and consequently $\tilde{u}$ and $\tilde{v}$, are unit orthogonal vectors. Then, with the previous notations, we have, using 2.109, $p(\tilde{C}_r) = C_r$ and $L(C_r) \leq L(\tilde{C}_r)$. Then our claim follows from 3.68. ■

3.69 bis Example. We have seen in 3.D.3 that there are very many compact homogeneous spaces which carry a metric with non-negative sectional curvature. We give now a more involved example. Take $S^3 \times S^2$, with

$$S^3 = \{Z = (z_0, z_1) \in \mathbf{C}^2, |z_0|^2 + |z_1|^2 = 1\}$$
$$S^2 = \{(u, r) \in \mathbf{C} \times \mathbf{R}, |u|^2 + r^2 = 1\},$$

and define a S^1 action by

$$e^{it}(Z, (u, r)) = (e^{it}Z, (e^{it}u, r)).$$

We have the Hopf action on S^3, and rotations with the same axis on S^2. If $S^3 \times S^2$ is equipped with the product of canonical metrics, this action is isometric, so that the quotient metric has non-negative sectional curvature. Topologically, the quotient of $S^3 \times S^2$ by this action is the space obtained from $S^3 \times [-1, 1]$ by quotienting $S^3 \times \{-1\}$ and $S^3 \times \{1\}$ by the Hopf action. * It is diffeomorphic to the connected sum $(P^2\mathbf{C}, +)\sharp(P^2\mathbf{C}, -)$ – that is the connected sum of two copies of $P^2\mathbf{C}$ with opposite orientation – and therefore does not carry any homogeneous metric. For details, cf. [SB].* This idea of using some "twisted" isometric action to obtain non trivial examples of manifolds with positive curvature is used systematically in [C-E].

3.E The behavior of length and energy in the neighborhood of a geodesic

3.E.1 Gauss lemma

We have seen (2.92, 2.97) that "short enough" geodesics provide local minima of length and energy. Jacobi fields will permit us to know what happens in the neighborhood of a "long" geodesic.

3.70 **Gauss lemma** (compare with 2.93). *Let u and v be tangent vectors at m in M, and c be the geodesic $\exp_m sv$. Then*

$$g_{c(r)}(T_{rv}\exp_m \cdot v, T_{rv}\exp_m \cdot u) = \langle u, v \rangle.$$

In particular, the geodesics from m are orthogonal to the distance spheres

$$S_r = \{m' \in M, d(m, m') = r\} = \exp_m S^e(0, r),$$

where $S^e(0, r)$ denotes the Euclidean sphere of center 0_m and radius r in T_mM.
Proof. We have seen in 3.46 that $T_{rv}\exp_m \cdot u = Y(r)$, where Y is the Jacobi field along c such that $Y(0) = 0$ and $Y'(0) = \frac{u}{r}$. If

$$\frac{u}{r} = \lambda v + u_1 \quad \text{where} \quad \langle u_1, v \rangle = 0,$$

then (cf. 3.43) $Y(s) = \lambda s c'(s) + Y_1(s)$, where Y_1 is a normal Jacobi field. Now

$$g(Y(r), c'(r)) = \lambda r = \langle u, v \rangle. \blacksquare$$

3.71 Lemma. *In T_mM, let ϕ be the radius $t \to tv$ (with $t \in [0,1]$) and $t \to \psi(t)$ be a piecewise C^1 curve such $\psi(0) = 0$ and $\psi(1) = v$. Then*

$$L(\exp_m \circ \psi) \geq L(\exp_m \circ \phi) = \mid v \mid \cdot$$

Proof. Suppose first that ψ is C^1, and set $\psi(t) = r(t)u(t)$ with $\mid u(t) \mid = 1$ (polar coordinates). Then

$$\psi'(t) = r'(t)u(t) + r(t)u'(t), \quad \text{with} \quad \langle u(t), u'(t) \rangle = 0,$$

and the Gauss lemma gives

$$\mid (\exp_m \circ \psi)'(t) \mid^2 = \mid T_{\psi(t)} \exp_m \psi'(t) \mid^2 = (r'(t))^2 + (r(t))^2 \mid T_{\psi(t)} \exp_m u'(t) \mid^2 .$$

Therefore

$$L(\exp_m \circ \psi) \geq \int_0^1 \mid r'(t) \mid dt \geq \mid r(1) - r(0) \mid = \mid v \mid .$$

If ψ is only piecewise C^1, just apply the same estimate for each interval where it is C^1.

3.E.2 Conjugate points

3.72 Definition. Let $c : [a, b] \to M$ a geodesic with ends p and q. The points p and q are said to be *conjugate along* c if there exists a non trivial Jacobi field along c such that $Y(a) = Y(b) = 0$.
If the vector space of Jacobi fields along c such that $Y(a) = Y(b) = 0$ is k-dimensional, it will be said that p and q are *conjugate of order (or multiplicity)* k.
Remarks. i) This property does not depend on the parameter of the geodesic.
ii) If $c(s) = \exp_p sv$ and $q = \exp_p rv$, the differential of $\exp_p$ at rv is *singular*, and the dimension of its kernel is the multiplicity (cf. 3.46).
iii) The vector field Y is forced to be normal to c (see 3.43).
3.73 Theorem. *Let $c : [a, b] \to M$ be a geodesic with ends p and q in a Riemannian manifold (M, g).*
i) *If there is no conjugate point to p along c, then there is a neighborhood V of c in Ω_{pq} (for the uniform topology) such that, for any γ in V,*

$$L(\gamma) \geq L(c) \quad \text{and} \quad E(\gamma) \geq E(c).$$

Moreover, as soon as $\gamma([a, b]) \neq c([a, b])$ these estimates are sharp.
ii) *If there exists s_0 in $]a, b[$ such that p and $c(s_0)$ are conjugate along c, there exists a variation c_t of c with fixed ends such that*

$$L(c_t) < L(c) \quad \text{and} \quad E(c_t) < E(c)$$

for t small enough.

iii)*If p and q are not conjugate along c, then c is a non-degenerate critical point. Its index is finite, and is equal to the number of conjugate points to p along c, each of them being counted with its multiplicity.*

Be careful: in case (i), we only know that c provides a *local* minimum of L and E. On a flat torus, there are no conjugate points. Any geodesic segment is then a local minimum for L. But there are infinitely many geodesics with given ends, and generically only one of them is minimizing.

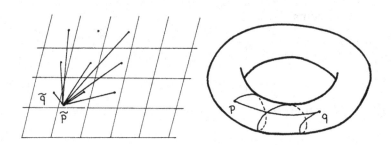

Fig. 3.4. Minimizing and non-minimizing geodesics on a torus

3.74 Exercise. Explain what happens for a geodesic of $P_n\mathbf{R}$ whose length is a little greater than $\frac{\pi}{2}$.

3.75 Proof of 3.73 i). We can assume that $a = 0$ and $b = 1$. Set $v = c'(0)$ and denote by ϕ the curve $s \to sv$ in T_pM. Our assumption implies that $\exp_p$ is a local diffeomorphism in a neighborhood of each $\phi(s)$. Cover $\phi([0,1])$ with a finite number of such neighborhoods $W_1,...,W_k$, and set $U_i = \exp_p W_i$. Take a subdivision

$$s_0 = 0 < s_1 < \cdots < s_k = 1$$

such that $\phi([s_{i-1}, s_i]) \subseteq W_i$. If H denotes a variation of c, for suitable positive ϵ_i and for any i

$$H\left([s_{i-1}, s_i]\times] - \epsilon_i, \epsilon_i[\right) \subseteq U_i.$$

Therefore $c_t([0,1]) \subseteq \bigcup U_i$ for t small enough.

The curve c_t can be lifted stepwise to a curve ψ_t from 0 to v. Indeed, suppose we have defined ψ_t up to s_{i-1} and that $\psi_t(s_{i-1})$ lies in W_i. Then we extend ψ_t to $[0, s_i]$ by setting

$$\psi_t(s) = \left(\exp_{p|_{W_i}}\right)^{-1}(c_t(s)).$$

Lemma 3.71 says that $L(c_t) \geq L(c)$. Moreover, since the differential of $\exp_p$ at $\psi_t(s)$ is invertible, the inequality is sharp as soon as c_t and c have different images. As for the energy, Schwarz inequality says that

$$E(c_t) \geq \frac{1}{2}(L(c_t))^2 \geq \frac{1}{2}(L(c))^2 = E(c).$$

Now, any curve γ in Ω_{pq} which is close enough to c for the uniform topology can be represented as a c_t. ■

Before proving ii), notice that the second variation formula can be extended to variations which are piecewise C^2.

3.76 Proof of 3.73 ii): Recall that $a = 0$, $b = 1$. By assumption, there is a normal Jacobi field such that $Y(0) = 0$ and $Y(s_0) = 0$. We shall produce a normal vector field Y_α such that

$$I(Y_\alpha, Y_\alpha) < 0.$$

Notice that $Y'(s_0) \neq 0$. Otherwise, using 3.43, Y should be identically zero. Now, take the parallel vector field Z_1 along c such that $Z_1(s_0) = -Y'(s_0)$, and a smooth function θ on $[0, 1]$ such that

$$\theta(0) = \theta(1) = 0 \quad \text{and} \quad \theta(s_0) = 1.$$

Set $Z(s) = \theta(s)Z_1(s)$, and define a new vector field Y_α by

$$Y_\alpha(s) = \begin{cases} Y(s) + \alpha Z(s) & \text{if} \quad s \in [0, s_0] \\ \alpha Z(s) & \text{if} \quad s \in [s_0, 1] \end{cases}$$

This vector field is not smooth at s_0. Nevertheless it defines a piecewise smooth variation of c. Using the second variation formula 3.34, we get

$$I(Y_\alpha, Y_\alpha) = I_1 + I_2 + I_3,$$

where

$$I_1 = \int_0^{s_0} \left(g(Y', Y') - R(Y, c', Y, c') \right) ds$$

$$I_2 = 2 \int_0^{s_0} \left(g(Y', Z') - R(Y, c', Z, c') \right) ds$$

and $I_3 = \alpha^2 I(Z, Z)$. Using the equalities

$$(g(Y', Y))' = g(Y', Y') + g(Y'', Y)$$
$$(g(Y', Z))' = g(Y', Z') + g(Y'', Z)$$

and taking the Jacobi equation for Y into account, we get

$$I_1 = [g(Y, Y')]_0^{s_0} \quad \text{and} \quad I_2 = 2\alpha[g(Z, Y')]_0^{s_0}.$$

Eventually, we obtain

$$I(Y_\alpha, Y_\alpha) = -2\alpha \mid Y'(s_0) \mid^2 + \alpha^2 I(Z, Z).$$

For small enough α, one has $I(Y_\alpha, Y_\alpha) < 0$, and for the corresponding variation of the geodesic c

$$\frac{d}{dt}L(c_t)_{|t=0} = 0 \quad \text{and} \quad \frac{d^2}{dt^2}L(c_t)_{|t=0} = I(Y_\alpha, I_\alpha) < 0.$$

Therefore $L(c_t)$ is strictly smaller than $L(c_t)$ for t small enough. The proof for E goes in the same way. Property iii), which is known as *the Morse index theorem*, will not be proved here. It will not be used in the sequel either. For a proof, see [Mi2], part III. ∎

3.77 Corollary. *Let Cut_m be the cut-locus of m in a complete Riemannian manifold (M, g). Consider the (star-shaped) open set U_m of T_mM with the following property: v lies in U_m if and only if there exists $\epsilon > 0$ such that the geodesic $\exp_m tv$ is minimizing up to $1 + \epsilon$. Then $\exp_m$ is a diffeomorphism of U_m onto $M \setminus \mathrm{Cut}_m$.*
Proof. We have seen in 2.111 that $\exp_m$ is injective on U_m. Part ii) of theorem 3.76 says that when v belongs to U_m, the points m and $\exp_m v$ are not conjugate, so that $\exp_m$ is a local diffeomorphism. ∎

3.78 Scholium.[1] *The point m' will belong to the cut-locus of m if and only if one of the two (not exclusive from each other) properties are satisfied:*
a) there exist two distinct minimizing geodesics from m to m';
b) there is a minimizing geodesic from m to m' along which m and m' are conjugate.

An important consequence is the following "symmetry" property of the cut-locus: m lies in $\mathrm{Cut}_{m'}$ if and only if m' lies in Cut_m.

3.79 Exercise. Revisit the examples of 2.114 and check whether a), b) (or both!) occur.

When r is not bigger $d(m, \mathrm{Cut}_m)$ the restriction of the exponential map to the open ball $B(0, r)$ in T_mM is a diffeomorphism. Recall that the infimum of $d(m, \mathrm{Cut}_m)$ is called the injectivity radius $i(g)$ of the manifold.

3.80 Klingenberg's lemma. *Let (M, g) be a compact manifold whose sectional curvature is smaller than a^2. Then*

$$i(g) \geq \min\left\{\frac{\pi}{a}, \frac{l}{2}\right\},$$

where l is the length of the shortest closed geodesic in M.
Proof. Let p and q be two points such that $d(p, q) = i(g)$. Suppose first that in 3.78 a) is realized but not b), and denote by c_1 and c_2 two minimal geodesics from p to q. If $c_1'(q) \neq -c_2'(q)$, there exists $u \in T_qM$ such that

$$g(v, -c_1'(q)) > 0 \quad \text{and} \quad g(v, -c_2'(q)) > 0.$$

---

[1] A scholium is basically a comment of some important statement. This term was used a long time ago in theology. We use it here as a homage to Spinoza. The scholia of [Sa] contain the best of his philosophy.

Let $\sigma : [0, \epsilon[\mapsto M$ be a curve such that $\sigma(0) = q$ and $\dot{\sigma}(0) = v$.
Since q is not conjugate to p along c_1 or c_2, there exist for small t one-parameter families of geodesics c_1^t and c_2^t from p to $\sigma(t)$ such that $c_1^0 = c_1$ and $c_2^0 = c_2$. Then, by the first variation formula, one has length$(c_i^t) < d(p, q)$ for t small enough. Suppose for instance that

$$\text{length}(c_1^t) \leq \text{length}(c_2^t) < \text{length}(c_1) = \text{length}(c_2).$$

Using 3.78 again, we see that the geodesic c_2^t is minimizing at most up to $\sigma(t)$, a contradiction.
We can use the same argument after exchanging p and q, and get a closed geodesic of length $2i(g)$.
If p and q are conjugate, then $d(p, q) \geq \frac{\pi}{a}$. In dimension 2, this comes from the Sturm-Liouville comparison theorem, applied to the differential equations of Jacobi fields. A similar but more involved argument works in higher dimension (cf. [C-E], [B-C] or [Kl]). ∎

3.E.3 Some properties of the cut-locus

3.81 The study of the cut-locus and of the injectivity radius may be quite delicate. Indeed, properties 3.78 a) and b) must be taken care of together, but the former is global while the latter is local. Let us give some examples.

i) It is not difficult to prove (see 2.113) that Cut_m is a deformation retract of $M \setminus m$. Hence, for any Riemannian metric on S^n and for any $m \in S^n$ the cut-locus is contractible. In particular, if $n = 2$, it is a tree, which is of finite type if g is real analytic. Moreover, for any $m' \in \text{Cut}_m$, the number of minimizing geodesics from m to m' is equal to the number of connected components of $(V \setminus m') \cap \text{Cut}_m$, where V is a small ball with center at m (we suppose that the cut-locus contains more than one point). In particular, the ends of the tree are conjugate points, therefore Cut_m always contains such points. All these results have been proved by Myers, cf. [My].

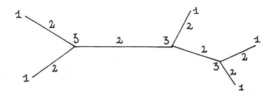

Fig. 3.5. The cut-locus of a (non round!) 2-sphere

ii) If $M \neq S^2$ there always exists a metric and a point m for which the cut-locus contains no conjugate points. If $\dim M = 2$, this property comes from the topological classification of compact surfaces. *Either $M = P^2\mathbf{R}$ and the canonical metric on the real projective plane answers the question, or $\chi(M) \leq 0$. In that case, M carries a metric with constant curvature (cf. [Sp], vol 4). Namely, the torus and the Klein bottle carry flat metrics, and the surfaces with $\chi(M) < 0$ carry metrics with constant negative curvature. In any case, 3.88 says there are no conjugate points.*

If $\dim M \geq 3$, the existence of such metrics comes from a direct construction due to Weinstein ([Wn]).

iii) In the opposite direction, the cut-locus of any point in the standard sphere or in the standard complex projective space is composed with conjugate points. *The same property is true for any simply connected Riemannian symmetric space (Crittenden, cf. [C-E], p.102). It can be deduced from that (cf. ibidem, or [Hn], VII Th. 4.7) that $\mathrm{codim}\mathrm{Cut}_m \geq 2$, and that $\mathrm{codim}\mathrm{Cut}_m \geq 3$ for any compact semi-simple Lie group G equipped with a bi-invariant metric. Therefore, using a dimension argument, one sees that any continuous map of S^2 in such a G is homotopic to a map with values in $G \setminus \mathrm{Cut}_m$, i.e. to a constant map since $G \setminus \mathrm{Cut}_m$ is contractible. In other words, $\Pi_2(G) = 0$, a classical result of Elie Cartan.*

3.F Manifolds with constant sectional curvature

From 3.20, we can suppose that $K = 0$, $+1$ or -1.

3.82 Theorem. *If (M, g) is a complete Riemannian manifold with constant sectional curvature K, its universal Riemannian covering is either*

i) $(\mathbf{R}^n, \mathrm{can})$ *if $K = 0$;*

ii) (H^n, can) *if $K = -1$;*

iii) (S^n, can) *if $K = +1$.*

Proof. Take $m \in M$, a unit tangent vector v at m, and the geodesic $c(t) = \exp_m tv$. Take also a normal vector u to v, the parallel vector field U along c such that $U(0) = u$, and the Jacobi field Y such that $Y(0) = 0$ and $Y'(0) = u$. Recall that (cf. 3.46)

$$Y(t) = T_{tv} \exp_m \cdot tu.$$

i) If $K = 0$, then $Y''(t) = 0$ so that $Y(t) = tU(t)$. Using Gauss lemma 3.70, we see that $T_{tv} \exp_m$ is an isometry. In other words, $\exp_m$ is a local isometry from $T_m M$ into M. It is a covering map since (M, g) is complete (cf. 2.106).

ii) If $K = -1$, one has $Y'' = Y$, therefore $Y(t) = \sinh t.U(t)$. Take $a \in H^n$ and an isometry f between $T_m M$ and $T_a H^n$. Let us transport all our constructions to H^n. More explicitly, take $\tilde{u}$ and $\tilde{v}$ to be $f(u)$ and $f(v)$ respectively, and the Jacobi field $\tilde{Y}$ along $\widetilde{\exp}_a t\tilde{v}$ with 0 and $\tilde{u}$ as initial data. Again $\tilde{Y}(t) = \sinh t.\tilde{U}(t)$. Since $\widetilde{\exp}_a$ is a diffeomorphism from $T_a H^n$ onto H^n, we can define a smooth map F of H^n into M by setting

$$F = \exp_m \circ f^{-1} \circ (\widetilde{\exp}_a)^{-1}.$$

Now $Y(t) = T_{c(t)}F \cdot \tilde{Y}(t)$, and using Gauss lemma again, we see that F is a local isometry, and consequently a covering map.

iii) If $K = 1$, then $Y(t) = \sin t.U(t)$. On the other hand, for a in S^n, we have seen in 3.77 that $\tilde{\exp}_a$ is a diffeomorphism of the ball $B(0, \pi)$ in S^n onto $S^n \setminus \{-a\}$. Therefore, by using the same procedure as in ii), we get a local isometry F of $S^n \setminus \{-a\}$ into M. Now, taking b in S^n different from a and $-a$, we get a local isometry of $S^n \setminus \{-b\}$ by setting

$$F_1 = \exp_{F(b)} \circ T_b F \circ (\tilde{\exp}_b)^{-1}.$$

It is clear that F and F_1 match together to give a local isometry of (S^n, can) into (M, g). Using the completeness argument again, we have a covering map, and we are done. ∎

Remark. A byproduct of this result is the corresponding *local* one: a manifold with constant curvature K is locally isometric to the Euclidean space, the hyperbolic space or the sphere.

3.83 Exercise. a) Show that any even dimensional complete manifold with constant positive sectional curvature is isometric to (S^{2n}, can) or to $(P^{2n}\mathbf{R}, \mathrm{can})$.
b) Take

$$S^3 = \{(z, z') \in \mathbf{C}^2, |z|^2 + |z'|^2 = 1\}.$$

Taking a p-th root of unit α (with p prime), and an integer k between 1 and $p - 1$, set

$$t_{p,k}(z, z') = (\alpha z, \alpha^k z').$$

Show that $t_{p,k}$ gives a free isometric action of $\mathbf{Z}/p\mathbf{Z}$ on S^3. Let $L_{p,k}$ be the quotiented Riemannian manifold. Show that $L_{p,k}$ and $L_{p,l}$ are isometric if and only if k and l are equal or inverse modulo p. These manifolds are called *lens spaces*. A complete classification of manifolds with constant positive curvature has been given by J. Wolf. Indeed, it is the main subject of the book [Wo 1]. *Using a very subtle topological invariant, called simple homotopy type (cf. [Mi 1]), it can be proved that $L_{p,k}$ and $L_{p,l}$ are not homeomorphic as soon as they are not isometric.*

c) Equip $SU(3)$ with the bi-invariant metric we introduced in 2.48. Show that the subgroup T of diagonal matrices of $SU(3)$ is a totally geodesic (cf. 5.8 c) flat 2-dimensional submanifold. More precisely, show that T is isometric to the *hexagonal* 2-torus.
More generally, the (flat) metric of maximal tori of a compact simple Lie group (all of them are conjugate, therefore isometric) can be read on the Dynkin diagram, cf. [Hn].

3.84 Remarks. We shall see in 3.L that any compact surface with negative Euler characteristic carries a metric with *constant negative curvature*.
Concerning flat compact manifolds, we have the following theorem:

Theorem (Bieberbach). *Let M^n be a flat compact Riemannian manifold, $\pi_1(M)$ be its fundamental group, and $\Gamma := \pi_1(M) \cap \mathbf{R}^n$ be the subgroup of* *translations in $\pi_1(M)$. Then, Γ is a free Abelian normal subgroup of rank n of $\pi_1(M)$, $G := \pi_1(M)/\Gamma$ has finite order, and $\mathbf{R}^n/\Gamma$ is a torus which covers M with deck group G.*

The proof of Bieberbach is explained in [Wo1]. See also [Bu1] for a particularly enlightening proof, discovered by M. Gromov.

3.G Topology and curvature: two basic results

The two theorems we present in this section are not difficult, but far reaching. Today, they still give rise to various improvements and generalizations.

3.G.1 Myers' theorem

Denote by $S^n(r)$ the radius r sphere in $\mathbf{R}^{n+1}$. Its metric is just $r^2 g_o$, where g_o is the standard metric on the sphere: the diameter is πr and the sectional curvature r^{-2}.

Let h and h' be two symmetric bilinear forms on a real vector space. It is said that $h \geq h'$ if $h - h'$ defines a non-negative quadratic form. Clearly, this definition can be extended to symmetric 2-tensors on a manifold. For a Riemannian manifold (M, g), we have $h \geq ag$ if and only if the (pointwise) eigenvalues of h are bigger than a.

3.85 Myers' theorem. *Let (M, g) be a complete Riemannian manifold such that*

$$Ric \geq (n-1)r^{-2}g \quad \text{where } r \text{ is real positive.}$$

Then $\mathrm{diam}(M, g) \leq \mathrm{diam}(S^n(r))$. In particular, M is compact and its fundamental group is finite.

Proof. For any L which is smaller than the diameter, we can find p and q such that $\mathrm{dist}(p, q) = L$. Hopf-Rinow's theorem says that there is a minimizing geodesic from p to q. For any vector field Y along c which vanishes at the ends, $I(Y, Y) \geq 0$.

Take an orthonormal basis $e_1, ..., e_n$ of $T_p M$ with $e_1 = c'(0)$. Using parallel transport, we get a field of orthonormal frames

$$t \to \{c'(t), X_2(t), ..., X_n(t)\}$$

along c. Set $Y_i(t) = \sin\frac{\pi t}{L} X_i(t)$ (where $2 \leq i \leq n$). Now

$$I(Y_i, Y_i) = -\int_0^L g\left(Y_i, Y_i'' + R(c', Y_i)c'\right) dt$$

$$= \int_0^L \sin^2\frac{\pi t}{L}\left[\frac{\pi^2}{L^2} - R(c', X_i, c', X_i)\right] dt.$$

All the $I(Y_i, Y_i)$ are nonnegative, and so is their sum

$$\sum_{i=2}^{n} I(Y_i, Y_i) = \int_0^L \sin^2 \frac{\pi t}{L} \left((n-1)\frac{\pi^2}{L^2} - Ric(c', c') \right) dt \geq 0.$$

In view of our assumption on the Ricci curvature, L must be smaller than πr, therefore $\text{diam}(M, g) \leq \pi r$.

From Hopf-Rinow's theorem we see that M is compact. Now, the same bound for Ricci curvature holds for the universal Riemannian covering of (M, g), which is also compact. Then $\pi_1(M)$ is finite. ∎

Be careful: a complete Riemannian manifold with strictly positive Ricci curvature may be non-compact (see 5.11).

3.85 bis Remark. Cheng's diameter theorem asserts that, if $Ric \geq (n-1)r^{-2}g$ and $\text{diam}(M, g) = \text{diam}(S^n(r))$, then (M, g) is isometric to $S^n(r)$. See [Pe], ch.9 for a proof.

3.86 Exercise. Take a Lie group G whose Lie algebra $\underline{G}$ has trivial center. Suppose that G carries a bi-invariant Riemannian metric. Show that G and its universal covering are compact.

Show that $O(n, 1)$ and $Sl(n, \mathbf{R})$ carry no bi-invariant Riemannian metric.

As a consequence of this exercise, the universal covering of compact semi-simple Lie group is compact (Weyl's theorem).

3.G.2 Cartan-Hadamard's theorem

3.87 Cartan-Hadamard's theorem. *Let (M, g) be a complete Riemannian manifold with non-positive sectional curvature. Then for any m in M, the exponential map is a covering map. In particular, M is diffeomorphic to $\mathbf{R}^n$ as soon as it is simply connected.*

The proof relies on the following elementary property.

3.88 Lemma. *Let c be a geodesic in a Riemannian manifold (M, g). If the sectional curvature is non-positive along c, then c carries no conjugate points.*
Proof. Take a parameter t of c such that $c(0) = m$, and let Y be a Jacobi field such that $Y(0) = 0$. The function

$$f(t) = g(Y(t), Y(t))$$

vanishes at $t = 0$, and so does its first derivative. Furthermore

$$f'' = g(Y'', Y) + g(Y', Y') = g(Y', Y') - R(c', Y, c', Y)$$

is non negative. Therefore f is convex, and if $f(t_o) = 0$ for some $t_o > 0$ then f vanishes identically on $[0, t_o]$. Then Y, as a solution of a linear differential equation, is forced to vanish everywhere. ∎

Proof of the theorem. The lemma says that $\exp_m$ is a local diffeomorphism. Then the symmetric 2-form $\tilde{g} = \exp_m^* g$ is positive definite, and defines a

Riemannian metric on $T_m M$. Using Gauss lemma, we see that the geodesics of $\tilde{g}$ from the origin are just the rays, and from 2.105 iii) this metric is complete. Therefore, theorem 2.106 shows that $\exp_m$ is a covering map. ■

Simply connected Riemannian manifolds with non positive curvature are called *Hadamard manifolds*. It is worth-noting that this property can be studied from a purely metric point of view. Namely, if γ and γ' are two geodesics of a Hadamard manifold, it can be easily proved, using the second variation formula, that the function $t \mapsto d\big(\gamma(t), \gamma'(t)\big)$ is convex. * This property can be taken as the definition of non positive curvature for the length-spaces of M. Gromov (cf. [Gr1], ch.2). They are, roughly speaking, metric spaces for which the notion of geodesic makes sense. See [BGS] and [B-H] for fascinating developments.*

3.89 Exercise. a) If (M, g) has non-positive curvature, any homotopy class of paths with fixed ends p and q contains a unique geodesic.
b)* The same property is true for *free* homotopy classes if (M, g) has strictly negative curvature.*

3.H Curvature and volume

3.H.1 Densities on a differentiable manifold

There is no "natural" measure on a smooth manifold M. What can be made intrinsic is the property to be absolutely continuous with respect to the Lebesgue measure.

3.90 Definition. Take a smooth structure on M defined by an atlas (U_i, ϕ_i). A *density* is the data, for each open set $\phi_i(U_i)$, of a measure μ_i with the following properties:
i) each μ_i is absolutely continuous and has strictly positive density with respect to the Lebesgue measure;
ii) for any continuous function f with compact support in $\phi_i(U_i \cap U_j)$ we have the compatibility condition

$$\int_{\phi_i(U_i \cap U_j)} f d\mu_i = \int_{\phi_j(U_i \cap U_j)} (f \circ \phi_i \circ \phi_j^{-1}) \mid J(\phi_i \circ \phi_j^{-1}) \mid d\mu_j$$

for any pair (i, j) such that $U_i \cap U_j \neq \emptyset$.

Example. Any volume form on an orientable manifold defines a density.

Any density defines a positive measure δ on the manifold. First, for a continuous function f with support contained in U_i, set

$$\delta(f) = \int_{\phi_i(U_i)} f \circ \phi_i^{-1} d\mu_i.$$

If supp$(f) \subseteq U_j$ for some other index j, the compatibility condition ensures that $\delta(f)$ is well defined. Now, using partitions of unit as explained in chapter I, we can define $\delta(f)$ for any continuous f.

3.91 Proposition. *If M is compact (or locally compact and countable at infinity), there always exists a density. Moreover, if δ and δ' are the measures associated with two densities, there exists a strictly positive continuous function f such that $\delta' = f\delta$.*

Proof. The first part comes from using partitions of unit like in 1.127. The second part is elementary. ∎

Thanks to that last proposition, we can speak of measurable sets and of sets with zero measure in a differentiable manifold.

3.92 Definition. A part A of M is *measurable* (resp. with *measure zero*) if it enjoys this property for some (and then from 3.91 for any) density on M.

3.H.2 Canonical measure of a Riemannian manifold

Recall that all the translation invariant measures on a given affine space are proportional. For the Euclidean space $\mathbf{R}^n$, we choose the measure for which the unit cube has volume 1. If (e_i) is an orthonormal basis and (a_i) any system of n vectors, the volume of the parallelepiped generated by the a_i is given by

$$\det((a_1, \cdots, a_n)/(e_1, \cdots, e_n)) = \sqrt{\det(\langle a_i, a_j \rangle)}$$

(Gram determinant).

Now, take an n-dimensional Riemannian manifold (M, g), and let

$$\sum_{i,j} g_{ij}^{(k)} dx^i dx^j$$

be the local expression of g in a local chart (U_k, ϕ_k). For m in U_k, the tangent vectors $\frac{\partial}{\partial x_i}$ generate a parallelepiped whose volume is

$$\sqrt{\det(g_{ij}^{(k)})}.$$

We are lead to define the *canonical measure* of (M, g), denoted v_g, as corresponding to the density which is given in our atlas by

$$\mu_k = \sqrt{\det(g_{ij}^{(k)})} L_n,$$

where L_n is the Lebesgue measure in $\mathbf{R}^n$.

If M is orientable, and if we take an atlas which is compatible with an orientation of M, then v_g can be given by a volume form. We shall still denote this volume form by v_g. Then the following property can be checked as an easy exercise.

3.93 Proposition. *The covariant derivative of v_g is zero.*

Remark. It is possible to define densities intrinsically, as nowhere vanishing sections of the orientation bundle of M (compare with the "odd forms" introduced by de Rham in the celebrated "Variétés Différentiables"). Then $Dv_g = 0$ makes sense in that general context.*

3.94 Example. Suppose that a submanifold V of $\mathbf{R}^3$ is given by a local parametrization $F(u, v)$. Then the measure associated with the induced metric is given by

$$| F'_u \times F'_v |^{1/2} \, dudv.$$

Here $\times$ is just the cross product in $\mathbf{R}^3$.

3.95 Definition. The *volume* of a Riemannian manifold is the (possibly infinite) integral $\int\limits_M v_g$.

Example. Let $g = dr^2 + f^2(r)d\theta^2$ be a smooth metric on $\mathbf{R}^2$ expressed in polar coordinates. This metric is complete (see 2.105 iii)), and the volume is

$$2\pi \int\limits_0^\infty f(r)dr.$$

It can be finite even though this manifold is not compact. To compute the volume of a Riemannian manifold, we express v_g in an exponential chart, and use the following property.

3.96 Lemma. *For any $m \in M$, the cut-locus Cut_m has measure zero.*

Proof. With the notations of 2.112 the cut-locus is the image by $\exp_m$ of ∂U_m. Since any ray from the origin in the tangent space meets ∂U_m once at most, ∂U_m has measure zero (use polar coordinates and Fubini theorem) and the result follows. ∎

Therefore $\mathrm{vol}(M, g) = \int\limits_{U_m} \exp_m^* v_g$. Now, using 3.45, we can use Jacobi fields to compute $\exp_m^* v_g$. Take a geodesic $c(t) = \exp_m tu$ from m, an orthonormal basis $\{u, e_2, ..., e_n\}$ of $T_m M$, and the Jacobi fields Y_i such that

$$Y_i(0) = 0 \quad \text{and} \quad Y'_i(0) = e_i.$$

Recall that $T_{tu} \exp_m \cdot u = c'(t)$ and $T_{tu} \exp_m \cdot e_i = \frac{1}{t} Y_i(t)$. Then, setting

$$J(u, t) = t^{-(n-1)} \sqrt{\det(g(Y_i(t), Y_j(t))},$$

we have

$$\exp_m^* v_g = J(u, t)dx^1...dx^n = J(u, t)t^{n-1}dtdu.$$

Here, du denotes the canonical measure of the unit sphere of $(T_m M, g_m)$. We see in particular that $J(u, t)$ does not depend on $\{e_2, ..., e_n\}$. If $\rho(u)$ is the (possibly infinite) distance to the cut-locus in the direction u, using Fubini's theorem we get

$$\text{vol}(M, g) = \int_{S^{n-1}} \int_0^{\rho(u)} J(u, t) t^{n-1} dt du. \tag{3.97}$$

3.H.3 Examples: spheres, hyperbolic spaces, complex projective spaces

Volume of the sphere (S^n, can).

Keeping the same notations, and taking the parallel vector field E_i whose value at 0 is e_i, recall that

$$Y_i(t) = \sin t . E_i(t).$$

Therefore

$$\text{vol}(S^n, \text{can}) = \int_{S^{n-1}} \int_0^\pi \left(\frac{\sin t}{t} \right)^{n-1} t^{n-1} du dt$$

$$= \text{vol}(S^{n-1}, \text{can}) \int_0^\pi \sin^{n-1} t dt.$$

We recover the well known formulas

$$\text{vol}(S^{2n}, \text{can}) = \frac{(4\pi)^n (n-1)!}{(2n-1)!} \quad \text{and} \quad \text{vol}(S^{2n+1}, \text{can}) = 2 \frac{\pi^{n+1}}{n!}.$$

In that case of course, any (good) elementary text-book on Analysis gives a better method, using the integral of $\exp -r^2$ over $\mathbf{R}^{n+1}$.

Volume of balls in the hyperbolic space (H^n, can)

Since $\exp_m$ is a global diffeomorphism, the method goes still more easily. We get, using 3.48,

$$\text{vol}(B_m(R)) = \text{vol}(S^{n-1}, \text{can}) \int_0^R \sinh^{n-1} r dr.$$

In particular, there is a constant c_n such that

$$\text{vol}(B_m(R)) \sim c_n \exp(n-1) R$$

when R goes to infinity.

Volume of the complex projective space $(P^n\mathbf{C}, \mathrm{can})$.
Take an orthonormal basis $\{u, e_2, ..., e_n\}$ of the tangent space at m with $e_2 = Ju$. We have seen in 3.58 that the corresponding Jacobi fields Y_i are given by

$$Y_2(t) = \sin t \cos t . E_2(t) \quad \text{and}$$

$$Y_i(t) = \sin t . E_i(t) \quad \text{for} \quad i \geq 3.$$

Since the distance from m to the cut-locus is $\frac{\pi}{2}$ (cf. 2.114), we get

$$\mathrm{vol}(P^n\mathbf{C}, \mathrm{can}) = \int_{S^{2n-1}} \int_0^{\pi/2} \sin^{2n-1} t \cos t \, du dt = \frac{\pi^n}{n!}.$$

Remark. In the three examples above, the computations are so easy because $J(u, t)$ only depends on t. This property is not surprising, since the isometry group acts transitively *on the unit tangent bundle*.
* A Riemannian manifold (M, g) is said to be *globally harmonic* if the function $J(u, t)$ is globally defined on the unit tangent bundle and only depends on t. It is conjectured that such a manifold is either covered by $\mathbf{R}^n$, or isometric to a rank one symmetric space (namely, in the compact case, the sphere, the projective spaces over $\mathbf{R}$, $\mathbf{C}$ or the quaternions, and the Cayley projective plane) equipped with its canonical Riemannian metric. This conjecture has been settled in the compact simply connected case by R. Szabó, cf. [Sz]. It is related with the study of manifolds all of whose geodesics are closed: cf. [B1], p.170 for a precise statement, and the whole chapter 6 of [B1] for a thorough study of harmonic manifolds.
In the non-compact case, there exist non symmetric harmonic manifolds: the Damek-Ricci spaces, cf. [D-R] or [Ro], which are solvable Lie groups equipped with some special left-invariant metric. In such spaces, as an easy consequence of Prop. 4.16 below, the Laplacian of a radial function (that is a function which only depends on the distance to a given point) is still a radial function: they are especially interesting in harmonic analysis. *

3.H.4 Small balls and scalar curvature

3.98 Theorem. *Let m be a point in (M, g). Then, if $\dim M = n$,*

$$\mathrm{vol}(B_m(r)) = r^n \mathrm{vol}(B^e(1)) \left(1 - \frac{s(m)}{6(n + 2)} r^2 + o(r^2)\right).$$

Proof. For r small enough

$$\mathrm{vol}(B_m(r)) = \int_{S^{n-1}} \int_0^r J(u, t) t^{n-1} du dt .$$

Take again the Jacobi fields Y_i of 3.96. From 3.67, we have the asymptotic expansion

$$Y_i(t) = tE_i - \frac{t^3}{6}R(c', E_i)c' + o(t^3).$$

The claimed result follows from the asymptotic expansion of $J(u,t)$. To get that expansion, we use the following lemmas.

3.99 Lemma. *Let $A(t)$ be a differentiable map from $I \subseteq \mathbf{R}$ into $Gl_n\mathbf{R}$. Then*

$$(\det A)' = (\det A)\mathrm{tr}(A^{-1}A').$$

The proof is left to the reader.

3.100 Lemma. *For any symmetric bilinear form ϕ on $\mathbf{R}^n$*

$$\int_{S^{n-1}} \phi(v,v)dv = \frac{1}{n}\mathrm{vol}(S^{n-1})\mathrm{tr}(\phi).$$

Proof. Just diagonalize ϕ with respect to an orthonormal basis. ∎

Remarks. i) As a by-product of this proof, we get the asymptotic expansion

$$(\exp_m^* v_g)(u) = \left(1 - \frac{1}{6}Ric(u,u) + o(|\,u\,|^2)\right)v_{\mathrm{eucl}}.$$

ii) A more general asymptotic expansion has been given by A. Gray (cf. [Gy]). The existence of such an expansion is not unexpected in view of the following result of Elie Cartan (see [B-G-M] for a proof in the spirit of this book, using Jacobi fields) : the coefficients of the Taylor expansion at 0 of $\exp_m^* g$ are universal polynomials in the curvature tensor and its covariant derivatives.

3.H.5 Volume estimates

The proof of 3.98 suggests that suitable curvature assumptions could give volume estimates. Denote by $V^k(r)$ the volume of a ball of radius r in the complete simply connected Riemannian manifold with constant curvature k. The following comparison theorem is due to Bishop (case i)) and Gunther (case ii)).

3.101 Theorem (Bishop-Gunther). *Let (M,g) be a complete Riemannian manifold, and $B_m(r)$ be a ball which does not meet the cut-locus of m.*
i) *If there is a constant a such that $Ric \geq (n-1)ag$, then*

$$\mathrm{vol}(B_m(r)) \leq V^a(r).$$

ii) *If there is a constant b such that $K \leq b$, then*

$$\mathrm{vol}(B_m(r)) \geq V^b(r).$$

Proof. Take a geodesic $c(t) = \exp_m tu$ from m, and an orthonormal basis $\{u, e_2, ..., e_n\}$ of the tangent space at m. Take also, as in the proof of Myers' theorem for example, the parallel vector fields E_i (with $2 \leq i \leq n$) along c such that $E_i(0) = e_i(0)$. Suppose that

$$0 \leq r \leq \rho(u).$$

For such an r, there exists a unique Jacobi field Y_i^r such that

$$Y_i^r(0) = 0 \quad \text{and} \quad Y_i^r(r) = E_i(r).$$

Indeed, since $T_{ru} \exp_m$ is an isomorphism from the tangent space at m onto the tangent space at $c(r)$, this Jacobi field is given by

$$Y_i^r(t) = T_{tu} \exp_m \cdot tv,$$

where v is the unique tangent vector at m such that

$$T_{ru} \exp_m \cdot rv = E_i(r).$$

Now,

$$J(u, t) = C_r t^{1-n} \det \left(Y_2^r(t), \cdots, Y_n^r(t) \right),$$
$$\text{where} \quad C_r^{-1} = \det \left(Y'^r_2(0), \cdots, Y'^r_n(0) \right).$$

For given u, set $f(t) = J(u, t)$.

3.102 Lemma. *Denoting by I the index form of energy, we have*

$$\frac{f'(r)}{f(r)} = \sum_{i=2}^{n} I(Y_i^r, Y_i^r) - \frac{(n-1)}{r}.$$

Proof of the lemma. First remark that

$$| \det(Y_2^r, ... Y_n^r) | = (\det g(Y_i^r, Y_j^r))^{1/2}.$$

In other words, denoting this last determinant by $D(t)$, we have

$$\frac{f'(t)}{f(t)} = \frac{D'(t)}{2D(t)} - \frac{n-1}{t}.$$

For $t = r$, the matrix $\left[g(Y_i^r, Y_j^r) \right]$ is just the unit matrix, and lemma 3.99 shows that

$$D'(r) = 2 \sum_{i=2}^{n} g((Y_i^r)', Y_i^r).$$

On the other hand, by the same argument as in 3.76, the second variation formula 3.34, when applied to a Jacobi field Y, gives

$$I(Y, Y) = \int_0^r \left(| Y' |^2 - R(Y, c', Y, c') \right) ds = [g(Y, Y')]_0^r.$$

The claimed formula is now straightforward. ∎

3.103 Lemma. *If $c : [a, b] \to M$ is a minimizing geodesic, Y is a Jacobi field and X is a vector field along c with the same values as Y at the ends, then $I(X, X) \geq I(Y, Y)$.*

Proof of the lemma. Since $X - Y$ vanishes at the ends, we have

$$I(X - Y, X - Y) \geq 0$$

because c is minimizing. On the other hand we have

$$I(Y, Y) = [g(Y', Y)]_a^b \quad \text{and} \quad I(X, Y) = [g(Y', X)]_a^b.$$

Therefore $I(X - Y, X - Y) = I(X, X) - I(Y, Y)$ and the result follows. ■

End of the proof of the theorem. i) We shall apply the above lemma to Y_i^r and to the vector field X_i^r given by

$$X_i^r(t) = \frac{s(t)}{s(r)} E_i(t),$$

where

$$s(t) = \sin \sqrt{a} t \quad \text{if} \quad a > 0$$

$$s(t) = t \quad \text{if} \quad a = 0$$

$$s(t) = \sinh \sqrt{-a} t \quad \text{if} \quad a < 0.$$

Lemma 3.103 gives

$$\sum_{i=2}^{n} I(Y_i^r, Y_i^r) \leq \sum_{i=2}^{n} I(X_i^r, X_i^r).$$

The right member of this inequality is just

$$\int_0^r \left(\frac{s(t)}{s(r)} \right)^2 ((n-1)a - Ric(c', c')) ds + \sum_{i=2}^{n} g(X_i^r, (X_i^r)')(r).$$

The assumption made on the curvature yields that the integral is negative. Then, using lemma 3.102 and the definition of X_i^r, we see that

$$\frac{f'(r)}{f(r)} \leq (n-1) \left(\sqrt{a} \cotan \sqrt{a} r - \frac{1}{r} \right) \quad \text{if} \quad a > 0$$

$$\frac{f'(r)}{f(r)} \leq 0 \quad \text{if} \quad a = 0$$

$$\frac{f'(r)}{f(r)} \leq (n-1) \left(\sqrt{-a} \cotanh \sqrt{-a} r - \frac{1}{r} \right) \quad \text{if} \quad a < 0.$$

In any case, if $f_a(r)$ denotes the function $J(u,r)$ for the "model space" with constant curvature a (recall that J does not depend on u in that case), we have

$$\frac{f'(r)}{f(r)} \leq \frac{f'_a(r)}{f_a(r)}.$$

By integrating, we get $f(r) \leq f_a(r)$, and the claimed inequality follows from a further integration, using 3.97. ∎

ii) Denoting by Y one of the Jacobi fields Y_i^r, we have (cf. the proof of lemma 3.102)

$$g(Y(r), Y'(r)) = \int_0^r (g(Y', Y') - R(Y, c', Y, c')) ds$$

$$\geq \int_0^r (g(Y', Y') - bg(Y, Y)) ds.$$

Write

$$Y(t) = \sum_{i=2}^n y^i(t) E_i(t).$$

On the simply connected manifold with constant curvature b, take a geodesic $\tilde{c}$ of length r, and define vector fields $\tilde{E}_i$ along $\tilde{c}$ in the same way as the vectors E_i. Set

$$\tilde{Y}(t) = \sum_{i=2}^n y^i(t) \tilde{E}_i(t).$$

Then

$$\int_0^r \left(|\tilde{Y}'|^2 - b |\tilde{Y}|^2 \right) dt = \int_0^r \left(|Y'|^2 - b |Y|^2 \right) dt = I(\tilde{Y}, \tilde{Y}).$$

Lemma 3.103, when applied to the simply connected manifold with constant curvature b, gives

$$I(\tilde{Y}_i^r, \tilde{Y}_i^r) \geq I(\tilde{X}_i^r, \tilde{X}_i^r),$$

where $\tilde{X}_i^r(t) = \frac{s(t)}{s(r)} \tilde{E}_i(t)$ is the Jacobi field which takes at the ends of $\tilde{c}$ the same values as $\tilde{Y}_i^r$. Using lemma 3.102, we see that

$$\frac{f'(r)}{f(r)} \geq \frac{f'_b(r)}{f_b(r)},$$

and the claim follows by integration. ∎

3.I Curvature and growth of the fundamental group

3.I.1 Growth of finite type groups

Let Γ be a group of finite type, and $S = \{a_1, ..., a_k\}$ be a system of generators of Γ. Any element s of Γ can be written as

$$s = \prod_i a_{k_i}^{r_i} \quad (r_i \in \mathbf{Z}),$$

with possible repetitions of the generators a_{k_i}. Such a representation is called a *word* with respect to the generators, and the integer

$$\sum_i |r_i|$$

is by definition the *length* of that word. For any positive integer s, the number of elements of Γ which can be represented by words whose length is not greater than s will be denoted by $\phi_S(s)$.

3.104 Exercise. i) Show that if Γ is the free Abelian group generated by the a_i,

$$\phi_S(s) = \sum_{i=0}^{k} 2^i \binom{k}{i} \binom{s}{i}.$$

In particular, $\phi_S(s) = O(s^k)$.
ii) Show that if Γ is the free group generated by the a_i, then

$$\phi_S(s) = \frac{k(2k-1)^s - 1}{k-1}.$$

In view of these examples, we are lead to introduce the following definitions.

3.105 Definitions. i) A group Γ of finite type is said to have *exponential growth* if for any system of generators S there is a constant $a > 0$ such that $\phi_S(s) \geq \exp(as)$.
ii) Γ is said to have *polynomial growth of degree $\leq n$* if for any system of generators S there is an $a > 0$ such that $\phi_S(s) \leq as^n$.
iii) Γ is said to have *polynomial growth of degree n* if the growth is polynomial of degree $\leq n$ without being of degree $\leq n-1$.
It is not difficult to show (cf. [VCN] for instance, and [Wo 2] for more details) that Γ has exponential, or polynomial growth of degree n, as soon as the above properties hold for *some* system of generators.

3.106 Theorem (Milnor-Wolf, cf. [Mi3] and [Wo2]). *Let (M, g) be a complete Riemannian manifold with nonnegative Ricci curvature. Then any subgroup of $\pi_1(M)$ with finite type has polynomial growth whose degree is at most $\dim M$. The same property holds for $\pi_1(M)$ if M is compact.*

Proof. The fundamental group acts isometrically on the universal Riemannian cover $(\tilde{M}, \tilde{g})$. Take $a \in \tilde{M}$. From the very definition of covering maps, we can find $r > 0$ such that the balls $B(\gamma(a), r)$ are pairwise disjoint. Take a finite system S of generators of the subgroup we consider, and set

$$L = \max d(a, \gamma_i(a)), \quad \gamma_i \in S.$$

Now, if $\gamma \in \pi_1(M)$ can be represented as a word of length not greater than s with respect to the γ_i, clearly

$$d(a, \gamma(a)) \le Ls.$$

Taking all such γ's, we obtain $\phi_S(s)$ disjoint balls $B(\gamma(a), r)$, such that

$$B(\gamma(a), r) \subseteq B(a, Ls + r).$$

Therefore

$$\phi_S(s) \le \frac{\mathrm{vol}(B(a, Ls + r))}{\mathrm{vol}(B(a, r))} \le C_M (Ls + r)^n$$

in view of Bishop's theorem 3.101.

The last claim is straightforward, since the fundamental group of a compact manifold has finite type (see [My]). ∎

3.107 Example. Take the Heisenberg group (cf. 2.90 bis) H, and the subgroup $H_{\mathbf{Z}}$ of H obtained by taking integer parameters. It can be proved (cf. [Wo 3]) that $H_{\mathbf{Z}}$ has polynomial growth of degree 4. Therefore, the compact manifold $H/H_{\mathbf{Z}}$ carries no metric with nonnegative Ricci curvature.

3.I.2 Growth of the fundamental group of compact manifolds with negative curvature

First recall some standard properties of the action of $\pi_1(M)$ on the universal covering $\tilde{M}$. First of all, if M is compact, there exists a compact K of $\tilde{M}$ whose translated $\gamma(K)$ cover $\tilde{M}$.

3.108 Proposition. *The covering* $(\gamma(K)), \gamma \in \pi_1(M)$, *is locally finite.*

Proof. Equip M with a Riemannian metric, and take the universal Riemannian covering $(\tilde{M}, \tilde{g})$ of (M, g). Let d be the distance which is given by $\tilde{g}$. As a consequence of the Lebesgue property for the compact K, there exists some $r > 0$ such that, for any ball B of radius r whose center lies in K, the balls $\gamma(B)$ are pairwise disjoint when γ goes through $\pi_1(M)$.

Now, we are going to show that for any x in $\tilde{M}$, the ball $B(x, \frac{r}{2})$ only meets a finite number of $\gamma(K)$. Since the γ's are isometries, we can suppose that x lies in K. Suppose there exists a sequence γ_n of distinct elements of Γ, and a sequence y_n of points of K, such that for any n

$$\gamma_n(y_n) \in B(x, \frac{r}{2}).$$

After taking a subsequence if necessary, we can suppose that y_n converges in K. Let y be the limit. Then, since the γ_n are isometries, $\gamma_n(y)$ belongs to $B(x, r)$ for n big enough, a contradiction. ∎

A direct consequence of this lemma is the following: for a given $D > 0$, the set

$$S = \{\gamma \in \pi_1(M), d(K, \gamma(K)) < D\}$$

is finite. Take D strictly bigger than the diameter δ of M.

3.109 Lemma. *Take $a \in K$, and $\gamma \in \pi_1(M)$ such that, for some integer s,*

$$d(a, \gamma(K)) \leq (D - \delta)s + \delta.$$

Then γ can be written as the product of s elements of S.

Proof. Take $y \in \gamma(K)$, a minimizing geodesic c from a to y, and points $y_1, y_2, \cdots, y_{s+1}$ such that

$$d(a, y_1) < \delta \quad \text{and} \quad d(y_i, y_{i+1}) \leq (D - \delta) \quad \text{for} \quad 1 \leq i \leq s.$$

Any y_i can be written as $\gamma_i(x_i)$, for some γ_i in $\pi_1(M)$ and some x_i in K, and we can take $\gamma_1 = Id$ and $\gamma_{s+1} = \gamma$. Then

$$\gamma = (\gamma_1^{-1}\gamma_2)(\gamma_2^{-1}\gamma_3)...(\gamma_s^{-1}\gamma_{s+1}).$$

On the other hand

$$d(x_i, \gamma_{i-1}^{-1}(\gamma_i(x_i))) = d(\gamma_{i-1}(x_i), y_i)$$

is smaller than

$$d(\gamma_{i-1}(x_i), \gamma_{i-1}(x_{i-1})) + d(\gamma_{i-1}(x_{i-1}), y_i)).$$

But this is just $d(x_{i-1}, x_i) + d(y_{i-1}, y_i)$, which is smaller than D, so that $\gamma_{i-1}^{-1}\gamma_i$ is in S.

3.110 Theorem (Milnor, cf. [Mi 3]). *If (M, g) is a compact manifold with strictly negative sectional curvature, then $\pi_1(M)$ has exponential growth.*

Proof. Take a system S of generators as in the preceding lemma. This lemma says that the ball

$$B(a, (D - \delta)s + \delta)$$

is covered by $\phi_S(s)$ compact sets $\gamma(K)$, so that

$$\text{vol}(B(a, (D - \delta)s + \delta) \leq \phi_S(s)\text{vol}(K).$$

On the other hand, if the sectional curvature is smaller than some $-b$, where $b > 0$, theorem 3.101 ii) gives

$$\text{vol}(B(a, (D - \delta)s + \delta) \geq V^{-b}((D - \delta)s + \delta) \approx c_n \exp((n - 1)Ds).$$

For s big enough, we get the lower bound we claimed for $\phi_S(s)$.

Theorem (Preissmann, [Pr]): *If (M, g) is a Riemannian manifold with strictly negative curvature, then $\pi_1(M)$ does not contain $\mathbf{Z}^2$.*

As a consequence of any of the two above theorems (Milnor or Preissmann), we obtain the following result.

Corollary. *The torus $\mathbf{T}^3$ does not carry metrics with strictly negative curvature.*

For a modern version of such results, using the notion of *simplicial volume*, see [Gr3].

Remarks. a) As soon as $n \geq 3$, there are n-dimensional compact manifold whose fundamental group has exponential growth and which carry no metric with negative sectional curvature (cf. [VCN]).

b) These results can be considered as the prehistory of a *Geometric Group Theory* which has known dramatic developments in the nineties of the last century, and is still very active today. See for instance [G-H] and [Gr6].

3.J Curvature and topology: some important results

This section is expository, no proofs are given.

3.J.1 Integral formulas

In dimension 2, the Gauss-Bonnet formula says everything about the relations between curvature and topology. Namely, if (M, g) is a compact Riemannian surface, its Euler-Poincaré characteristic, that is the alternate sum of Betti numbers, is given by

$$\chi(M) = \frac{1}{4\pi} \int_M Scal(g)v_g = \frac{1}{2\pi} \int_M K_g dv_g$$

(recall that the topological type of a compact surface, once known whether it is orientable or not, is entirely given by $\chi(M)$).

In higher dimension, the Gauss-Bonnet formula has been extended by Chern as follows. Suppose the dimension n is even, and take the $\frac{n}{2}$-th exterior power of the curvature tensor: we get a field of endomorphisms of $\Lambda^n TM$, therefore a scalar field. In that way, we have obtained a polynomial $P_n(R)$ of degree $\frac{n}{2}$ with respect to the curvature tensor. Then, for some universal constant c_n (which can be computed by taking the standard sphere) we have (see for example [Sp],t.V)

$$\chi(M) = c_n \int_M P_n(R)v_g.$$

If $n = 4$ it can be proved (Chern-Milnor theorem, see [SB]) that if the sectional curvature has constant sign, then $P_n(R) \geq 0$, so that $\chi(M) \geq 0$. Unfortunately, hardly anything is known when dimension $n > 4$.

* There are more involved integral formulas, which involve Pontriagin classes and Pontriagin numbers. In particular, the signature of a $4k$-dimensional manifold can be expressed in terms of curvature. See [Sp], t.5 for elementary properties of Pontriagin classes, and [Gi] for the signature theorem and related topics. It turns out that Pontriagin forms, that represent Pontriagin classes once a metric has been chosen, only depend on the Weyl tensor, that we shall meet in 3.126 (cf. [K-P]).*

3.J.2 (Geo)metric methods

We have already seen some of them in 3.G and 3.I. They have also been very successful in the study of manifolds with positive δ-pinched curvature.

Namely, a complete Riemannian manifold (M, g) with positive curvature is said to be δ-pinched if there is a constant $A > 0$ such that $\delta A < K < A$ (after rescaling the metric, A can be taken equal to 1). Successive efforts of H. Rauch, M. Berger and W. Klingenberg have given the following result.

3.111 **Theorem** *If M is simply connected, and if $\delta > 1/4$, then M is homeomorphic to S^n. Furthermore, if $\delta \geq 1/4$, either M is homeomorphic to S^n, or M is isometric to *a compact rank one symmetric space*, namely a projective space.*

The basic tool in the proof of this result is Toponogov's theorem, which compares metric properties of geodesic triangles in (M, g) and in a constant curvature space (see [Br1] for a review, and [C-E] for detailed proofs).

If δ is close enough to 1, it has been proved that M is *diffeomorphic* to S^n (J. Cheeger, E. Ruh, cf. [C-E] again).

The case of pinched manifolds with negative curvature is completely different. M. Gromov and W. Thurston ([G-T]) have produced examples of δ-pinched manifolds of negative curvature, with δ as close to 1 as you like, which carry no metric with constant curvature. This construction supports in a striking way the general feeling that there are many more manifolds with negative than with positive curvature. It is also worth-pointing out that, for these examples, the diameter goes to infinity as δ goes to one: the compactness results we shall see below are not relevant.

One can also ask about the intermediate case, when the absolute value of the sectional curvature is small. For homogeneity reasons, some normalization is necessary. Consider the following situation.

3.112 **Exercise.** Take $\epsilon > 0$. Show that there is on the Heisenberg group H a left-invariant metric for which the quotient metric on $H/H_{\mathbf{Z}}$ satisfies $|K(g)| < \epsilon$ and $\mathrm{diam}(g) < 1$.

This example is interesting in that $H/H_{\mathbf{Z}}$ carries no flat metric (* otherwise, using Bieberbach theorem (3.84 bis), $H_{\mathbf{Z}}$ would contain a subgroup isomorphic to $\mathbf{Z}^3$ *). A compact manifold which carries for any ϵ a metric with diam $< \epsilon$ and $| K | < 1$ is said to be *almost flat*. The example of exercise 3.112 can be easily generalized. Namely, any compact quotient of a nilpotent Lie-group by a discrete cocompact subgroup, or in other words any *nilmanifold,* is almost flat (cf. [B-K], 1.4). This property has a converse, which may be one of the most difficult results in Riemannian geometry.

3.113 Theorem (M. Gromov, cf. [B-K]). *There is a constant ϵ_n such that any compact n-dimensional Riemannian manifold (M,g) with*

$$| K(g) | \, (\mathrm{diam}(g))^2 \leq \epsilon_n$$

is covered by a nilmanifold.

In particular, for any n, there is but a finite number of differential manifolds which satisfy the assumptions of this theorem.

It may be said that purely geometric methods culminate with a result of M. Gromov, who gave an *a priori* bound of the sum of the Betti numbers of a compact manifold of non-negative curvature, by applying Morse theory to the distance function (cf. [Pe], ch.11).

3.J.3 Analytic methods

They seem to be necessary as soon as a weaker invariant than sectional curvature is involved.

Linear methods have been known for a long time. Basically, they consist in proving, under various positivity assumptions, vanishing theorems for suitable differential operators, whose kernels carry geometric or topological information. This is the Bochner technique, which is still fruitful today. We'll give some examples in the next chapter.

Non linear methods are more recent. A typical result is that of R.S. Hamilton ([Ha]).

3.114 Theorem. *Any compact, 3-dimensional simply connected Riemannian manifold with* strictly positive Ricci curvature *is diffeomorphic to S^3.*

The basic idea is to deform the metric along the Ricci curvature, that is to integrate the partial differential equation

$$\frac{\partial g_t}{\partial t} = -\mathrm{Ric}_{g(t)},$$

which can be viewed a differential equation on the space of Riemannian metrics. The solution develops singularities whose control is fairly delicate.

Using powerful refinements of this technique, G. Perel'man (2003) has announced he has solved Poincaré's conjecture, which asserts that any compact simply connected 3-dimensional manifold is homeomorphic to S^3.

We have seen that the conditions Ric > 0 or Ric ≥ 0 have strong topological consequences. It turns out that the condition Ric < 0 gives no information.

It was proved by L.Z. Gao and S.T. Yau in dimension 3 ([G-Y]) and by
J. Lohkamp in any dimension ≥ 3, that *any compact manifold carries a metric
with strictly negative Ricci curvature*. See [Lo] for an expository paper.

The weakest curvature invariant is of course the scalar curvature. However,
it has rather subtle properties. Of course, the results of J. Lohkamp say in
particular that any compact manifold carries a metric with negative scalar
curvature, but this way already known (T. Aubin) in the seventies. See [B2],
ch.IV for an easy proof.

But there are topological obstructions to positive scalar curvature. One of
them was first pointed out by A. Lichnerowicz, by using a subtle version of
the Bochner technique, involving spinors. In spite of decisive break-throughs
by R. Schoen, S.T. Yau, M. Gromov and B. Lawson, J. Rosenberg, J. Stolz,
the situation remains mysterious. A good reference is [L-M].

3.J.4 Coarse point of view: compactness theorems

Taking much weaker assumptions, we get for instance finiteness results for the
topological type of the manifolds that are considered. The first result in that
direction were due (independently) to J. Cheeger and A. Weinstein.

3.115 **Theorem** *If a, b, c are positive real numbers, there is but a finite number
of topological types of compact n-dimensional manifolds which carry a metric
g such that*

$$\mid K(g) \mid \leq a, \quad \mathrm{diam}(g) \leq b, \quad \mathrm{vol}(g) \geq c.$$

This result lies over a lower bound of the injectivity radius involving a, b, c
only.

M. Gromov has endowed the isometry classes of compact metric spaces with
a distance which generalizes the well-known Hausdorff distance between compact parts of $\mathbf{R}^n$. Namely, the Hausdorff distance of X and Y is just the
infimum of the ordinary Hausdorff distance of $f(X)$ and $f(Y)$ for all possible
isometric embeddings (with disjoint images of course) in a third metric space.
See [Gr1], chapter 2 for details. Using a refinement of Bishop's theorem (that
we shall meet in 4.19) he proved the following spectacular property.

3.116 **Theorem.** *The set of isometry classes of compact manifolds such that*

$$\mathrm{diam}(M, g) \leq D \quad and \quad Ric(g) \geq ag$$

(with a real, D real positive) is precompact for the Hausdorff distance.

For more refined compactness theorems, see [Gr1], ch.8. Although Hausdorff distance is a very rough tool, this result provides a powerful guide
for guessing suitable statements about curvature and topology. For example Myers' theorem (Ric $> 0 \implies b_1(M) = 0$) and Bochner's theorem
(Ric $\geq 0 \implies b_1(M) \leq n$, cf. ch.IV) have the following generalization.

3.117 **Theorem** (Gromov [Gr1] p.73, Gallot [Ga2,3], Bérard-Gallot [Be-G]).
*There is an explicit function $f(n, a, D)$, with $f(n, 0, D) = n$ and $f(n, a, D) = 0$
for $a > 0$, such that the first Betti number of any n-dimensional Riemannian
manifold with diameter smaller than D and Ricci curvature greater than ag
is smaller than $f(n, a, D)$.*

Since the eighties of the past century, there has been intensive work addressing
this kind of continuity and finiteness results. The last chapters of P. Petersen's
book [Pe], that we already quoted, give a good account until 1995.

Dramatic progresses have still occurred afterwards, mainly due to J. Cheeger
and T. Colding. Let us quote a typical sample of their results.

3.117 bis **Theorem** *Let M be a compact n-dimensional manifold with
$b_1(M) = n$. There exists an $\epsilon(n) > 0$ such that, if M carries a metric g for
which*

$$\mathrm{diam}_g(M)^2 \mathrm{Ric}_g \geq -\epsilon(n),$$

then it is diffeomorphic to T^n.

See the expository article [Ga6] for more information.

3.K Curvature tensors and representations of the orthogonal group

3.K.1 Decomposition of the space of curvature tensors

According to ideas which go back to H. Weyl, we are going to look a little closer
at the algebraic properties of the curvature tensor. Let (E, q) be a real vector
space of dimension $n > 1$, equipped with a non degenerate quadratic form q.
Indeed, we have in mind the case $(E, q) = (T_m M, g_m)$ for some Riemannian
manifold (M, g), but everything we shall see works as well in the pseudo-
Riemannian case. There is a natural action of the linear group $Gl(E)$ on each
tensor space $\otimes^k E \otimes^l E^*$: just set, for $x_i \in E$ ($1 \leq i \leq k$) and $y_j^* \in E^*$
($1 \leq j \leq l$),

$$\gamma(x_1 \otimes \cdots \otimes x_k \otimes y_1^* \otimes \cdots \otimes \gamma_l^*)$$
$$= \gamma x_1 \otimes \cdots \otimes \gamma x_k \otimes^t \gamma^{-1} y_1 \otimes \cdots \otimes^t \gamma^{-1} y_l.$$

The quadratic form q permits us to identify E and E^*. More explicitly, if $\gamma \in
O(q)$, since $\gamma = {}^t\gamma^{-1}$, we see that E and E^* are isomorphic as representations
spaces for $O(q)$. In other words, we can identify the $O(n)$-modules E and E^*:
from now on, only tensor powers of E^* will be considered. We shall denote by
$\circ$ the symmetric product (with the convention $x \circ x = x \otimes x$), and by $S^k E^*$
the k-th symmetric tensor power of E^*.

The $O(q)$-module E^* is of course irreducible. A basic question of representation theory is to find the irreducible components of its tensor powers. The case
of $\otimes^2 E^*$ is easy. Denote by $S_o^2 E^*$ the space of traceless symmetric 2-tensors,
and notice that the scalar product associated with q has a natural extension

to tensors. Then (cf. [W1], ch.5 and [SB], exp. IX), the decomposition is as follows.

3.118 Proposition. *The $O(q)$-module $\otimes^2 E^*$ admits the irreducible orthogonal decomposition*

$$\otimes^2 E^* = \Lambda^2 E^* \oplus S_o^2 E^* \oplus \mathbf{R} \cdot q.$$

Sketch of the Proof. The existence of such a decomposition is clear. Irreducibility is a direct consequence of classical invariant theory for the orthogonal group: it can be checked directly that the space of $O(n)$-invariant quadratic forms over $\otimes^2 E^*$ is generated by

$$\sum t_{ij} t^{ij}, \quad \sum t_{ij} t^{ji}, \quad \text{and} \quad (\text{trace}(t))^2,$$

and therefore is 3-dimensional. Hence there are 3 irreducible components at most. ∎

More generally, for studying $\otimes^k E^*$, we must look at its $Gl(E)$-module structure. The elements of the symmetric group S_k and of the group algebra $\mathbf{R}[S_k]$ give rise to $Gl(E)$-morphisms of $\otimes^k E^*$. Two simple and important examples are given by the symmetrisation and the antisymmetrisation operators. In fact, the $Gl(E)$-irreducible components of $\otimes^k E^*$ appear as kernels (or images) of some idempotents of the algebra $\mathbf{R}[S_k]$. They are the famous Young symmetrizers (cf. [F-H]). Here we are only making an implicit use of them. We have seen in 3.5. that the curvature tensor lies in the subspace $S^2 \Lambda^2 E^*$ of $\otimes^4 E^*$. It can be checked that the only idempotent of $\mathbf{R}[S_4]$, whose restriction to $S^2 \Lambda^2 E$ is non trivial, is the map b defined as follows:

3.119 Definition. The *Bianchi map* is the endomorphism of $S^2 \Lambda^2 E^*$ given by

$$b(T)(x, y, z, t) = \frac{1}{3} \left(T(x, y, z, t) + T(y, z, x, t) + T(z, x, y, t) \right),$$

where $T \in S^2 \Lambda^2 E^*$ and $x, y, z, t \in E$.

Clearly, b is $Gl(E)$-equivariant and idempotent. Hence we have the $Gl(E)$-invariant decomposition

$$S^2 \Lambda^2 E^* = \text{Ker}\, b \oplus \Im\, b \tag{3.120}.$$

Using elementary properties of the symmetric group S_4, it is easy to see that $\Im b$ is $Gl(E)$-isomorphic to $\Lambda^4 E^*$. By the way, this proves that $b = 0$ if $\dim E = 2$ or 3.

3.121 Definition. The space $C(E) = \text{Ker}\, b$ is the space (and the $O(q)$-module) of *curvature tensors*.

To see how $C(E)$ can be decomposed as an $O(q)$-module, two elementary remarks will be useful. First, the *Ricci contraction* defined by

$$c(R)(x, y) = \sum_{i=1}^{n} R(x, e_i, y, e_i)$$

(where $(e_i)_{1 \leq i \leq n}$ denotes an orthonormal basis) is $O(q)$-equivariant. Secondly, there is a natural way to make curvature tensor with symmetric 2-tensors.

3.122 Definition. The *Kulkarni-Nomizu product* of the symmetric 2-tensors h and k is the 4-tensor $h \cdot k$ given by

$$(h \cdot k)(x, y, z, t) = h(x, z)k(y, t) + h(y, t)k(x, z) - h(x, t)k(y, z) - h(y, z)k(x, t).$$

Remarks: i) We just used suitable symmetrizations to make an element of $S^2 \Lambda^2 E^*$ from $h \otimes k$.
ii) We have $g \cdot g = 2R_o$ (cf. 3.9).

3.123 For the scalar products on $S^2 E^*$ and $S^2 \Lambda^2 E^*$ which are given by q, we take the following normalizations. Via q, we have $O(q)$-equivariant embeddings of $S^2 E^*$ and $S^2 \Lambda^2 E^*$ into $\text{End}(E)$ and $\text{End}(\Lambda^2(E))$ respectively. Then the scalar product of two tensors will be the trace of the product of the corresponding isomorphisms.

3.124 Lemma. *If $n > 2$, the map $h \to h \cdot q$ from $S^2 E^*$ into $C(E)$ is injective, and the transposed map is just the Ricci contraction c.*
Proof. Everything can be checked by a direct computation, after having noticed that

$$c(h \cdot q) = (n - 2)h + (\text{tr}_q h)q.$$

3.125 Theorem. *If $n > 4$, the $O(q)$-module $C(E)$ admits the orthogonal irreducible representation*

$$C(E) = \mathbf{R} \oplus S_o^2 E^* \oplus W(E),$$

where $W(E) = \text{Ker}\, c \cap \text{Ker}\, b$. The factor $\mathbf{R}$ is realized by the line $\mathbf{R} \cdot q \cdot q$, and the factor $S_o^2 E^*$ by $q \cdot S_o^2 E^*$.
Sketch of the proof. The existence of factors $\mathbf{R}$ and $S_o^2 E$ is a consequence of 3.118 and 3.124. On the other hand, using 3.121 and 3.124 again, one sees that the orthogonal of $q \cdot S_o^2 E^*$ in $C(E)$ is $\text{Ker}\, c \cap \text{Ker}\, b$. Now, the vector space of $O(q)$-invariant quadratic forms on $C(E)$ is generated by

$$T \to |T|^2, \quad T \to |c(T)|^2, \quad \text{and} \quad T \to \left(\text{tr}(c(R))\right)^2.$$

It is 3-dimensional, hence $C(E)$ has at most three irreducible $O(q)$-invariant components, and the decomposition we obtained is irreducible. ∎

3.126 Definition. The space $W(E)$ is the space of *Weyl tensors* of (E, q).

Now, we come back to Riemannian geometry. Namely, we take $T = R$, the curvature tensor of a Riemannian metric. Then $c(R) = \text{Ric}$ and $\text{tr}(c(R)) = \text{Scal}$. The decomposition of R given by theorem 3.118 is just

$$R = \frac{s}{2n(n-1)} g \cdot g + \frac{1}{n-2}(\text{Ric} - \frac{\text{Scal}}{n}g) \cdot g + W. \tag{3.127}$$

The tensor W, which denotes the $W(E)$-component of R, can be viewed as a "remainder" after two successive divisions of R by g.

This formula is still in principle valid if $n = 3$. However, for dimension reasons, the space $W(E)$ is then reduced to zero, and the decomposition reduces itself to

$$R = (\text{Ric} - \frac{1}{4}\text{Scal} \cdot g) \cdot g. \qquad (3.128)$$

In particular, any Einstein 3-dimensional manifold has constant sectional curvature. To finish this overview, let us mention that many interesting properties occur when $n = 4$, related to the non-irreducibility of $W(E)$ as an $SO(4)$-module. See [SB] for basic ideas, [L2] and [F-U] for fascinating developments about the topology of 4-manifolds.

3.129 Exercise. Let (M, g) be a Riemannian manifold, and $f \in C^\infty(M)$. Use the above formalism to compute the curvature tensor of $g_1 = \exp(2f)g$. If the Weyl component is denoted by $W(g)$, deduce from that the equality $W(g_1) = \exp(2f)W(g)$.

3.K.2 Conformally flat manifolds

3.130 Definition. A Riemannian manifold (M, g) is said to be *conformally flat* if for any $m \in M$ there is an open set U containing m and a function $f \in C^\infty(U)$ such that the metric $\exp(2f)g$ is *flat*.

Example. Any manifold with constant sectional curvature is conformally flat. If $K = 0$, it is clear. If $K \neq 0$, using 3.82, we see that it has just to be checked for (S^n, can) and (H^n, can), since the property is local. Now, we actually proved it in exercise 2.11. It amounts to the same to say (with a little extra subtlety in the Lorentzian case), that stereographic projections are angle-preserving.

3.131 Exercise. Show that $(S^p, \text{can}) \times (H^q, \text{can})$ is conformally flat. (We set $(S^1, \text{can}) = (S^1, dt^2)$ and $(H^1, \text{can}) = (\mathbf{R}, dt^2)$).

Now, if we want to detect conformal flatness, exercise 3.129 gives us an obvious necessary condition, namely that $W(g)$ should vanish. In dimensions 2 and 3, this condition is empty: not unexpectedly, a particular treatment is necessary for those cases. We just state the results.

3.132 Theorem. i) *Any 2-dimensional Riemannian manifold is conformally flat.*

ii) it *A 3-dimensional Riemannian manifold is conformally flat if and only if the covariant derivative of the tensor* $\text{Ric} - \dfrac{\text{Scal}}{2(n-1)}g$ *is a symmetric 3-tensor.*

iii) *If* $\dim M \geq 4$, *a Riemannian manifold* (M, g) *is conformally flat if and only if* $W(g) = 0$.

Assertion i) is just the existence theorem of isothermal coordinates, cf. [Sp] for example. The condition in ii) is precisely the Frobenius integrability condition for the equation $R(\exp(2f)g) = 0$ in dimension 3. For the general case, see [W 2].

It has been proved (Kuiper, [Kr]) that any compact simply connected conformally flat manifold is globally conformal to (S^n, can). A reference for conformally flat manifolds is [K-P].

3.133 **Exercises:** a) Recover the results of 3.131 by using theorem 3.132.

b) Let (M, g) be an hypersurface of $\mathbf{R}^{n+1}$ equipped with the induced metric. If $n \geq 4$, the manifold (M, g) is conformally flat if and only if the second fundamental form of M (cf. 5.2) has one eigenvalue of multiplicity at least $n - 1$ (Elie Cartan). See [dCa-Da] for a nice global consequence of this local result.

3.K.3 The Second Bianchi identity

Let DR be the covariant derivative of the curvature tensor. It is a tensor of type $(0, 5)$, and $(DR)_m \in T_m^* M \otimes C(T_m^* M)$. Remember that R has been introduced as an antisymmetrized second covariant derivative (cf. 3.2). This fact explains the following property, which can be viewed as a Riemannian version of the identity $d \circ d = 0$.

3.134 **Proposition** (second Bianchi identity):

If R is the curvature tensor of a Riemannian manifold (M, g) then, for any tangent vectors x, y, z, t, u at m, we have

$$D_x R(y, z, t, u) + D_y R(z, x, t, u) + D_z R(x, y, t, u) = 0.$$

Proof. We make the computations for the curvature of type $(1, 3)$. Take vector fields X, Y, Z in a neighborhood of m whose values at m are respectively x, y, z. Then we have the following equalities in $\Gamma(\text{End}(TM))$:

$$D_X R(Y, Z) = [D_X, R(Y, Z)] - R(D_X Y, Z) - R(Y, D_X Z)$$

(from the very definition of covariant derivatives). Then, using 3.4

$$D_X R(Y, Z) = [D_X, D_{[Y,Z]}] - [D_X, [D_Y, D_Z]] - R(D_X Y, Z) - R(Y, D_X Z)$$

$$= D_{[X,[Y,Z]]} - [D_X, [D_Y, D_Z]] - R(X, [Y, Z]) - R(D_X Y, Z) - R(Y, D_X Z).$$

When summing up with terms obtained by cyclic permutation with respect to X, Y, Z, we see that the terms involving D disappear because of Jacobi identity. Using the identity $D_X Y - D_Y X = [X, Y]$, we see that the terms involving R also disappear. ∎

3.135 **Corollary.** i) *For any Riemannian metric (M, g) one has*

$$\delta(\text{Ric}) + \frac{1}{2} d \, \text{Scal} = 0,$$

where δ denotes the linear map of $\Gamma(S^2 M^*)$ in $\Gamma(T^* M)$ given by

$$\delta h = -tr_{12} Dh.$$

ii) *If there is a smooth function f such that* $\mathrm{Ric} = fg$ *and if* $\dim M \geq 3$, *the function f is constant.*

Proof. i) is obtained from the second Bianchi identity by taking traces with respect to (y, t) and (z, u) (recall that the covariant derivative commutes with traces).

ii) On one hand, $\delta(\mathrm{Ric}) = \delta(fg) = -df$. On the other hand, $d\,\mathrm{Scal} = ndf$. Then use i). ∎

3.L Hyperbolic geometry

3.L.1 Introduction

In this section, we come back to the study of manifolds with *constant* negative curvature. We shall see that the hyperbolic space has some kind of boundary, which consists in points at "infinity". In the hyperboloid model, this boundary is represented by the generatrices of the asymptotic cone. In the Poincaré disk model, it is just the boundary of the disk, and the Poincaré half-plane model, it is the boundary of the half-plane, with an extra ideal point. In fact, even if the hyperboloid model is conceptually more satisfactory, Poincaré models present an overwhelming advantage: they allow drawings. Moreover, working with the disk model corresponds to the choice of a point in the hyperbolic space (see Proposition 3.136 below), and working with the half-plane model corresponds to the choice of a point at infinity.

We shall focus on the 2-dimensional case, and give elementary but useful formulas involving angles and distances. Then we shall give several proofs -two of them are geometric and elementary (3.152, 3.159), one (sketchy) uses complex analysis (3.162), and one uses non-linear analysis (3.163)- that *any compact surface with negative Euler characteristic carries a metric with constant negative curvature.*

3.L.2 Angles and distances in the hyperbolic plane

We begin with the following characterization of the Poincaré disk model, which is valid in any dimension.

3.136 **Proposition.** *Let g be a complete Riemannian metric of constant negative curvature -1 on the unit ball of $\mathbf{R}^n$. Suppose that g is both pointwise conformal to the Euclidean metric g_0 and invariant under the linear action of $O(n)$. Then*

$$g = \frac{4g_0}{(1 - |x|^2)^2} \cdot$$

Proof. Any two-plane through the origin is the set of fixed points for some isometry of g. Hence the two-planes are totally geodesics, and it is enough to make the proof for $n = 2$. Note also that lines through the origin are geodesics.

Using both polar coordinates (r, θ) and normal coordinates (ρ, θ) with respect to the origin, and using 3.51, we get

$$g = f^2(r) \left(dr^2 + r^2 d\theta^2\right) = d\rho^2 + \sinh^2 \rho \, d\theta^2 \,.$$

Therefore $\dfrac{d\rho}{\sinh \rho} = \dfrac{dr}{r}$, and by integrating we get

$$r = \frac{\exp \rho - 1}{\exp \rho + 1}, \quad \rho = \log\left(\frac{1+r}{1-r}\right) \quad \text{hence} \quad f(r) = \frac{2}{(1-r^2)} \cdot \blacksquare$$

3.137 Exercise. Take a $(n-1)$−sphere tangent to the boundary of the ball. Show that the metric induced by the Poincaré metric is *flat*. (These spheres are called *horospheres*). They admit a nice generalization to the case of negative variable curvature, see [BGS] pp. 21.

3.138 Corollary. *The hyperbolic space is two-points homogeneous. Namely, given two pairs (p, q) and (p', q') such that $d(p, q) = d(p', q')$, there exists an isometry I such that $I(p) = p'$ and $I(q) = q'$. Moreover, in dimension 2, there is exactly one such isometry which preserves (resp. reverses) the orientation.*
Proof. Using homogeneity, we can suppose that $p = p'$. Now, suppose that p is the center of the ball in the Poincaré model. The preceding proof shows that, for transforms which fix p, Euclidean and hyperbolic isometries are the same. Moreover, points which are equidistant to p for the hyperbolic distance are also equidistant for the Euclidean distance. $\blacksquare$
Now, we look at the geodesics in these Poincaré models.

3.139 Proposition. *In the Poincaré half-plane model, the geodesics are half-circles whose center belongs to the real axis, and half-lines orthogonal to this axis.*
In the Poincaré disk model, the geodesics are arcs of circles meeting the unit circle orthogonally, or segments through the origin.
Proof. We have seen the first part already (exercise 2.83 d)). For the second one, notice that the inversion of pole $i = (0, 1)$ and modulus 2 yields an isometry between the two models (one can take also the transform $z \to \dfrac{z-i}{z+i}$ from the half-plane onto the disk).
We have got all the geodesics by the standard uniqueness argument. $\blacksquare$

3.140 Remark. The hyperbolic plane (or space) is homogeneous, and for our "Euclidean" eyes, this homogeneity seems to be lost in both Poincaré models. Indeed, taking the disk (or ball) model amounts to choose a point in H^n, and taking the half-plane (or space) model amounts to choose a point at infinity (or a direction of geodesics, as will be explained now).

3.141 Proposition. *Given two distinct geodesics γ and γ' in the hyperbolic plane, there are three possibilities:*
i) they meet in exactly one point in the plane;
ii) they do not meet, but their distance is zero (in other words, they meet at infinity);

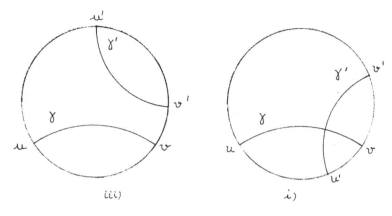

Fig. 3.6. Geodesics in the Poincaré disk

iii) their distance is strictly positive. In that case, the distance is realized by a unique geodesic segment which is orthogonal to both.

Proof. Take the Poincaré disk model. Any geodesic meets the unit circle in two points, and conversely two points on this circle define one and only one geodesic. Now, take two distinct geodesics γ and γ', and let (u, v) and (u', v') be the (unordered) pairs of points in S^1 defining them. Suppose first that the four points u, v, u', v' are distinct. Then either u' and v' lie in two different connected components of $S^1 - \{u, v\}$, and γ and γ' meet, this is case i), or they lie in the same component, and γ and γ' do not meet. See figure 3.6.

What happens in that case is better seen in the half plane model, see figure 3.7. All that we have said about the disk model is still valid in that case: one agrees that the set of points at infinity is $\mathbf{R} \cup \infty$. The half-line which is orthogonal to the real axis at the point u' is the geodesic defined by the pair (u', ∞). After using a suitable isometry, we can suppose that γ' for instance is a half-line. We are in the case when $u' \notin [u, v]$. Then the half-circle whose center is u' and radius the tangent from a to the circle of diameter $[u, v]$ yields the unique geodesic which is orthogonal to both γ and γ'. This is case iii).

Now, suppose that $\{u, v\}$ and $\{u', v'\}$ have a common point, say $v = v'$. By using a suitable isometry, we can suppose that $v = \infty$. Take the points (u, y) and (u', y) on γ and γ' respectively. Their distance is certainly lower than $\dfrac{|u - u'|}{y}$. Taking y arbitrarily large, we see that we are in case ii). ∎

3.142 Exercise. Extend this result to the n-dimensional hyperbolic space. Hint: using the second variation formula, first prove the following:
Let γ and γ' be two geodesic lines in a simply connected manifold of non-positive curvature. Take any arc-length parametrization of these geodesics.

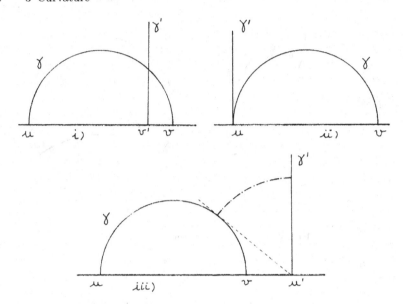

Fig. 3.7. Geodesics in the Poincaré half-plane

Then the function

$$(t, t') \rightarrow d(\gamma(t), \gamma'(t'))$$

is convex.

To summarize the previous discussion:
i) a geodesic is entirely determined by its two points at infinity;
ii) according to whether two geodesics intersect or not, they make some angle, or have some shortest distance (asymptotic geodesics have both an angle and a shortest distance, which are both zero!).

This angle, or this distance, only depends on the data of four points at infinity. We are going to compute them explicitly in terms of these four points. To do that, we shall need the following notion.

3.143 Definition. Let z_1, z_2, z_3, z_4 be four real or complex numbers, such that three of them are distinct. Their *cross − ratio* is the (possibly infinite) number

$$(z_1, z_2, z_3, z_4) = \frac{z_1 - z_3}{z_1 - z_4} : \frac{z_2 - z_3}{z_2 - z_4}.$$

This definition naturally extends to the case when some z_i is infinite. For instance

$$(z_1, z_2, z_3, \infty) = \frac{z_1 - z_3}{z_2 - z_3}.$$

Remark. This notion comes from projective geometry. Namely, it is easily checked that the projective group $PGl_2\mathbf{K}$ is three times transitive on the projective line $P^1\mathbf{K}$. Two quadruples of points are mapped into each other if and only if their cross ratio is the same. See [Br2], ch.6 for details. We shall only use the down-to-earth version of this property:

The homographic transforms $x \rightarrow \dfrac{ax+b}{cx+d}$ *leave the cross-ratio invariant.*

To see that, write

$$\frac{ax+b}{cx+d} = \frac{a}{c} + \frac{bc-ad}{c(cx+d)}.$$

Since translations and homotheties clearly preserve the cross-ratio, we are reduced to check that it is preserved by $x \mapsto \frac{1}{x}$, a straightforward computation. From now on, we shall work with one of the Poincaré models, and the *complex* cross-ratio, which does not depend on the model we choose, in view of the preceding remark.

3.144 Proposition. *Let p and q be two distinct points of the hyperbolic plane, u and v the points at infinity of the geodesic containing them. Then, in either Poincaré model,*

$$d(p,q) = |\log|(p,q,u,v)||.$$

Proof. We use the Poincaré half-plane model. Recall that in that case the direct isometries are the homographic transforms

$$z \rightarrow \frac{az+b}{cz+d} \qquad \text{with} \quad a,b,c,d \in \mathbf{R}, \ ad-bc=1.$$

Therefore, using transitivity, we can suppose that $v = \infty$, which just means that

$$u = x, \quad p = x + iy_1, \quad q = x + iy_2.$$

Now

$$d(p,q) = |\int_{y_1}^{y_2} \frac{dt}{t}| = |\log|\frac{y_1}{y_2}||. \ \blacksquare$$

For a variant of this result using the hyperboloid model, see [Br2], ch.19.

3.145 Proposition. *Let γ and γ' be two distinct geodesics of the hyperbolic plane, whose points at infinity are respectively (u,v) and (u',v'). Let r be the cross-ratio (u,v,u',v'). Then r is real, and γ and γ' meet (resp. do not meet) if and only if r is negative (resp. positive). Furthermore, one has*

$$\left|\frac{1+r}{1-r}\right| = \begin{cases} \cos\angle(\gamma,\gamma') & \text{if} \quad r < 0 \\ \cosh d(\gamma,\gamma') & \text{if} \quad r > 0 \end{cases}$$

Proof. Take the Poincaré half plane model. After performing suitable isometries, we can suppose that $v = -u$ and $v' = \infty$. Then the cross-ratio (u,v,u',v') is just $\dfrac{u'-u}{u'+u}$. Then γ, that is the circle of center 0 and radius u, meets γ', that is the line $y = u'$, if and only if $|u'| < u$. Since the Poincaré model is pointwise conformal to the Euclidean metric, we can compute this angle using elementary geometry (see the second picture of Proposition 3.141). We have

$$\cos \angle(\gamma, \gamma') = \left|\frac{u'}{u}\right| = \frac{1 + \frac{u'-u}{u'+u}}{1 - \frac{u'-u}{u'+u}} \, .$$

If γ and γ' do not meet, the geodesic which is orthogonal to both is just the circle of center u' and radius $\sqrt{u'^2 - u^2}$. Take the parametrization

$$t \rightarrow u' + \sqrt{u'^2 - u^2} \, \exp it \, .$$

Then

$$d(\gamma, \gamma') = \int_\phi^{\frac{\pi}{2}} \frac{dt}{\sin t} \qquad \text{where} \qquad \sin \phi = \left|\frac{u}{u'}\right|,$$

so that

$$d(\gamma, \gamma') = - \log \tan \frac{\phi}{2} \qquad \text{and}$$

$$\cosh d(\gamma, \gamma') = \frac{1}{2}\left(\tan \frac{\phi}{2} + \cot \frac{\phi}{2}\right) = \frac{1}{\sin \phi} = \left|\frac{u}{u'}\right|.$$

The result follows in the same way as in the former case. ∎

3.L.3 Polygons with "many" right angles

Take three points p, q, r in the hyperbolic plane such that the geodesics from p to q and from p to r are orthogonal (we shall use the vocabulary of elementary Geometry and say *perpendicular*). Such a configuration will be called a *right-angled triangle*, and is clearly determined by the lengths $d(p, q)$ and $d(p, r)$ of the sides of the right angle. Take the perpendiculars through q and r to these sides. Either they meet, and we get a quadrilateral with *three* right angles, or they do not meet, and taking their common perpendicular, we get a *right-angled pentagon*, that is a pentagon whose all angles are right. (We neglect the intermediate case when these perpendicular meet at infinity, and give a quadrilateral with three right angles and one null angle). More precisely:

3.146 Proposition. *Let a and b be two positive real numbers.*
i) If $\sinh a \sinh b < 1$, there exists a unique (up to isometry) quadrilateral Q such that:
a) Q has three right angles and one acute angle ϕ;
b) the lengths of the sides which are not contiguous to ϕ are a and b.
ii) If $\sinh a \sinh b > 1$, there is a unique right-angled convex pentagon which has two contiguous sides of length a and b.
Moreover, denoting by c the length of the side which is not contiguous to the sides of lengths a and b, we have

$$\sinh a \sinh b = \begin{cases} \cos \phi & \text{in case} \quad i) \\ \cosh c & \text{in case} \quad ii) \end{cases}$$

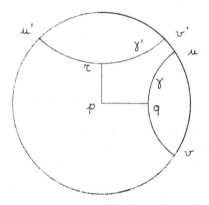

Fig. 3.8. Constructing a right-angled pentagon

Proof. Take the right-angled triangle we considered just above, and consider the geodesics γ and γ' through q and r respectively, which are perpendicular to the sides pq and pr. If u, v and u', v' are their points at infinity (see figure 3.8), we are reduced to compute the cross ratio (u, v, u', v') in terms of a and b. Now, we work in the Poincaré disk model, and suppose that the center of the disk is p, that q is real positive and r imaginary.

If x is the Euclidean distance from p to q, we have

$$a = d(p, q) = \int_0^x \frac{2dt}{1 - t^2} \quad \text{so that} \quad x = \tanh \frac{a}{2}.$$

Then the geodesic γ is carried by the circle whose diameter is $[x, x']$, where $x' = \coth \frac{a}{2}$. Its points at infinity are

$$u = \exp i\phi, \quad v = \exp -i\phi \quad \text{where} \quad \sin \phi = \tanh a.$$

In the same way, the points at infinity of γ' are $\exp \pm i\psi$, where $\sin \psi = \tanh b$, so that

$$(u, v, u', v') = \frac{\cos(\phi - \psi)}{\cos(\phi + \psi)}.$$

Taking into account that for positive a and for $\phi \in [0, \frac{\pi}{2}]$,

$$\sin \phi = \tanh a \iff \tan \phi = \sinh a$$

we eventually get

$$(u, v, u', v') = \frac{1 - \sinh a \sinh b}{1 + \sinh a \sinh b}.$$

Then our claims are direct consequences of Proposition 3.145. ∎

3.147 Corollary. *In the hyperbolic plane, there exist right-angled hexagons (that is hexagons all of whose angles are right).*

Proof. Take two right angled pentagons having one side of same length. Since H^2 is two-points homogeneous, we can glue them together along this side. ∎

Remark. The argument in the proof of Lemma 3.160 below shows that for any $n \geq 5$, there exists a ("regular") right-angled n-gon. But our present argument is more geometric, and allows to show (with some extra-work), that there exists a unique right-angled hexagon, when prescribing the lengths of three non-adjacent sides.

Now, we can begin the proof the existence of metrics of negative constant curvature on any surface of genus $\gamma > 1$. It will be done by a two-steps gluing procedure. To see that, we begin with the following.

3.148 Definition. A *pant P* is a compact Riemannian manifold with boundary which satisfies the following properties:

i) it is homeomorphic to a sphere which three disks taken off;

ii) it carries a Riemannian metric with constant curvature -1 such that the boundary consists in three disjoint closed geodesics.

Now, if we take two copies of a given right-angled hexagon, and glue them together along three non-adjacent sides, we obtain a pant in view of the following elementary property.

3.149 Lemma. *Let γ be a piece of geodesic in H^2. Then any tubular neighborhood of γ can be parametrized by the (algebraic) distance u to γ and the curvilinear abciss s of the orthogonal projection. The hyperbolic metric is then given by*

$$\cosh^2 u \, ds^2 + du^2 \,.$$

In particular, the orthogonal symmetry with respect to γ provides a (local) isometry.

Proof. Take the half-plane Poincaré model, and suppose that γ is carried by the line $x = 0$. Let

$$s \rightarrow (0, \exp s)$$

be an arc-length parametrization of γ. We can parametrize a tubular neighborhood of γ by

$$(s, t) \rightarrow (\exp s \sin t, \exp s \cos t) \,,$$

and in terms of t, the arc-length u on the geodesics $s = cte$ is given by

$$du = \frac{dt}{\cos t} \,.$$

Therefore, we get the parametrization

$$(s, u) \rightarrow (\exp s \tanh u, \frac{\exp s}{\cosh u}) \qquad \text{of } H^2 \,,$$

and the claimed formula is now straightforward. ∎

3.150 Exercise. Classify the two-dimensional complete Riemannian manifolds with constant negative curvature whose fundamental group is isomorphic to $\mathbf{Z}$.

3.L.4 Compact surfaces

Before going on, let us recall basic facts about compact two-dimensional manifolds, which are also called *compact surfaces*. For simplicity, we shall deal mainly with orientable surfaces, although the non-orientable case is not much more difficult. The sphere and the connected sum T_n of n tori (or in other words tori with $n \geq 1$ holes) clearly provide examples of compact surfaces. These surfaces are topologically distinct when n goes through **N**. *Indeed, using Mayer-Vietoris exact sequence (cf. [Sp], t.1), one checks that the Euler characteristic $\chi(T_n)$ is equal to $2 - 2n$.*

A basic result of algebraic topology (see [Gr] for an elementary and elegant account, using Morse theory) is that *any orientable surface is diffeomorphic to some T_n* (we set $T_0 = S^2$).

3.151 **Definition.** The integer n is the *genus* of the surface.

An important alternative description of T_n is the following. Take a polygon with $4n$ sides in the plane, and label the consecutive sides

$$a_1, b_1, a_{-1}, b_{-1}, \cdots, a_n, b_n, a_{-n}, b_{-n}.$$

Orient a_i and b_i clockwise if $i > 0$, anti-clockwise if $i < 0$ and identify a_i with a_{-i}, b_i with b_{-i}. Clearly, we get $T_1 = T^2$ for $n = 1$. If $n = 2$, the identification for $i = 1$ or $i = 2$ gives a torus with one disk removed (see the figure of 3.159 below if necessary). These two manifolds have the same boundary, so that the complete identification yields T_2. The general case follows by induction. Now, we come to the main result of this section.

3.152 **Theorem.** *Any compact orientable surface of genus $\gamma > 1$ carries a metric with constant negative curvature.*

Proof. Take two copies of the same pant, and glue them along a boundary geodesic. We get a surface whose boundary consists in four closed geodesics, which has the shape of an "X". (We could also glue two pairs of boundary geodesics simultaneously, and obtain a torus with 2 disks removed). Gluing pairwise the components of same length, which is made possible by Lemma 3.149, we obtain a surface of genus 2 with constant curvature -1 (see figure 3.9). For higher genus, we proceed in the same way: first we glue $\gamma - 1$ "X"together, and afterwards we identify pairwise the components of the boundary. ∎

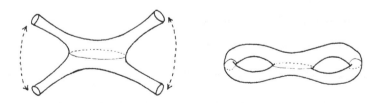

Fig. 3.9. From pants to surfaces

3.153 Exercise. Extend this result to non-orientable surfaces of negative Euler characteristic.

3.154 Remark. The above proof makes it clear that on a given surface of genus $\gamma > 1$, there are "many" metrics with curvature -1. Namely, it can be proved (cf. [FLP], ch.3) that right-angled hexagons are determined by the lengths of non-adjacent sides, so that pants depend on three real parameters. Moreover, two boundary geodesics of two given pants can be glued as soon as they have the same length, and when performing this gluing, we can choose arbitrarily one point of each of them and identify these points. Therefore, metrics of constant curvature -1 depend on 6 real parameters for a surface of genus 2, and $6\gamma - 6$ real parameters for a surface of genus γ. The elaboration of these remarks is the celebrated Teichmüller theory, which is still very lively. The reference [FLP] gives a geometric view-point which should be easy of access when you have read this section. A very complete reference is [Bu2]. See also [J2] for an account using analytic methods.

3.L.5 Hyperbolic trigonometry

We shall give another proof, less elementary but more classical of theorem 3.152. This proof uses the alternative description of orientable surfaces given by gluing the sides of a $4n$-gon. We need elementary properties of hyperbolic triangles. For completeness, we begin with the following "fundamental formula of hyperbolic trigonometry" which, strangely enough, was not needed before.

3.155 Proposition. *Let an hyperbolic triangle have side-lengths a, b, c. Denote by A (resp. B, C) the opposite angle to the side of length a (resp. b, c). Then*

$$\cosh a = \cosh b \cosh c - \sinh b \sinh c \cos A .$$

Proof. We come back to the hyperboloid model. Let x,y,z be the vertices corresponding to the angles A,B,C, and u,v be the unit tangent vectors at x to the oriented geodesics xy and xz. Then

$$y = x \cosh c + u \sinh c$$
$$z = x \cosh b + v \sinh b,$$

whereas $\cos A = (u, v)$ and $\cosh a = -(y, z)$. The claimed formula is now straight-forward. ∎

3.156 Proposition. *The area of any hyperbolic triangle is equal to*

$$\pi - A - B - C .$$

The formula is still valid if one (or more) vertex is at infinity, in which case the corresponding angle is zero.

Proof. It is enough to prove this result for triangles with one vertex at infinity, since any hyperbolic triangle is the difference of two such triangles (see figure 3.10).

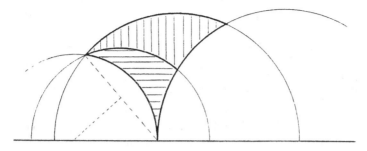

Fig. 3.10. Triangles and ideal triangles

Now, we work in the Poincaré half-plane model and suppose that the infinite sides are vertical lines. The area form is

$$\frac{dx \wedge dy}{y^2} = -d(\frac{dx}{y})$$

and the result follows from the Stokes formula. ∎

Remark. We get a very nice particular case when all the vertices are at infinity. (The triangle is called an *ideal triangle* in that case). The area is then equal to π. Indeed, more can be said: since the Möbius group of the circle, which is just the isometry group of H^2, is triply transitive (see 3.143), all ideal triangles are isometric.

3.157 Corollary. *The area of an n-sided polygon of the hyperbolic plane whose angles are $(\alpha_i)_{1 \leq i \leq n}$ is*

$$(n-2)\pi - \sum_{i=1}^{n} \alpha_i .$$

Example: The area of a right-angled hexagon is equal to π. It follows that the area of a pant is 2π.

3.158 Theorem. *The area of a compact orientable surface of genus γ and curvature -1 is $4\pi(\gamma - 1)$.*

Proof. This is clearly true if the metric is constructed by gluing pants as in the proof of Theorem 3.152. It is true indeed (intuitively clear using 2.98, and not too difficult to prove, cf. [FLP], appendix of ch.4) that any metric with curvature -1 can be obtained in that way.

*Another argument uses elementary algebraic topology. First we take a triangulation of the manifold (M, g). Then we refine it (using barycentric subdivision for example). If the vertices of the new triangulation are close enough, they can be taken as the vertices of a geodesic triangulation. Then, using 3.157, we get

$$\text{Area}(M, g) = \pi \times \sharp(\text{faces}) - 2\pi \times \sharp(\text{vertices}) .$$

Since $3\sharp(\text{faces}) = 2 \times \sharp(\text{wedges})$, this yields

$$\text{Area}(M,g) = 2\pi\big(\sharp(\text{wedges}) - \sharp(\text{faces}) - \sharp(\text{vertices})\big) = -2\pi\chi(M),$$

where $\chi(M) = 2(1 - \gamma)$ is the Euler characteristic.* ∎

Remark: What we have seen is just a particular case of Gauss-Bonnet theorem.

3.159 Existence of metrics with curvature -1: an alternative proof.
Take the 4γ-gon we described in 3.151. We suppose (this possibility shall be justified below) that this polygon P has the following properties:
i) the sides are geodesic segments;
ii) $L(a_i) = L(a_{-i})$ and $L(b_i) = L(b_{-i})$ for any i;
iii) the sum of interior angles is 2π.
iv) P is geodesically convex.
Recall the $4n$-gon we described in 3.151, and denote by A_i (resp. B_i) the unique orientation preserving isometry of H^2 which transforms the (oriented!) side a_i (resp. b_{-i}) into a_{-i} (resp. b_i).

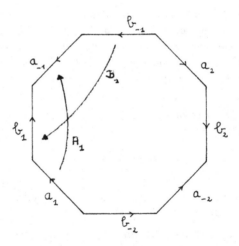

Fig. 3.11. Identifying the sides of an octagon

Performing successively the transforms

$$A_1, B_1, A_1^{-1}, B_1^{-1}, \cdots, A_\gamma^{-1}, B_\gamma^{-1}$$

taking the angle relation into in mind and following the story of b_γ, we see that

$$B_\gamma^{-1}.A_\gamma^{-1}.B_\gamma.A_\gamma \cdots B_1.A_1 = \text{Id}.$$

Moreover, the 4γ polygons

$$B_\gamma(P), A_\gamma(B_\gamma(P)), B_\gamma^{-1}(A_\gamma(B_\gamma(P))), \cdots$$

yield a tiling of a neighborhood of the origin of a_1. Of course, we get a tiling around any vertex of P by the same procedure: just change the labels of the sides. Repeating the process for the polygons around P, we see that H^2 is tiled by the polygons $\tau(P)$, where τ runs through the group Γ generated by the A_i and the B_i. This construction shows that Γ acts freely on H^2. Moreover, for any $\tau \neq Id$, $\tau(P) \cap P$ is a point, a geodesic segment or the empty set. Therefore Γ is a discrete group of isometries which acts properly, so that the quotient H^2/Γ is diffeomorphic to the surface T_γ. Now, the hyperbolic metric of H^2 goes down to the quotient, and we are done.

It remains to prove the existence of a polygons which satisfy the assumptions i), ii), iii) and iv) above. This is a consequence of the following stronger property.

3.160 Lemma: *For any integer $n \geq 5$, there exists a convex n-gon in H^2 whose sides have equal length and whose angles are equal to $\frac{2\pi}{n}$.*
Proof. Take a point $a \in H^2$, and n geodesics rays from a whose unit tangent vectors at a form a regular polygon in $T_a H^2$. Let P_r be the convex n-gon whose vertices are the points on the rays at distance r from a. The homogeneity property 3.138 shows that the sides of P_r are equal, and so are the interior angles. When r goes to zero, these angles converge to $\frac{(n-2)\pi}{n}$, that is their value in the Euclidean case. When r goes to the infinite, they converge to zero, since we get an *ideal* polygon. Therefore, provided that

$$\frac{2\pi}{n} < \frac{(n-2)\pi}{n} \qquad \text{i.e.} \quad n > 4,$$

there is an $r > 0$ such that the common value of the interior angles is just $\frac{2\pi}{n}$. ∎

3.161 Remarks. i) The above continuity argument uses (in a very weak way indeed) the existence of explicit trigonometric formulas for triangles of the hyperbolic plane.

ii) As a by-product of the proof of 3.159, we get that the fundamental group of a surface of genus γ is the group with 2γ generators A_i, B_i $(1 \leq i \leq \gamma)$ and the relation

$$\prod_{i=1}^{n}[A_i, B_i] = e.$$

(The bracket denotes the commutator). This property is also a consequence of van Kampen theorem (cf. [Gr]).

3.162 The existence of metrics with curvature -1 on surfaces of high genus can also be obtained using complex analysis. Since the arguments are much less geometric, we just sketch them.

Picking a Riemannian metric and an orientation on T_γ, we get a complex structure (in other words a Riemann surface), which is given by the rotation of angle $\frac{\pi}{2}$ on each tangent space. Since the (topological) universal covering of T_γ is a plane, the lifted Riemann surface is either $\mathbf{C}$ or the unit disk. If $\gamma > 1$, the first case is ruled out. Indeed, the only subgroups of complex automorphisms of the complex plane which yield compact quotients are lattices, while the fundamental group of T_γ is not Abelian. Now, the complex automorphisms of the disk are isometries for the Poincaré hyperbolic metric and we are done.

3.L.6 Prescribing constant negative curvature

The last approach for existence of hyperbolic surfaces will be analytic. We only sketch the proof, which relies on material to be introduced in Chapter IV. The result will be a consequence of the following more general result.

3.163 **Theorem.** *Let (M, g_0) be a compact Riemannian surface with negative Euler characteristic, and $K : M \to \mathbf{R}$ with $\sup K < 0$. Then there exists a unique smooth conformal metric $g = e^{2u} g_0$ on M with curvature K.*

Remarks. All data are smooth. We are actually interested in the case $K = -1$, but the proof works the same when K is everywhere negative.

For further results concerning prescribed curvature on surfaces, see [KW1], [KW2], [T], [H-T]. For the relevant Analysis results, see [G-T].

Proof. a) It can be checked, using 3.129, that the Gaussian curvature K of a conformal metric $e^{2u} g_0$ is $(\Delta_0 u + K_0) e^{-2u}$, where Δ_0 and K_0 are the Laplace operator and the curvature of g_0. We have to solve the non-linear partial differential equation

$$\Delta_0 u = K e^{2u} - K_0 \tag{3.1}$$

It appears that the appropriate space to construct a solution is the Sobolev space $H^1(M)$, which is defined as the completion of $C^\infty(M)$ for the norm

$$u \to \left(\| u \|_2^2 + \| \nabla_0 u \|_2^2 \right)^{1/2}.$$

We shall prove regularity afterwards.

For convenience, we will set $H = H^1(M)$ and $H' = \{u' \in H, \int u' v_{g_0} = 0\}$.

Existence of a solution: We introduce the functionals $\mathcal{F}$ and $\mathcal{G}$ defined on H as

$$\mathcal{F}(u) = \int_M |\nabla_0(u)|^2 + 2K_0 u \, v_{g_0}, \quad \mathcal{G}(u) = \int_M e^{2u} K \, v_{g_0}.$$

Observing that any (weak) solution $u \in H$ of (3.1) satisfies

$$\mathcal{G}(u) = \int_M K_0 v_{g_0} = 2\pi \chi(M)$$

(Gauss-Bonnet formula, cf. 3.J.1), we are lead to minimize the functional $\mathcal{F}$ on the hypersurface of H defined by $\mathcal{G}(u) = 2\pi\chi(M) := \gamma$.

Actually, for such a minimum u, the theory of Lagrange multipliers will ensure the existence of $\lambda \in \mathbf{R}$ with $\nabla\mathcal{F}(u) = \lambda\nabla\mathcal{G}(u)$ (for $\mathcal{G}$ has no critical points), that is

$$\Delta_0 u + K_0 = \lambda K e^{2u}.$$

Integrating over M yields

$$\gamma = \int K_0 v_{g_0} = \lambda \int K e^{2u} v_{g_0} = \lambda \int K_g v_g = \lambda\gamma,$$

(Gauss-Bonnet again) so that $\lambda = 1$ and u satisfies 3.1.

b) The key analytical argument to exhibit a minimum u for $\mathcal{F}$ is Trudinger inequality which yields the following property as a corollary.

3.164 Proposition. *The embedding $H \to L^2$ defined by $u \to e^{2u}$ is compact.* See [KW1] for details and references. This result ensures that $\mathcal{G}$ is weakly continuous on H. We also state Poincaré's inequality (compare with 4.57).

3.165 Proposition. *There exists a constant $c = c(M, g_0)$ such that for $u' \in H'$, $\| u' \|_2^2 \leq c \| \nabla_0 u' \|_2^2$.*

3.166 Corollary. *The functional $\mathcal{F}$ is bounded below on H'.*

Proof. Just apply Cauchy-Schwarz inequality.■

c) Since $\sup K < 0$, there exists a constant $a \in \mathbf{R}$ such that $\mathcal{G}(a) = \gamma$. Let $\mathcal{F}(a) = m$, set

$$B = \{u \in H, \quad \mathcal{F}(u) \leq m, \quad \mathcal{G}(u) = \gamma\},$$

and observe the following.

3.167 Proposition. *a) B is not empty;*
b) $\mathcal{F}$ is bounded below on B: let $\mu := \inf_B \mathcal{F}$;
c) B is bounded in H.

Sketch of proof. a) is clear.

b) Use for any $u \in H$ the decomposition $u = u' + \bar{u}$, where $u' \in H'$ and $\bar{u} = (\int u\, v_{g_0})/(\int v_{g_0})$. Now for $u' \in H'$, Jensen inequality[2] yields

$$|\mathcal{G}(u')| \geq \inf |K| \int_M e^{2u'} v_{g_0} \geq \inf |K| \operatorname{Area}(g_0)$$

while $\mathcal{G}(u) = e^{2\bar{u}}\mathcal{G}(u')$: thus $\bar{u}$ is bounded above on B. Since $\mathcal{F}(u) = \mathcal{F}(u' + \bar{u}) = \mathcal{F}(u') + 2\gamma\bar{u}$, we are done by 3.166.

c) The proof is in the same spirit, using Poincaré inequality (one must show that $\mathcal{F}$ is bounded below on H' and $\bar{u}$ is bounded below on B) and is left to the reader (see [T] for details).■

---

[2] if φ is a convex function, $\varphi(\int f/ \int 1) \leq \int \varphi(f)/ \int 1$

c) Let now $u_n \in B$ be a minimizing sequence for $\mathcal{F}$. Since B is bounded in H, there exists a subsequence which converges weakly to $v \in H$. Since $\mathcal{G}$ is weakly continuous (3.164), we have $\mathcal{G}(v) = 2\pi\chi(M)$. Also $\mathcal{F}(v) \leq \liminf \mathcal{F}(u_n)$ (for $u \mapsto \| \nabla u \|^2$ is weakly continuous on H), so that $\mathcal{F}(v) = \mu$ by minimality, and $v \in H$ is a weak solution for (E).

d) **Regularity:** this is standard elliptic regularity (see [GT]). First note that, for a weak solution $u \in H$ for (E),

$$\Delta_0 u = He^{2u} - K_0 \in L^2,$$

hence u is Hölder continuous. Thus Δu itself is Hölder continuous, so that $u \in C^{2,\alpha}$. Conclude by bootstrapping that u is actually smooth. ∎

3.L.7 A few words about higher dimension

In higher dimension, we have the same description for geodesics, and the discussion of 3.141 is still valid. The set points at infinity can be identified with $\mathbf{R}^{n-1} \cup \infty$ or S^{n-1}. It is clear that an isometry of H^n has a unique extension to infinity. This extension will be studied in detail in the next section. Here we just make some remarks:

a) like in the Euclidean case, regular polyhedra are more difficult to find than regular polygons. As a consequence, the existence of compact quotients of H^n, that is compact manifolds with constant negative curvature, is not a trivial property.

For instance, adapting 3.147 and 3.160, one can prove the existence of a regular dodecahedron whose faces are right-angled pentagons.

b) we shall see in M. that the extension of $\mathrm{Isom}(H^n)$ to the sphere at infinity is still 3-transitive, but no more.

The notion of *ideal simplex* still makes sense. They are the $(n+1)$-simplices whose vertices lye on the infinity sphere. But these simplices are no more isometric. It can be proved that there is a maximum volume, which is achieved for *regular* simplices. See [Mi4] for the computation in the 3-dimensional case, and other very nice things about higher dimensional hyperbolic geometry.

M. Gromov has pointed out that this fact can be used to give a proof of Mostow's rigidity theorem for hyperbolic manifolds: *two compact manifolds with constant curvature -1 and dimension strictly bigger than 2 are isometric as soon as they are diffeomorphic.* (Something much stronger is true, but we do not care here.) See [B-P] and [Ra], for a proof along those lines, and also for a very detailed account of higher dimensional hyperbolic geometry.

See also [Th] for the role of hyperbolic geometry in 3-dimensional topology.

Lastly, the notion of infinity sphere still makes sense for simply connected manifolds with negative curvature. It is obtained by identifying oriented geodesics γ and γ' such that

$$\lim_{t \to +\infty} d\big(\gamma(t), \gamma'(t)\big) = 0$$

There is still a notion of points at infinity in the non-positive curvature case, but the situation is much more involved. For these questions, see [BGS].

3.M Conformal geometry

3.M.1 Introduction

In this section, we shall see that the points at infinity of the hyperbolic space enjoy beautiful geometric properties.

In fact, we shall take the other side of the picture: we begin with the so called Möbius group, that is the group of conformal transformations of the Euclidean space $\mathbf{R}^n$ ($n \geq 3$). This formulation is not completely correct: the natural setting turns out to be S^n. We shall see that the conformal geometry of S^n "generates" elliptic, hyperbolic and Euclidean geometry.

In fact, speaking of conformal geometry has a double meaning. On one hand, it can be viewed as the geometry of the Möbius group, and more generally of Möbius or conformally flat manifolds, that are manifolds equipped with an atlas $(U_i, \phi_i)_{i \in I}$, such that the transition diffeomorphisms are Möbius maps (compare with 3.K.2, and see [K-P]). On the other hand, given a Riemannian manifold (M, g), one can study its conformal group or its conformal invariants. These two points of view connect each other. A basic theorem of J. Lelong-Ferrand and M. Obata (see [K-P]) says that, *if a compact Riemannian manifold (M, g) is not conformal to the standard sphere, there is a conformal scale u such that* $\mathrm{Conf}(M, g) = \mathrm{Isom}(M, e^u g)$. Another important result is the solution of the Yamabe conjecture, due to the successive efforts of many mathematicians (H. Yamabe, N. Trudinger, T. Aubin, R. Schoen), see [L-P] for an illuminating account: *the metric g being given, there is a conformal scale such that* $\mathrm{Scal}(e^u g)$ *is constant.* It is a highly non trivial result in non-linear analysis, and it turns out that conformally flat manifolds require special efforts. Moreover, the non-compactness of the Möbius group permit to use it in renormalization techniques, not only for this problem, but also for instance in Yang-Mills theory (cf. [L2]).

3.M.2 The Möbius group

We begin with some elementary geometry.

3.168 Definition. In the Euclidean space $\mathbf{R}^n$, the *inversion of pole a and modulus k* ($k \neq 0$), is given by

$$i_{a,k}(x) - a = k\frac{x - a}{|x - a|^2} \, .$$

Clearly, $i_{a,k}$ is a self-inverse diffeomorphism of $\mathbf{R}^n - \{a\}$.

3.169 Definition. A smooth map ϕ from a Riemannian manifold (M, g) to (M', g') is *conformal* if there is a function f such that

$$\phi^\star g' = fg \, .$$

In other words, the differential of ϕ is angle-preserving. Holomorphic functions on open sets of the complex plane provide nice examples of conformal maps.

Another example is given by the inversion. Indeed, the differential of $i_{a,k}$ is just

$$T_x i_{a,k} \cdot \xi = \frac{k}{|x-a|^2}(\xi - 2\frac{\langle x-a, \xi \rangle}{|x-a|^2}(x-a)).$$

Up to a dilation, $T_x i_{a,k}$ is the orthogonal symmetry with respect to $(x-a)^\perp$. Now, a classical result of Liouville says that any conformal diffeomorphism between open sets of $\mathbf{R}^n$, *if the dimension n is strictly bigger than 2*, is a product of inversions (see [Br2], 9.5 or [K-P], p.12, and enjoy yourself in comparing the proofs.)

This setting for Liouville theorem is not so pleasant, since an inversion or a product of inversions is very seldom a diffeomorphism of $\mathbf{R}^n$. However, inversions become everywhere defined and bijective if we add to $\mathbf{R}^n$ a point ω at infinity, and set

$$i_{a,k}(a) = \omega \quad \text{and} \quad i_{a,k}(\omega) = a.$$

This "conformal compactification" of $\mathbf{R}^n$ can be made more precise as follows. Let K be the map of $\mathbf{R}^n$ into $P^{n+1}\mathbf{R}$ defined as follows

$$K(x) = [(\frac{1}{2}, x, \frac{1}{2}|x|^2)]. \tag{3.170}$$

Denoting by $X_0, X_1, \cdots, X_{n+1}$ the homogeneous coordinates of $P^{n+1}\mathbf{R}$, we see that K maps $\mathbf{R}^n$ into the n-dimensional quadric Q_n whose homogeneous equation is

$$\sum_{p=1}^{n} X_p^2 - 4X_0 X_{n+1} = 0,$$

and that K is a diffeomorphism of $\mathbf{R}^n$ onto $Q_n - [(0,0,\cdots,0,1)]$. Moreover, the similarities and inversions of $\mathbf{R}^n$ extend naturally to projective automorphisms of $P^{n+1}\mathbf{R}$ which preserve Q_n. Such maps can be represented in $Gl_{n+2}\mathbf{R}$ by maps which preserve the quadratic form

$$q = \sum_{p=1}^{n} X_p^2 - 4X_0 X_{n+1},$$

whose signature is $(n+1, 1)$.

For example, a linear orthogonal map A gives the projective map whose matrix in $Gl_{n+2}\mathbf{R}$ is just

$$\begin{pmatrix} 1 & 0 & \cdots & 0 \\ 0 & A & & 0 \\ 0 & 0 & \cdots & 1 \end{pmatrix}.$$

The dilation $x \to tx$ and the translation $x \to x + v$ respectively yield the matrices

$$\begin{pmatrix} 1 & 0 & \cdots & 0 \\ 0 & tI_n & & 0 \\ 0 & 0 & \cdots & t^2 \end{pmatrix} \quad \text{and} \quad \begin{pmatrix} 1 & 0 & \cdots & 0 \\ 2v & I_n & & 0 \\ |v|^2 & v & & 1 \end{pmatrix}.$$

As for the inversion $i_{0,1}$, it yields the matrix

$$\begin{pmatrix} 0 & 0 & 1 \\ 0 & I_n & 0 \\ 1 & 0 & 0 \end{pmatrix}.$$

More generally, it can be checked that the inversions and the hyperplane orthogonal symmetries of $\mathbf{R}^n$ exactly yield the projective transforms which can be represented in $O(n + 1, 1)$ by *reflections* (recall that a reflection in an orthogonal group is an involutive automorphism which leaves some hyperplane point-wise invariant). But the orthogonal group of a non-degenerate quadratic form is generated by reflections (Cartan-Dieudonné theorem, cf. [Br2], ch. 13). Eventually we get:

3.171 Theorem. *The inversions of $\mathbf{R}^n$ generate a group which is isomorphic to $PO(n + 1, 1)$. Its neutral component is isomorphic to $O_0(n + 1, 1)$.*

Proof. It just remains to see that the connected component of the identity in $PO(n + 1, 1)$ is isomorphic to $O_0(n + 1, 1)$. But the restriction to $O_0(n + 1, 1)$ of the covering map

$$O(n + 1, 1) \to PO(n + 1, 1)$$

is injective. ∎

Remark: More precisely, Cartan-Dieudonné theorem says that any element of $O(q)$ is the product of at most N reflections, where N denotes the dimension of the vector space. Therefore any conformal map of $\mathbf{R}^n$ (with $n > 2$) is the product of at most $n + 2$ inversions.

3.172 Definition. The *Möbius group* in dimension n is the group $PO(n+1, 1)$ acting on the quadric Q_n.

3.173 Remark. All this discussion makes sense even if $n = 1$. Indeed, in that case the "inversions" of $\mathbf{R}$ generate the group of homographic transforms $x \to \dfrac{ax + b}{cx + d}$, that is the projective group $PGl_2\mathbf{R}$. We get the isomorphisms

$$PGl_2\mathbf{R} \simeq PO(2, 1) \quad \text{and} \quad PSl_2\mathbf{R} \simeq O_0(2, 1).$$

Now, look at *complex* homographic transforms. The same method gives the isomorphism

$$PGl_2\mathbf{C} \simeq O(3, \mathbf{C})/\{\pm\mathrm{Id}\}.$$

Moreover, the transform $z \to \frac{1}{z}$ is the product of line-symmetry with a (real) inversion. So the group $PGl_2\mathbf{C}$ (which by the way is isomorphic to $PSl_2\mathbf{C}$), yields a subgroup of $PO(3, 1)$. Since it is connected and 6-dimensional, we get the isomorphism

$$PSl_2\mathbf{C} \simeq O_0(3, 1).$$

It should also be noted that, although the Liouville theorem is not true in dimension 2 (holomorphic transforms give a host of counter-examples), any conformal diffeomorphism of S^2 is still a Möbius map. Indeed, such a map is an holomorphic or anti-holomorphic isomorphism of $P^1\mathbf{C}$.

In any dimension, we have the following.

3.174 Theorem. *The Möbius group is triply transitive.*

Proof. Identifying Q_n with $\mathbf{R}^n \cup \omega$, and using elementary geometry, we see that it is transitive. Then, the Möbius transforms which fix ω are the (affine) similarities. Using elementary geometry again, it is easy to check that the affine similarity group is transitive on pairs of distinct points of $\mathbf{R}^n$. ∎

Another proof. The algebraic-minded reader will easily build a more conceptual proof, applying Witt theorem (cf. [Se], or [Br2], 13.7) to $\mathbf{R}^{n+2}$ equipped with the $(n + 1, 1)$ quadratic form q. ∎

3.M.3 Conformal, elliptic and hyperbolic geometry

Clearly, the quadric Q_n is diffeomorphic to S^n. An explicit diffeomorphism can be given as follows. It is (more or less) well-known that the complement of any hyperplane in the projective space is diffeomorphic to the affine space (cf. [Br2], ch.4). The removed hyperplane is the so called "infinite hyperplane". Pick an hyperplane which does not meet Q_n, for instance the hyperplane P whose equation is $X_0 + X_{n+1} = 0$. Then the formula

$$[(X_0, \cdots, X_{n+1})] \rightarrow \left(\frac{X_0 - X_{n+1}}{X_0 + X_{n+1}}, \frac{X_1}{X_0 + X_{n+1}}, \cdots, \frac{X_n}{X_0 + X_{n+1}}\right), \quad (3.175)$$

gives a diffeomorphism A of $P^{n+1}\mathbf{R} - P$ onto $\mathbf{R}^{n+1}$ and the image of Q_n is just the unit sphere

$$\sum_{p=0}^{n} y_p^2 = 1.$$

Now, $P^{n+1}\mathbf{R} - Q_n$ has two connected components. They are distinguished by the sign of q, which makes sense in a real projective space. Each component is diffeomorphic to the $(n + 1)$-dimensional ball, and is acted on transitively by the Möbius group. Pick a point p in the *negative* component. Such a point is the image of a line L in $\mathbf{R}^{n+2}$ on which the quadratic form q is strictly negative. Then q is positive definite on $L^\perp$. The subgroup of $O(q)$ which preserves L (and therefore $L^\perp$) is isomorphic to $O(n) \times \{\pm 1\}$. Therefore, as an homogeneous space, the negative component of $P^{n+1}\mathbf{R} - Q_n$ is just

$$O(n + 1, 1)/O(n + 1) \times \{\pm 1\} \simeq SO_0(n + 1, 1)/SO(n + 1).$$

This is precisely the hyperbolic space H^{n+1}, viewed as an isotropy irreducible homogeneous space (cf. 2.10 and 2.40). Therefore, we get the following

3.176 Theorem. *Let q be a quadratic form of signature $(n+1,1)$ on $\mathbf{R}^{n+2}$ and Q the image of $q^{-1}(0)$ in the projective space $P^{n+1}\mathbf{R}$. Then:*

i) Q is diffeomorphic to S^n;

ii) the connected component B_n of $P^{n+1}\mathbf{R} - Q$ for which q is negative, acted on by the group $PO(q)$, is naturally isometric to the $(n+1)$-dimensional hyperbolic space.

iii) Q has a conformal structure for which $PO(q)$ acts conformally. More precisely, for any point a in B_n, the stabilizator of a in $PO(q)$ is the isometry group of a positive constant curvature metric on Q, and $PO(q)$ is the conformal group of this metric. Moreover, for two different choices of points in B_n, the corresponding metrics are pointwise conformal.

Proof. We have already seen the first two parts, and also that the map A used in (3.143) sends Q onto the standard sphere in $\mathbf{R}^{n+1}$. Moreover, A sends B_n onto the standard unit ball, and $A \circ K$ is just the map

$$x \to \left(\frac{1 - |x|^2}{1 + |x|^2} , \frac{2x}{1 + |x|^2} \right) .$$

This is just the stereographic projection of $\{0\} \times \mathbf{R}^n$ with pole $(-1, 0, \cdots, 0)$ onto the unit sphere of $\mathbf{R}^{n+1}$, which is conformal, as we have seen in the beginning of this paragraph. Thus, in a first step, using K, we have conjugated the group generated by the inversions of $\mathbf{R}^n$ and the group $PO(n+1,1)$, which acts transitively both on the quadric Q_n of $P^{n+1}\mathbf{R}$ and its "inside" B_n. Then, using A, that is taking an affine subspace of this projective space, we get a transitive action of $PO(n+1,1)$ on both S^n and the unit ball B^{n+1}, which is conformal on S^n. The subgroup which fixes the center of the ball is just $O(n+1)$, acting linearly on $\mathbf{R}^{n+1}$. This forces B^{n+1} to appear as the homogeneous space

$$PO(n+1,1)/O(n+1) \simeq SO_0(n+1,1)/SO(n+1).$$

Here, $O(n+1)$ (resp. $SO(n+1)$) is the stabilizator of $A[(1,0,\cdots,0,1)]\cdot$ in $PO(n+1,1)$ (resp. $SO_0(n+1,1)$). We have proved i), ii), and the first part of iii) for the point $a_0 = [(1,0,\cdots,0,1)]$.

Now, take another point a. The stabilazator of a (or of $A(a)$, which amounts to the same) is conjugate to the stabilizator of a_0 (or of $A(a_0)$). Therefore, we have another transitive action of $O(n+1)$ on S^n, which yields another constant curvature metric on the sphere, say g_a. Take a Möbius map ϕ such that $\phi(a_0) = a$. On one hand

$$\phi^\star g_a = g_{a_0}$$

since the two metrics are completely defined by their isometry group. On the other hand, we know that ϕ is conformal (for both g_a and g_{a_0}). This forces g_{a_0} and g_a to be pointwise conformal. ∎

3.177 Remarks. a) If we look at the model provided by the map A, we see that the choice of the constant curvature metric on S^n, that is the choice of a conformal scale, amounts to the choice of a point inside the sphere. This point is just the "center" for this scale. The existence of this family of pointwise metrics, parametrized by the unit ball plays a very important role in many places:

i) the Yamabe problem (cf. [L-P], § 3). Propositon 4.6 of the same paper gives a typical example of a "renormalization" technique;

ii) the study of self-dual Yang-Mills connections (cf. [L2], and in particular p.45-46 for an elementary example related to our discussion).

b) The component for which q is positive, that is the *outside* of the ball in the Poincaré model, has also a nice interpretation: it can be identified with the de Sitter space (in fact, with a $\mathbf{Z}_2$ quotient of $(n+1)$-dimensional de Sitter space.

3.178 Exercises. a) Show that any compact subgroup of $SO_0(n+1,1)$ is conjugate to $SO(n+1)$ (so that points of the unit ball B^{n+1} correspond to maximal compact subgroups).

b) Compute the conformal scale of the preceding remark.

c) Prove that any conformal transform of (H^n, can) is an isometry (use Liouville theorem; the result is still true for $n=2$, it is then a classical theorem in complex analysis).

d) What can be said of the subgroup of $PO(n+1,1)$ that fixes a point of $P^{n+1}\mathbf{R}$ for which q is positive? And what for a point on the quadric Q_n itself?

e) The one-parameter group $x \to e^t x$ of homotheties of $\mathbf{R}^n$ yields a one-parameter group of Möbius transforms, with two fixed points, one attractive and one repulsive. As in the proof of theorem 3.143, represent Q_n as the unit sphere of $\mathbf{R}^{n+1}$, and suppose that these fixed points are the north and south poles. What is the infinitesimal generator of this one-parameter group?

3.179 Remark One could equip $\mathbf{R}^n$ with a non degenerate quadratic form of any signature (p, q), make the same construction, and obtain a "pseudo Möbius group", isomorphic to $PO(p+1, q+1)$. (In fact, we have already met "Lorentz inversions" in exercise 2.11, as a way of yielding the Poincaré disk model from the hyperboloid model).

The conformal compactification of $\mathbf{R}^n$ is a quadric in $P^{n+1}\mathbf{R}$ defined by a quadratic form of signature $(p+1, q+1)$. This quadric is diffeomorphic to the quotient of $S^p \times S^q$ by the antipody.

For this quadric, the discussion we made in the Euclidean case remains valid. Namely, the flat $\mathbf{R}^n$, the de Sitter and the Anti de Sitter space (or their analogues in higher signature) can be realized as suitable homogeneous open sets.

See the thesis of C. Frances ([Fr]) for details, and for rich and amazing examples of conformally flat Lorentz manifolds.

Analysis on Riemannian manifolds and Ricci curvature

Analysis on Riemannian manifolds stems from the following simple fact: the classical Laplace operator has an exact Riemannian analog. Indeed, the properties of the Laplacian on a bounded Euclidean domain and on a compact Riemannian manifold are very similar, and so are the proofs. It can be said that the difficulties of the latter case, compared with the former, are essentially conceptual.

We begin with some basic facts about manifolds with boundary. In fact, using them is as crucial as using relative homology or cohomology in Algebraic Topology. In section **B**, we explain how the Laplace operator gives a less involved proof of the Bishop inequality (compare with **III.H**), with the equality case as a bonus. This method also yields another volume comparison theorem, concerning tubes around hypersurfaces, namely the Heintze-Karcher inequality.

In **C**, we present the celebrated Hodge-deRham-Weyl theorem (a neat proof of which can be found in [Wa]) and apply it to our favorite examples. The next sections are devoted to a presentation of some salient facts of Spectral Geometry.

Lastly, we come back to geometric inequalities in **H**.

Non-linear Analysis, which presents some specific and difficult aspects in Riemannian Geometry (Yamabe and Calabi conjectures, Yang-Mills theory, cf. [A] and [L2]) is beyond the scope of this book.

4.A Manifolds with boundary

4.A.1 Definition

Even though we are mainly concerned with the so called closed manifolds, it is necessary to introduce manifolds with boundary. On one hand, we cannot ignore mathematical beings which locally behave like domains on $\mathbf{R}^n$. On the other hand, when doing Analysis on manifolds, it may be useful to cut them into small pieces (cf. for example 4.65 and 4.68 below). These pieces are no more manifolds, but they will be manifolds with boundary.

4.1 Definition. A *smooth manifold with boundary* is a Hausdorff topological space M, countable at infinity, and equipped with coordinate charts (U_i, ϕ_i) such that:

i) the U_i give an open covering of M;

ii) the map ϕ_i is an homeomorphism of U_i onto an open set of

$$\mathbf{R}^n_+ = \left\{ (x^1, \cdots, x^n) \in \mathbf{R}^n, x^1 \geq 0 \right\};$$

iii) for any pair (i, j) of indices such that $U_i \cap U_j \neq \emptyset$, the coordinate change

$$\phi_i \circ \phi_j^{-1} : \phi_j(U_i \cap U_j) \to \phi_i(U_i \cap U_j)$$

is a smooth diffeomorphism.

To understand this definition, some explanations are necessary. Indeed, manifolds with boundary are modeled on $\mathbf{R}^n_+$, in the same way as ordinary manifolds are modeled on $\mathbf{R}^n$. More explicitly, set $\partial \mathbf{R}^n_+ = \{(0, x^2, \ldots, x^n)\}$, and for any open set $V \subseteq \mathbf{R}^n_+$, let ∂V be $V \cap \partial \mathbf{R}^n_+$: we get an open set ∂V of $\partial \mathbf{R}^n_+ = \mathbf{R}^{n-1}$.

Now, we shall say that that $f : U \to V$ is a diffeomorphism if f is an homeomorphism and if moreover f (resp. f^{-1}) can be extended to a smooth map on some open set of $\mathbf{R}^n$ containing V (resp. W). In particular, f induces (check it!) a diffeomorphism (in the usual meaning) between ∂U and ∂V.

4.2 Definitions. The *boundary* of M, denoted by ∂M, is the (possibly empty) set of points such that, if $m \in U_i$, then $\phi_i(x) \in \partial \mathbf{R}^n_+$, i.e. the union of the $\phi_i^{-1}(\partial \mathbf{R}^n_+ \cap \phi_i(U_i))$.

The restrictions of the ϕ_i to ∂M are the charts of an $(n-1)$-dimensional smooth structure, in the usual meaning. Clearly, ∂M is closed in M. The complement $M \backslash \partial M$, which is an n-dimensional manifold in the usual meaning, is called the *interior* of M. A *closed* manifold is a compact manifold without boundary.

Simple examples show that ∂M need not be connected, even if M is.

Now, suppose that $M \backslash \partial M$ is *oriented*, and take an orientation atlas (U_i, ϕ_i). It can be checked easily that the Jacobians of the coordinate changes of the atlas

$$(U_i \cap \partial M, \phi_{i|U_i \cap \partial M})$$

of ∂M are positive. Therefore, ∂M *is also orientable*. To give an orientation of ∂M amounts to choose an orientation of $\partial \mathbf{R}^n_+$. Namely, if e_1 denotes the vector $(1, 0, \ldots, 0)$ of $\mathbf{R}^n$, take vectors $e_2, \ldots, e_n$ of $\mathbf{R}^n_+$ such that, for the orientation of $\mathbf{R}^n$ given by $e_1 \wedge e_2 \wedge \ldots \wedge e_n$, the charts ϕ_i are positive. Then we shall orient $\partial \mathbf{R}^n_+$ by $e_2 \wedge \ldots \wedge e_n$.

4.A.2 Stokes theorem and integration by parts

The famous Stokes theorem (see [Wa] or [Sp], vol.1 for a proof) can be viewed as a generalization of the fundamental theorem of calculus.

4.3 Theorem. *Let M be a compact oriented n-dimensional manifold with boundary ∂M, and $i : \partial M \to M$ be the inclusion map. Then, for any differential form α of degree $n - 1$ on M,*

$$\int_M d\alpha = \int_{\partial M} i^*\alpha.$$

In particular, if $\partial M = \emptyset$, then $\int_M d\alpha = 0$.

Now, we shall give Riemannian versions of this theorem. We need some preliminary definitions.

4.4 Definition. In a Riemannian manifold (M, g) the *divergence* of a vector field X (resp. the *codifferential* of a 1-form α) is the function given by

$$\operatorname{div} X = -\operatorname{tr} DX \qquad \text{resp.} \qquad \delta\alpha = -\operatorname{tr} D\alpha.$$

Here DX is the endomorphism $u \mapsto D_u X$, while $D\alpha$ is a covariant two-tensor, whose trace is computed with respect to the Riemannian metric.

4.5 Remark. Using musical isomorphisms (cf. 2.66), these two notions can be viewed as equivalent. More generally, we shall define the *divergence* of a tensor field T of type $(0, q)$ as the $(0, q - 1)$-tensor given by

$$\delta T = -\operatorname{tr}_{12} DT,$$

where the notation $_{12}$ just means that the trace is taken with respect to the first two variables. This formula generalizes both definitions 4.4 and 3.135.

4.6 Example. It is easy to check that $\delta(fT) = f\delta T - i_{\nabla f} T$.

4.7 Definition. The *Laplacian* of f is the function Δf given by

$$\Delta f = \operatorname{div} \nabla f = \delta df.$$

The Euclidean Laplacian on $\mathbf{R}^n$ is merely

$$\Delta = -\sum_{i=1}^{n} \frac{\partial^2}{\partial x_i^2}.$$

The minus sign is chosen to make the associated quadratic form positive. Indeed, 4.6 tells us that

$$\delta(f df) = f d\Delta f - |df|^2,$$

so that, if M is closed,

$$\int_M f \, \Delta f \, dv_g = \int_M |df|^2 \, dv_g \geq 0$$

(we shall see in a moment that the integral of the divergence of a 1-form always vanishes on a closed manifold).

Detailed information about this very important operator will be given later on. Now, we use the divergence and the Laplacian to give Riemannian versions of Green and Ostrogradski formulas which shall be useful.

4.8 Lemma. *Let (M, g) be an oriented Riemannian manifold, and let v_g be its canonical volume form. Then, for any vector field X*

$$d(i_X v_g) = -(\operatorname{div} X) v_g.$$

Proof. For $m \in M$, take an orthonormal basis $(e_1, \cdots, e_n)$ of $T_m M$. Since the n-form v_g is parallel, we have (see 2.58)

$$(D_{e_1}(i_X v_g))_m (e_2, \cdots, e_n) = (v_g)_m (D_{e_1} X, e_2, \cdots, e_n)$$

and (2.61) gives

$$(d(i_X v_g))_m (e_1, \cdots, e_n) = \sum_{i=1}^n v_g(e_1, \cdots, D_{e_i} X, \cdots, e_n)$$

$$= \sum_{i=1}^n g(e_i, D_{e_i} X) v_g(e_1, \cdots, e_n). \blacksquare$$

Now, take an oriented Riemannian manifold (M, g) with boundary ∂M. With the choice of an orientation for ∂M we explained above, we define on ∂M *the normal vector field pointing inside*, denoted by ν, as follows: if $m \in \partial M$, take for ν_m the unit normal vector to $T_m \partial M$ such that $i_{\nu_m} v_g$ is the canonical volume form of $(\partial M, g_{|\partial M})$. It is convenient to set

$$\omega_n = v_g \quad \text{and} \quad \omega_{n-1} = i_\nu \omega_n.$$

Notice that, the canonical volume form of S^{n-1} being usually defined as $i_x dx^1 \wedge \cdots dx^n$, the volume form ω_{n-1} obtained by considering S^{n-1} as ∂B^n is just the opposite.

4.9 Proposition. *For any vector field X on a compact oriented Riemannian manifold (M, g) with boundary ∂M, we have*

$$\int_M \operatorname{div} X \cdot \omega_n = -\int_{\partial M} g(X, \nu) \, \omega_{n-1}.$$

Moreover, on any closed manifold, orientable or not,

$$\int_M \operatorname{div} X v_g = 0.$$

Proof. Using the Stokes theorem and the preceding lemma, we see that

$$\int_M \operatorname{div} X \cdot \omega = -\int_M d(i_X \omega_n) = -\int_{\partial M} i_X \omega_n.$$

But if $(\nu_m, e_2, \cdots, e_n)$ is a direct orthonormal basis of $T_m M$ for $m \in \partial M$, then

$$(i_X \omega_n)_m(e_2, \cdots, e_n) = (\omega_n)_m(X, e_2, \cdots, e_n)$$

$$= g_m(X, \nu).(\omega_n)_m(\nu_m, e_2, \cdots, e_n) = g_m(X, \nu).(\omega_{n-1})_m(e_2, \cdots, e_n). \blacksquare$$

If ∂M is empty, the second claim is now trivial if M is oriented. ** If M is closed and non-orientable, the same conclusion follows by going up to the orientable covering of M.**

4.10 Definition. Let (M, g) be a Riemannian manifold with boundary. The *normal derivative* of a smooth function f on M is the smooth function on ∂M defined by

$$\frac{\partial f}{\partial \nu} = g(\nabla f, \nu).$$

4.11 Proposition (Green's formula). *Let f_1 and f_2 be smooth functions on a compact Riemannian manifold (M, g) with boundary. Then*

$$\int_M (f_1 \Delta f_2 - f_2 \Delta f_1) \omega_n = -\int_{\partial M} (f_1 \frac{\partial f_2}{\partial \nu} - f_2 \frac{\partial f_1}{\partial \nu}) \omega_{n-1}.$$

Proof. Just notice that $\delta(f_1 df_2) = f_1 \Delta f_2 - g(df_1, df_2)$, and use 4.8 and the Stokes formula. $\blacksquare$

Using the same technique, we get the following basic property of the Laplacian.

4.12 Proposition. *Let f be a smooth function on a compact manifold (M, g) with boundary. Suppose that Δf identically vanishes on the interior of M and that one of the following boundary conditions holds:*
a) $f_{|\partial M} \equiv 0$ (Dirichlet's condition),
b) $\frac{\partial f}{\partial \nu} \equiv 0$ (Neumann's condition).
Then $f \equiv 0$ in case a), or f is constant in case b).
Proof. If $\Delta f \equiv 0$, then $\delta(f df) = - \mid df \mid^2$. Now Proposition 4.8 and the Stokes formula give

$$\int_{\partial M} f \frac{\partial f}{\partial \nu} \omega_{n-1} = \int_M \mid df \mid^2 \omega_n. \blacksquare$$

4.B Bishop inequality revisited

4.B.1 Some commutation formulas

A basic tool of tensor Analysis is the so-called Ricci identity.

4.13 **Lemma.** *Let X, Y, Z be vector fields on (M, g), and α be a differential one-form. Then*

$$((D_X D_Y - D_Y D_X)\alpha)(Z) = \alpha(R(X,Y)Z).$$

Proof. First note that in the formula above,

$$D_X D_Y \alpha(Z) = D^2_{X,Y} \alpha(Z) = D^2 \alpha(X, Y; Z)$$

denotes the second covariant derivative of α (see our conventions 2.58). From the very definition of the curvature, we have

$$
\begin{aligned}
((D_X D_Y - D_Y D_X)\alpha)(Z) &= R(Y, X, \alpha^{\#}, Z) \\
&= R(X, Y, Z, \alpha^{\#}) = \alpha(R(X,Y)Z). \blacksquare
\end{aligned}
$$

Remark: There is a natural action of $\Gamma(\text{End}(TM))$ on one-forms, given by

$$(A, \alpha) \to \alpha \circ A,$$

where $A \in \Gamma(\text{End}(TM))$ is a field of endomorphisms. This action can be extended as a derivation on the space of tensors on M (cf. 1.110). Denote by $A \cdot T$ the result of this derivation when applied to the tensor T. Then for any tensor T on M, we have

$$(D_X D_Y - D_Y D_X) \cdot T = R(X, Y) \cdot T. \tag{4.14}$$

Indeed, the operators $D_X D_Y - D_Y D_X$ and $R(X, Y)$ are derivations, and the preceding lemma says that they coincide on $\Gamma(T^*M)$.

4.15 **Proposition (Bishop's formula).** *For any smooth function on (M, g), we have*

$$g(\Delta(df), df) = |\, Ddf \,|^2 + \frac{1}{2}\Delta(|\, df \,|^2) + \text{Ric}(\nabla f, \nabla f).$$

Proof. Applying 4.13 to $\alpha = df$ gives

$$DDdf(X, Y, Z) - DDdf(Y, X, Z) = df(R(X,Y)Z).$$

Taking the trace with respect to X and Z and noticing that

$$DDdf(X, Y, Z) = DDdf(X, Z, Y)$$

(since df is closed) we see that

$$\text{tr}_{12} DDdf(Y) = -d\Delta f(Y) + \text{Ric}(\nabla f, Y).$$

On the other hand, $D_X \mid df \mid^2 = 2g(D_X df, df)$, so that

$$D_{X,Y}^2 \mid df \mid^2 = 2g(D_X df, D_Y df) + 2g(D_X D_Y df, df).$$

Taking the trace, we get

$$-\frac{1}{2}\Delta \mid df \mid^2 = \mid Ddf \mid^2 + \mathrm{tr}_{12} DDdf(\nabla f).$$

The claimed formula follows by eliminating $DDdf(\nabla f)$. ∎

Remarks. i) The Bochner (Bochner-Weitzenbock) formula is indeed a generic name for a multitude of formulas obtained by applying commutation laws and taking traces. We shall meet other examples later on.

ii) The Bochner formula above is actually valid for any 1-form α. The proof is easier when α is closed and we will use this formula only in case $\alpha = dr$, where r is a distance function.

4.B.2 Laplacian of the distance function

4.16 Proposition. *Let m be a point in a complete Riemannian manifold (M,g), and let $r(p) = d(m,p)$. Then for any point which $p \neq m$ which is not on the cut-locus of m, we have*

$$\Delta r = -\frac{n-1}{r} - \frac{J'(u,r)}{J(u,r)},$$

where $p = \exp_m ur$ and J denotes, as in 3.96, the density of the volume form in normal coordinates.

Proof. Set $c_u(s) = \exp_m su$, and let $\big(c_u'(r), e_2, \cdots, e_n\big)$ be an orthonormal basis of $T_p M$. Then

$$\Delta r(p) = -\sum_{i=2}^{n} Ddr(e_i, e_i).$$

But $Ddr(e_i, e_i)$ is the second variation of the length of the geodesic c_u associated with the Jacobi field Y_i^r such that $Y_i^r(0) = 0$ and $Y_i^r(r) = e_i$. The claimed formula then follows from lemma 3.102. ∎

Another proof. Let $\Phi : S^{n-1} \times]0, \infty[\longrightarrow M$ be the inverse of the normal chart at m. Namely, $\Phi(r, u) = \exp_m(ru)$. Set $\Phi^* v_g = a(u,r)dr \cdot du$ (indeed, $a(u,r) = r^{n-1}J(u,r)$). For any open set $U \subset S^{n-1}$, set $\Omega_{r,\epsilon} = \Phi(U \times [r, r+\epsilon])$. Now, if for any $t \in [r, r+\epsilon[$, tU lies in the open set of $T_m M$ where $\exp_m$ is a diffeomorphism,

$$\int_{U \times [r, r+\epsilon]} a(u,r)\Delta r\, dr \cdot du = \int_{\Omega_{r,\epsilon}} \Delta r\, v_g,$$

which by Proposition 4.11 is equal to

$$-\int_{\partial\Omega_{r,\epsilon}} \frac{\partial r}{\partial\nu}\,\omega_{n-1} = \int_U a(u,r)du - \int_U a(u,r+\epsilon)du$$

(for this last change of sign, see the remark preceding proposition 4.9). But this is just $\int_{U\times[r,r+\epsilon]} -\frac{\partial a}{\partial r}\,dudt$. Therefore, for any U, r, ϵ, the functions $(a\Delta r)$ and $(-\frac{\partial a}{\partial r})$ have the same integral on $U \times [r, r+\epsilon[$. It follows that they are equal. ∎

4.17 Exercise. Show that on (S^2, can) we have

$$\Delta_y\big[\log\big(1 - \cos(d(x,y))\big)\big] = 1 - 4\pi\delta_x\,,$$

where both members of this inequality are distributions. *Hint:* imitate the proof of the equality $\Delta\log r = -2\pi\delta_o$ in $\mathbf{R}^2$.

4.B.3 Another proof of Bishop's inequality

Applying Bochner's formula, we shall give a proof which uses more directly comparison theorems for differential equations.

4.18 Lemma. *With the notations of* 4.16, *at any point lying outside the cut-locus of* m,

$$-\frac{a''}{a} = \mid Ddr \mid^2 -(\Delta r)^2 + \text{Ric}(\frac{\partial}{\partial r}, \frac{\partial}{\partial r}).$$

Proof. Just make $f = r$ in 4.15, and notice that $\mid dr \mid = 1$. Therefore

$$g\big(d(\Delta r), dr\big) = \frac{\partial}{\partial r}\Delta r = -\frac{\partial}{\partial r}\frac{a'}{a} = -\frac{a''}{a} + (\Delta r)^2. \blacksquare$$

4.19 Theorem. [Gr1] *Let* (M,g) *be a complete Riemannian manifold such that* $\text{Ric} \geq (n-1)kg$, *where* k *is a given real number. Denote by* $V^k(R)$ *the volume of the ball of radius* r *in the complete space with constant curvature* k. *Then, for any* $m \in M$, *the function*

$$\frac{\text{vol}\big(B(m,R))\big)}{V^k(R)}$$

is non-increasing. In particular $\text{vol}\big(B(m,R)\big) \leq V^k(R)$.

4.20 Lemma. *Set* $b = a^{\frac{1}{n-1}}$. *Then (without any curvature assumption):*

$$(n-1)\frac{b''}{b} + \text{Ric}(\frac{\partial}{\partial r}, \frac{\partial}{\partial r})b = -\left(\mid Ddr \mid^2 -\frac{1}{n-1}(\Delta r)^2\right)b \leq 0.$$

Proof. Just replace in the result of lemma 4.18 a by b^{n-1}, using that

$$\frac{a'}{a} = (n-1)\frac{b'}{b} = -\Delta r.$$

Now, Schwarz inequality forces

$$| \, Ddr \, |^2 - \frac{1}{n-1}(\Delta r)^2$$

to be positive, since $\frac{\partial}{\partial r}$ is in the null-space of the bilinear form Ddr. ∎

Proof of the theorem. With our curvature assumptions, the preceding lemma says that

$$b'' + kb \leq 0.$$

Using Sturm-Liouville theory (namely, using the fact that the previous inequality implies that $(b'\overline{b} - b\overline{b}')' \leq 0$), we can compare b with the solution $\overline{b}$ of the differential equation

$$\overline{b}'' + k\overline{b} = 0$$

which satisfies the same initial condition, namely $\overline{b}(0) = b(u,0) = 0$ and $\overline{b}'(0) = b'(u,0) = 1$. We get that

$$\frac{b'}{b} \leq \frac{\overline{b}'}{\overline{b}}.$$

Therefore, for any given u in the unit sphere of $T_m M$, the ratio $\frac{b}{\overline{b}}$ is non-increasing with respect to r. By integration, we already get Bishop's theorem in the form which was given in 3.101. To get our new statement, in which the cut-locus of m is ignored, just replace $b(u,r)$ by 0 if $\exp_m ru$ is beyond the cut-locus of m in the direction u. Denote this new function by b^+. Clearly, $\frac{b^+}{\overline{b}}$ is still non-increasing. Therefore, if $r > \epsilon$ we have

$$\frac{b^+(r)}{b^+(\epsilon)} \leq \frac{\overline{b}(r)}{\overline{b}(\epsilon)}.$$

Eventually we get that

$$\frac{\int_0^R (b^+)^{n-1}dr}{\int_0^\epsilon (b^+)^{n-1}dr} = 1 + \frac{\int_\epsilon^R [b^+/b^+(\epsilon)]^{n-1}\,dr}{\int_0^\epsilon [b^+/b^+(\epsilon)]^{n-1}\,dr}$$

is smaller than the same expression where b^+ is replaced by $\overline{b}$. The claimed estimate follows by integration on the unit sphere of $T_m M$. ∎

4.20 bis **Theorem.** *With the same curvature assumption, suppose that, for some positive R,*

$$\mathrm{vol}\big(B(m,R)\big) = V^k(R)$$

or that, for some couple (r, R), $0 < r < R$,

$$\frac{\text{vol}\big(B(m, R)\big)}{\text{vol}\big(B(m, r)\big)} = \frac{V^k(R)}{V^k(r)}.$$

Then the ball $B(m, R)$ *is isometric to the ball of radius* R *in the model space. In particular, if* $\text{Ric} \geq (n - 1)g$ *and* $\text{vol}(M, g) = \text{vol}(S^n, \text{can})$, *then* (M, g) *is isometric to* (S^n, can).

Proof. If one of these equalities holds, then $\text{vol}\big(B(m, r)\big) = V^k(r)$ for *any* $r < R$. Moreover, we know that we are before the cut-locus. Now, let us trace back the equality cases in the proof of 4.19 and 4.20. We first see that $b = \bar{b}$. Next, the equality case for Schwarz inequality in lemma 4.20 gives

$$Ddr = -\frac{\Delta r}{n - 1}g_r,$$

where g_r denotes the induced metric on the sphere $S(m, r)$. Therefore, for any tangent X vector to $S(m, r)$,

$$D_X \frac{\partial}{\partial r} = -\frac{\Delta r}{n - 1}X.$$

This means that the spheres $S(m, r)$ are totally umbilical submanifolds (cf. 5.10); in particular their mean curvature is constant, so that in normal coordinates

$$g = dr^2 + f^2(r)\sigma,$$

where σ is the canonical metric of S^{n-1}. If we write down that $\text{vol}\big(B(m, r)\big) = V^k(r)$ for such a metric, our claim is now straightforward. ∎

4.B.4 Heintze-Karcher inequality

The same method yields estimates for volumes of hypersurfaces. Let H be a closed hypersurface of a Riemannian manifold, and take a domain $D \subset H$ for which we can define a unit continuous normal vector field N. Define

$$\Phi : D \times] - \infty, +\infty[\longrightarrow M$$

by $\Phi(x, r) = \exp_x rN_x$ and set

$$\Phi^* v_g = h(x, r)^{n-1}dx.$$

(We have denoted by dx the Riemannian measure of H.) The same method as just above proves that, for any regular value of Φ (i.e. for *non-focal* points, see [Mi]) the Laplacian of the distance to H is given by

$$\Delta r = -(n - 1)\frac{h'}{h}.$$

Therefore, lemma 4.20 is still valid when b is replaced by h and the distance function to a point by the distance function to H. It will give the

4.21 Theorem (Heintze-Karcher, cf. [H-K]): *Let (M, g) be a complete Riemannian manifold such that* $\mathrm{Ricci} \geq (n-1)g$. *Let H be a closed hypersurface of M with mean curvature η. Then for any compact domain $D \subset H$ we have*

$$\mathrm{vol}\big(\Phi(D \times \{r\})\big) \leq \int_D \overline{h}(x, r)^{n-1} dx,$$

where $\overline{h}(x, r)$ is the solution of the differential equation $\overline{h}'' + k\overline{h} = 0$ which satisfies the initial conditions $\overline{h}(x, 0) = 1$ and $\overline{h}'(x, 0) = \eta(x)$.

Proof. If $\mathrm{Ricci} \geq (n-1)g$, the same computation as in 4.19 shows that

$$h'' + kh \leq 0.$$

Now, we must check the initial conditions satisfied by h. Denote by g_r the restriction of the metric g to $\Phi(H \times \{r\})$. Clearly $v_{g_r} = h(x, r)^{n-1} dx$, so that $h(x, 0) = 1$. On the other hand, we shall see in 5.19 that

$$\frac{d}{dr}(g_r)_{|r=0} = l \quad \text{second fundamental form of H.}$$

and that

$$\frac{d}{dr}(v_{g_r})_{|r=0} = (n-1)\eta.$$

(Notice that $l = Ddr$ and $(n-1)\eta = -\Delta r$). As in 4.19, our claim follows from the Sturm-Liouville comparison theorem. The explicit formula for $\overline{h}$ is given by

$$\overline{h}(x, r) = s'_k(r) + \eta(x)s_k(r),$$

where

$$s_k(r) = \begin{cases} \sin(\sqrt{k}r)/\sqrt{k}, & \text{if } k > 0; \\ r, & \text{if } k = 0; \\ \sinh(\sqrt{-k}r)/\sqrt{-k}, & \text{if } k < 0. \end{cases} \blacksquare$$

4.C Differential forms and cohomology

4.C.1 The de Rham complex

In this paragraph, many proofs will be omitted, for which we send back to [Wa]. We shall focus ourselves on vanishing theorems provided by the Bochner method.

The Poincaré lemma (cf. [Wa]) says that, on a contractible open set, any closed differential form is exact. Hence, if M is a smooth manifold, we have a *complex* (this simply means that $d \circ d = 0$), called the *de Rham complex*

$$\Omega^0 M \to \Omega^1 M \to \cdots \to \Omega^n M \to 0,$$

which is locally exact. This complex is not generally globally exact. Denote by $F^p(M)$ the vector space of smooth closed p-forms.

4.22 Definition. The *p-th de Rham cohomology group*, denoted by $H^p_{DR}(M)$ is the quotient

$$F^p(M)/d\Omega^{p-1}(M).$$

4.23 Example. Let M be a compact *oriented* manifold. Take $\omega \in \Omega^n M$ and $\alpha \in \Omega^{n-1} M$. The Stokes theorem says that

$$\int_M \omega = \int_M \omega + d\alpha.$$

In other words, the map $\omega \to \int_M \omega$ goes to the quotient and gives a linear map of $H^n_{DR}(M)$ into $\mathbf{R}$. This map is surjective (take for ω a volume form), which proves by the way that the de Rham complex is not globally exact. Moreover, it can be proved (cf. [Wa]) that this map is an isomorphism.

*We have just seen a particular case of the celebrated *de Rham theorem* which says that $H^p_{DR}(M)$ is isomorphic to the p-th singular cohomology group with real coefficients (cf. [Wa]).* If you don't know any algebraic topology, just remember that the de Rham cohomology has a topological meaning.

Now, if M is compact and equipped with a Riemannian metric, we are going to see that any cohomology class in the de Rham groups can be represented by a canonical differential form. Some more developments on Analysis will be necessary for that.

4.C.2 Differential operators and their formal adjoints

4.24 Definition. A *scalar differential operator* on a manifold M is a $\mathbf{R}$-linear map P from $C^\infty(M)$ into itself whose local expression with respect to any chart (U, ϕ) of M is a differential operator from $C^\infty(\phi(U))$ into itself in the usual sense.

Namely, $(\phi^{-1})^* \circ P \circ (\phi^{-1})^*$ has the form

$$\sum_\alpha a_\alpha(x) D^\alpha,$$

where the D^α are partial derivatives of index $(\alpha_1, \cdots, \alpha_n)$. The maximum of $|\alpha|$ is the *order* of P.

4.25 Exercise. Give the expression of the Laplacian in local coordinates.

When working with manifolds, the more general notion of a differential operator on vector bundles is necessary.

4.26 Definition. Let E and F be vector bundles over a compact manifold M. A *differential operator* from E to F is is a $\mathbf{R}$-linear map P from $\Gamma(E)$ into $\Gamma(F)$ such that every point $m \in M$ has a neighborhood U diffeomorphic to an open set Ω in $\mathbf{R}^n$, such that E and F are trivial over U and that the local expression of P over U is given by a $s \times r$ matrix of ordinary differential operators (r and s denote the ranks of E and F respectively).

Using distribution theory, differential operators have a simple characterization (cf. [N], p. 175 for a proof).

4.27 Theorem. *A linear map from $\Gamma(E)$ onto $\Gamma(F)$ such that*

$$\text{supp}(Ps) \subset \text{supp}(s)$$

for any $s \in \Gamma(E)$ is a differential operator. (The converse is clear).

4.28 Examples: the exterior differential d is a differential operator of degree 1 of $\Omega^p M$ into Ω^{p+1}. We have seen in 1.121 the following *naturality property:* for any smooth map ϕ of a smooth manifold M into another smooth manifold n,

$$d_M \circ \phi^* = \phi \circ d_N.$$

R. Palais proved (cf. [Pa]) that d, as a differential operator, is characterized by this naturality property.

On the other hand, if we are interested in naturality with respect to local isometries between Riemannian manifolds, we find many more natural differential operators. Namely, all the powers of the covariant derivative D are natural in that sense, and so are the operators which are obtained from those by performing traces and symmetrizations. With a mild regularity assumption, it can be proved (cf. [St]) that all the natural Riemannian differential operators are obtained in that way.

Among such operators, we have already met the divergence, and the Laplacian for functions. Another important operator is the following:

4.29 Definition. The *Hodge Laplacian* is the order two differential operator of $\Omega^p M$ into itself given by

$$\Delta^p = d^{p-1} \circ \delta^p + \delta^{p+1} \circ d^p.$$

Since there is no ambiguity about the spaces on which these operators are acting, we shall most often write $d\delta + \delta d$. To apply the Stokes theorem to these operators, two more definitions are necessary.

4.30 Definition. On a compact Riemannian manifold (M, g), the *integral scalar product* of two (p, q) tensors s and t is given by

$$\langle s, t \rangle = \int_M g_m(s, t) v_g.$$

This scalar product endows each tensor space $\Gamma(T_q^p M)$ with a prehilbertian structure. The corresponding norm will be denoted by $\| \ \|_g$ or just $\| \ \|$ if there is no ambiguity. The reader will check as an easy exercise that the norms associated with different metrics are *equivalent*.

4.31 Definition. Let E and F be two tensor bundles on a Riemannian manifold (M, g), and let P (resp. Q) be a differential operator from $\Gamma(E)$ into

$\Gamma(F)$ (resp. from $\Gamma(F)$ into $\Gamma(E)$). These operators are said to be *formally adjoint* if, for any $s \in \Gamma(E)$ and $t \in \Gamma(F)$ with compact support,

$$\langle Ps, t \rangle = \langle s, Qt \rangle.$$

It can easily be proved that any differential operator P has a unique formal adjoint (of the same order), which shall be denoted by P^*. We shall not need this result: we deal with natural Riemannian differential operators, whose formal adjoints are given explicitly.

4.32 Examples. The formal adjoint of d^p is δ^{p+1}. Hence the Hodge Laplacian is formally auto-adjoint.

Remark: The word *formal* is introduced to avoid confusion with the theory of self-adjoint unbounded operators in Hilbert spaces, which lies much deeper. Anyhow, this theory is crucial for proving theorem 4.34 below and the basic facts of spectral theory for the Laplacian (cf. [Ta]).

4.C.3 The Hodge-de Rham theorem.

4.32 bis Definition. A differential form α on a Riemannian manifold (M, g) is said to be *harmonic* if $\Delta\alpha = 0$.

4.33 Lemma. *If M is compact, a differential form is harmonic if and only if it is closed and coclosed (i.e. if its codifferential vanishes).*
Proof. Just notice that

$$\langle \Delta^p \alpha, \alpha \rangle = \langle d^p \alpha, d^p \alpha \rangle + \langle \delta^p \alpha, \delta^p \alpha \rangle. \blacksquare$$

Now, we have the following deep theorem, proved independantly by G. de Rham and H. Weyl during the Second World War (cf. [Wa] for an accessible proof).

4.34 Theorem. *For a compact Riemannian manifold (M, g), we have the following decompositions, which are orthogonal with respect to the integral scalar product:*

$$\Omega^p M = \mathrm{Ker}\,\Delta^p \oplus \mathrm{Im}\,d^{p-1} \oplus \mathrm{Im}\,\delta^{p+1}.$$

In particular, $\mathrm{Ker}\,\Delta^p$ is isomorphic to $H^p_{DR}(M)$.

4.35 Examples. i) Any *parallel* differential form is clearly harmonic. Therefore, if M is oriented, the space of harmonic forms of maximum degree is generated by the volume form v_g.

ii) It can be proved using elementary methods (cf.[B-G]) that $H^p_{DR}(S^n) = 0$ if $0 < p < n$. Therefore, for any Riemannian metric on S^n, there are no harmonic p-forms unless $p = 0$ or $p = n$.

iii) Any differential p-form on the torus T^n can be written as

$$\alpha = \sum_{i_1 \cdots i_p} \alpha_{i_1 \cdots i_p} dx^{i_1} \wedge \cdots \wedge dx^{i_p}.$$

Therefore, if T^n is equipped with a flat metric, since the forms $dx^{i_1} \wedge \cdots \wedge dx^{i_p}$ are then parallel, a straightforward computation shows that α is harmonic if and only if the functions $\alpha_{i_1 \cdots i_p}$ are harmonic, i.e. constant by 4.12. In particular,

$$\dim H^p_{DR}(T^n) = \binom{n}{p}.$$

iv) The 2-form ω on $(P^n\mathbf{C}, \mathrm{can})$ which is associated with the almost complex structure J (cf. 3.57) is parallel, hence harmonic. Clearly, we also have $D(\bigwedge^k \omega) = 0$ for any exterior power of ω, which proves that $H^{2k}_{DR}(P^n\mathbf{C}) \neq 0$ if $0 \leq k \leq n$. In fact, it can be proved that

$$H^{2k}_{DR}(P^n\mathbf{C}) \simeq \mathbf{R} \quad \text{and} \quad H^{2k+1}_{DR}(P^n\mathbf{C}) = 0.$$

4.C.4 A second visit to the Bochner method

4.36 Proposition. *For any differential one-form α on (M, g) one has*

$$\Delta\alpha = D^*D\alpha + \mathrm{Ric}(\alpha),$$

where Ric *is the Ricci curvature viewed as a field of endomorphisms which acts on one-forms, –that is with* $\big(\mathrm{Ric}(\alpha)\big)(X) = \alpha(\mathrm{Ric}(X))$.
Proof. Let $(E_i)_{1 \leq i \leq n}$ be a local field of orthonormal frames. Then we have

$$\delta\alpha = -\sum_i D_{E_i}\alpha(E_i) \quad \text{and} \quad d\delta\alpha(X) = -\sum_i D_X D_{E_i}\alpha(E_i).$$

On the other hand,

$$d\alpha(X, Y) = D_X\alpha(Y) - D_Y\alpha(X)$$

so that

$$\delta d\alpha(X) = -\sum_i D_{E_i}D_{E_i}\alpha(X) + \sum_i D_{E_i}D_X\alpha(E_i).$$

Therefore,

$$\Delta\alpha(X) = -\sum_i D_{E_i}D_{E_i}\alpha(X) + \sum_i (D_{E_i}D_X - D_X D_{E_i})\alpha(E_i)$$

$$= D^*D\alpha(X) + \alpha\,(\mathrm{Ric}(X))$$

from the Ricci identity (4.13). ∎

4.37 Theorem. *Let (M, g) be a compact n-dimensional Riemannian manifold. Then*
i) *if the Ricci curvature is strictly positive, $H^1_{DR}(M) = 0$;*
ii) *if it is non-negative, $\dim H^1_{DR}(M) \leq n$;*
iii) *if it is non-negative and if $\dim H^1_{DR}(M) = n$, then (M, g) is isometric to an n-dimensional flat torus.*

Proof. i) Using Proposition 4.25, we see that for any 1-form α

$$g(\Delta\alpha, \alpha) = g(D^*D\alpha, \alpha) + \mathrm{Ric}(\alpha, \alpha),$$

and by integrating on M we get

$$\langle \Delta\alpha, \alpha \rangle = \langle D^*D\alpha, \alpha \rangle + \int_M \mathrm{Ric}(\alpha, \alpha)v_g$$

$$= \langle D\alpha, D\alpha \rangle + \int_M \mathrm{Ric}(\alpha, \alpha)v_g.$$

If the Ricci curvature is strictly positive everywhere, the second member of this inequality is strictly positive as soon as the form α is not zero. Then there are no non trivial harmonic forms, and theorem 4.23 shows that $H^1_{DR} = 0$.
ii) If the Ricci curvature is non-negative, then $\Delta\alpha = 0$ if and only if $D\alpha = 0$. But a form whose covariant derivative vanishes is invariant by parallel transport, and therefore determined by its value at a point. Hence

$$\dim H^1_{DR}(M) = \dim \mathrm{Ker}\, D \leq n.$$

iii) Suppose that $\dim H^1_{DR}(M) = n$. It follows from ii) that there exist n linearly independent parallel 1-forms or, which amounts to the same, n linearly independent parallel vector fields. Using these vector fields to compute the curvature, we see that the manifold is *flat*. Our claim follows from the following lemma.

4.38 Lemma. *A compact flat manifold whose first Betti number is equal to the dimension is a flat torus.*

Proof. *This is just a consequence of Bieberbach's theorem, which gives the structure of compact flat Riemannian manifolds (cf. [Wo]). Let us give a more direct argument. For n linearly independent harmonic one-forms $(\alpha_i)_{1 \leq i \leq n}$, introduce the *Jacobi map* J

$$m \rightarrow \left(\int_{m_0}^m \alpha_i \right)_{1 \leq i \leq n},$$

where $m_0 \in M$ is given. Since two paths from m to m_0 differ by a Z-cycle, J maps M into a torus. Since the α_i are linearly independent at each point, J is a local diffeomorphism, hence a covering map since the source and the target are both compact (cf. 1.84). Now, a finite covering of a torus is a torus. ∎*
Another proof of i) and ii): * i) From Myers' theorem (cf. 3.85), we know that the fundamental group $\pi_1(M)$ is finite. Hence, so is $H_1(M, \mathbf{Z}) = \pi_1/[\pi_1, \pi_1]$ (cf. [Gg]). Using Poincaré duality and the universal coefficient theorem (ibidem), we see that $H^1(M, \mathbf{R}) = 0$.*
ii) From the Milnor-Wolf theorem (cf. 3.106), $\pi_1(M)$ has polynomial growth of degree at most n, and so has $H_1(M, \mathbf{Z})$ from the very definition of growth. Using 3.104, we see that this forces the rank of $H_1(M, \mathbf{R})$ to be lower than n. ∎

4.39 Remarks. a) On one hand, the geometric method of chapter III and the Bochner method *roughly* give the same result. On the other hand, some refinements are accessible by one method only: Myers' theorem is purely geometric, while the flatness result in 4.37 iii) requires an analytic proof.

b) The method of proof of theorem 4.37, known as Bochner (or Bochner-Weitzenbock method) can be used in a fairly general context. Namely, we write a differential operator whose kernel gives a geometric information (the Hodge Laplacian in that case) as the sum of a *positive* operator (D^*D) and of an algebraic operator which involves curvature. Then, if a suitable geometric assumption ensures that this last operator is positive, we get geometric information about the manifold (cf. [SB] exp.XVI, or [F-U] appendix C for details, [G-M], [Hu], [La1], for further Riemannian geometric results, [G-H] for complex geometry. Here we just state the following generalization of 4.37, due to D. Meyer (cf.[G-M]), and [L-M])

If (M, g) is a compact Riemannian manifold with positive curvature operator, *then $H_{DR}^k(M) = 0$ for $1 \leq k \leq \dim M$.*

In the book [L-M] of B. Lawson and M.L. Michelsohn, the relevant algebra for the Bochner method is presented in a very elegant and efficient way, using Clifford algebras and spinors. They also explain how the Bochner method, combined with spinorial methods, gives topological obstructions to positive scalar curvature.

Another point of view has been successfully developed by P. Li and S. Gallot (cf. [Be-G] and [Li] for instance): controlling the negativeness of the Ricci curvature (and modulo some extra geometric information, such as an upper bound for the diameter, compare with 3.116), they obtain upper bounds for $\dim H_{DR}^1(M)$. In fact, their method works in any context where the classical Weitzenbock method applies. The reader can consult the expository paper [Be4] for these and related problems.

4.40 Exercise. i) Let S be the differential operator of T^*M into S^2M defined by $S(\alpha) = L_{\alpha\#}g$, and let δ be the restriction of D^* to $\Gamma(S^2(M))$ (cf. 3.135). Show that for any $\alpha \in \Omega^1M$ and $h \in \Gamma(S^2(M))$

$$\langle S\alpha, h \rangle = 2\langle \alpha, \delta h \rangle.$$

ii) Show that $\delta S = D^*D - \text{Ric}$. Use this formula to prove that if (M, g) is compact with strictly negative Ricci curvature, there are no Killing vector fields, and therefore that $\text{Isom}(M, g)$ is finite.

iii) What can be said if Ric is only non-positive?

4.D Basic spectral geometry

4.D.1 The Laplace operator and the wave equation

Consider an homogeneous vibrating membrane, whose equilibrium position is represented by a compact domain $\Omega \subset \mathbf{R}^2$, and position at the instant t by the graph of a function $x \to u_t(x) = u(t, x)$ of Ω into $\mathbf{R}$. In other words, $u(t, x)$ is the height at the instant t of the point $x \in \Omega$. The potential energy is proportional to the increase of area, namely

$$E_t = \int_\Omega \left(1 + | \nabla u_t |^2\right)^{\frac{1}{2}} dx - \int_\Omega dx.$$

For "small" vibrations, the potential energy can be replaced by

$$E_t = \frac{1}{2} \int_\Omega | \nabla u_t |^2 dx$$

and the corresponding Lagrangian (kinetic energy minus potential energy) is just

$$L(u_t) = \frac{1}{2} \int_\Omega \left[\left(\frac{\partial u}{\partial t}\right)^2 - | \nabla u |^2 \right] dx.$$

The Lagrange-Hamilton principle says that any motion $t \to u_t$ of the membrane (say with fixed boundary, which gives the initial conditions $u(0, x) = 0$ and $\frac{\partial u}{\partial t}(0, x) = 0$ for $x \in \partial\Omega$), is a critical point of the map

$$u \to \int_{t_1}^{t_2} L(u_t)dt \qquad (\forall t_1, t_2).$$

Taking variations v which satisfy the same conditions and integrating by parts, we get, for any such v

$$\int_{t_1}^{t_2} \int_\Omega \left[\frac{\partial u}{\partial t} \frac{\partial v}{\partial t} - \nabla u \nabla v \right] dx dt = 0$$

and integrating by parts we get

$$\frac{\partial^2 u}{\partial t^2} + \Delta u = 0.$$

This equation is known as the *wave equation*. The method of separation of variables yields for solutions of type $f(t)\phi(x)$ the conditions

$$f'' + \lambda f = 0 \quad \text{and} \quad \Delta\phi = \lambda\phi$$

where λ is some constant.

4.41 Definition. The *spectrum of the Dirichlet problem* for Ω is the set of real numbers λ such that the equation

$$\Delta\phi = \lambda\phi$$

with the boundary condition $\phi_{|\partial\Omega} = 0$ has non trivial solutions. Such a λ is called an *eigenvalue*, the corresponding solutions are the *eigenfunctions* for λ.

They constitute the *eigenspace* E_λ, whose dimension is the *multiplicity* of the eigenvalue λ.

4.42 Had the boundary of the membrane not been fixed, we would have been lead to the same equation, with the *Neumann condition* at the boundary, namely

$$\frac{\partial u}{\partial \nu} = 0$$

Now, if Ω is replaced by a compact Riemannian manifold with boundary, the above equations, and therefore the above definitions still make sense, and we shall speak of the spectrum of (M, g) for the Dirichlet or the Neuman problem. If M has no boundary, we shall simply speak of the spectrum of (M, g), since there is no distinction between the Dirichlet and the Neumann problem in that case.

Another relevant physical problem is heat conduction. Indeed, Fourier series, which provide the most basic example of eigenfunctions expansion for the Laplacian, were first introduced for that. The temperature $u(t, x)$ at a point x of a solid $\Omega \subset \mathbf{R}^3$ and at the instant t satisfies the equation

$$\frac{\partial u}{\partial t} + \Delta u = 0 \,,$$

and decomposable solutions are given by $e^{-\lambda t}\phi(x)$, where ϕ is again an eigenfunction with eigenvalue λ for the Dirichlet problem (if the temperature at the boundary is fixed), or the Neumann problem (in case of free exchange of heat with the outside). The heat equation also makes sense in a compact Riemannian manifold. It turns out to have a *fundamental solution* $k(t, x, y)$. Namely, the solution of the heat equation on (M, g) such that $u(0, x) = f(x)$ is given by

$$u(t, x) = \int_M k(t, x, y) f(y) v_g \,,$$

where k is given, at least formally, by

$$k(t, x, y) = \sum_i e^{-\lambda_i t} \phi_{\lambda_i}(x) \phi_{\lambda_i}(y).$$

The study of the heat kernel k provides much information about the spectrum. The reader is invited to look at the text-books [BGM] (a classical reference), and [Be2], [Cha], [Ga4] (for more recent Riemannian geometric results).

4.D.2 Statement of basic results on the spectrum

A basic result (which is indeed a particular case) of elliptic operator theory is the following. It is concerned with both Dirichlet's and Neumann's problems for a *compact* Riemannian manifold with boundary. We denote by $C_a^\infty(M)$ the space of smooth functions which satisfy one of the boundary conditions.

4.43 Theorem. i) *The eigenvalues of Δ_a can be arranged into a discrete sequence*

$$0 \leq \lambda_0 < \lambda_1 \leq \lambda_2 \leq \cdots \leq \lambda_n \leq \cdots$$

converging to $+\infty$.

ii) *For any eigenvalue λ_i, the eigenspace E_{λ_i} is finite-dimensional, and the eigenspaces are pairwise orthonormal.*

iii) *The algebraic sum of the eigenspaces is dense in $C_a^\infty M$ for the uniform topology. In particular, the Hilbert space $L^2(M, v_g)$ has a basis of eigenfunctions.*

Our display of signs $<$ and $\leq$ in i) is implicitly contained in the following important statement.

4.44 Theorem. *The first eigenvalue λ_0 is simple, and the corresponding eigenfunctions never vanish on the interior of M.*

Of course, for the Neumann problem or for closed manifolds this is not very interesting: we have already seen (cf. 4.12) that $\lambda_0 = 0$ and $\phi_0 = const$ in that case).

Coming back to the physical interpretation, the $\sqrt{\lambda_i}$ are the eigenfrequencies of the membrane, and $\phi(x)$ is the corresponding maximal amplitude at x. In particular, the lowest significant eigenvalue (namely λ_0 in the Dirichlet case, λ_1 in the Neumann or boundaryless case) is very important: it corresponds to the sound you hear the best !

Two basic questions are the following: a) Can we compute the spectrum of a given Riemannian manifold (M, g); b) conversely, is a Riemannian manifold determined up to isometry by the data of its spectrum. In other words, that we borrow from the title of a very stimulating article of M. Kac [Ka], *can one hear the shape of a drum ?*

Although quite a few partial answers to these questions are known (cf. for example [B-G-M]), their formulation turns out to be very optimistic, as we shall soon explain. Therefore, mathematicians who are concerned with spectral geometry mainly deal with the following problems:

i) What estimates on the spectrum can be deduced from (suitably chosen) geometric data of the manifold ?

ii) *Conversely, what geometric data can be read form spectral ones ?*

These problems can be considered from two different points of view: if one is interested in *the first eigenvalue* then techniques and results related to isoperimetric inequalities occur (cf. [Ga1] for an expository article), and 4.71 below). On the other hand, if one is interested in the asymptotic behavior of the eigenvalues, heat and wave equation methods are used. One of the most beautiful results in that direction is the relation between the spectrum and the lengths of closed geodesics in a closed Riemannian manifold. (See Colin de Verdiere [CV], Duistermaat-Guillemin [D-G], or [B1] for a shorter account).

4.E Some examples of spectra

4.E.1 Introduction

The effective computation of the spectrum is impossible but in a very few cases. Even in the apparently simple case of domains of $\mathbf{R}^2$, the only known explicit examples of such a computation are the rectangle (the simple minded method of separation of variables works), the disk (use the separation of variable in polar coordinates: not unexpectedly, the spectrum appears to be related with the Bessel functions), the circular sector (use the same method), and the equilateral, isocel-rectangle and half-equilateral triangles (see [Bel] for the general method in that case, which uses in a nice way the group generated by the reflections with respect to the sides). Indeed, there are presently two cases where the computation is possible:

i) fundamental domains of a group which acts by isometries on a space whose spectrum is known, or Riemannian quotients of a manifold whose spectrum is known;

ii) manifolds or domains which have so many symmetries that the problem can be reduced to a one-dimensional differential equation problem.

Notice that the spectrum of (M, g) is known as soon as we know a family E_λ of eigenfunctions vector spaces whose algebraic sum is uniformly dense in $C^\infty(M)$. It may be useful to consider the space $C^\infty_{\mathbf{C}}(M)$ of complex-valued smooth functions on M.

4.E.2 The spectrum of flat tori

Parseval's theorem just says that the (complex-valued) eigenfunctions and corresponding eigenvalues of the Laplacian for the circle T^1_L of length L are

$$f_k(x) = \exp\left(\frac{2ik\pi x}{L}\right) \quad \text{and} \quad \lambda_k = \frac{4\pi^2 k^2}{L^2} \quad k \in \mathbf{Z}.$$

Each non-zero eigenvalue has multiplicity 2.

The same method works for an n-dimensional flat torus $T^n = \mathbf{R}^n/\Gamma$. Namely, consider the *dual lattice* of Γ, i.e. the set Γ^* of vectors $\lambda^* \in \mathbf{R}^n$ such that

$$\langle \lambda, \lambda^* \rangle \in \mathbf{Z}, \quad \forall \lambda \in \Gamma.$$

Clearly, the functions

$$f_{\lambda^*}(x) = e^{2i\langle \lambda^*, x \rangle}$$

are eigenfunctions (since they are Γ-invariant eigenfunctions for the Laplacian in $\mathbf{R}^n$), and the corresponding eigenvalues are the real numbers $4\pi^2 \mid \lambda^* \mid^2$. Since the f_{λ^*} form a Hilbertian basis of $L^2(\mathbf{T}^n)$, we have got the whole spectrum. Although this example looks very simple, it is related with several important non trivial questions. Namely

i) The Poisson formula

$$\sum_{\lambda \in \Gamma} e^{-\pi |\lambda|^2} = \frac{1}{\text{vol}(\Gamma)} \sum_{\lambda^* \in \Gamma^*} e^{-\pi |\lambda^*|^2} \qquad (4.1)$$

says that the spectrum determines the lengths of periodic geodesics (here, we have followed the notations of Number-theorists: $\text{vol}(\Gamma)$ denotes the measure of a fundamental domain of Γ, that is the volume of the flat torus T^n/Γ. Indeed, the general result as presented in [D-G] can be viewed as a deep generalization of the Poisson formula.

ii) A classical result in Geometry of Numbers, due to Minkowski, says that (cf.[Se] for instance)

$$\text{card} \left\{ \lambda \in \Gamma, \, |\lambda| \le t \right\} \sim \left(\frac{\Gamma(\frac{n}{2}+1)}{(4\pi)^{n/2}} \right) \text{vol}(\Gamma) t^{\frac{n}{2}} \qquad (4.2)$$

This is just a particular case of the *Weyl asymptotic formula* (cf.[Be2] for a proof), which says that the same property holds for a Riemannian manifold, possibly with boundary, when replacing $\text{vol}(\Gamma)$ by the volume of the manifold. Therefore the volume (and the dimension) are determined by the asymptotic behavior of the spectrum.

iii) Cleverly choosen flat tori have provided the first example of isospectral non-isometric Riemannian manifolds (J. Milnor, cf. [BGM] for a detailed account).

4.46 Exercise. Compute the spectrum of flat Klein bottles.

4.47 About isospectral non-isometric manifolds.

In J. Milnor's example,the manifolds are two 16-dimensional flat tori, and the arguments have a strong arithmetic flavor. Further examples were found afterwards: surfaces and n-dimensional compact manifolds with constant curvature -1, lens spaces of dimension 5, compact domains of $\mathbf{R}^2$ (see [Be3] for credits and references). In the eighties, C. Gordon and E. Wilson have developed methods for finding *continuous families* of isospectral non isometric Riemannian metrics on suitable nilmanifolds. In all these examples, the manifolds are non-simply connected.

However, overthrowing the believes of many people (even though nobody, up to our knowledge, ventured any conjecture), Dorothy Schüth (cf. [Sc]) has found isospectral deformations in the simply connected case.

4.E.3 Spectrum of (S^n, can)

Now, we are in a case where strong homogeneity properties make the computation possible. Let F be a smooth function on $\mathbf{R}^{n+1}$, and let f be its restriction to S^n. Denote by $\tilde{D}$ and $\tilde{\Delta}$ the covariant derivative and the Laplacian of $\mathbf{R}^{n+1}$, and by $\frac{\partial}{\partial r}$ the radial derivative.

4.48 Proposition. *On S^n, we have*

$$\tilde{\Delta} F = \Delta f - \frac{\partial^2 F}{\partial r^2} - n \frac{\partial F}{\partial r}.$$

Proof. Let $N = \frac{\partial}{\partial r}$ be the normal vector field to S^n. If X and Y are vector fields of $\mathbf{R}^{n+1}$ which are tangent to S^n, from 2.56 we have

$$\tilde{D}_X Y = D_X Y - \langle X, Y \rangle N.$$

Let $(e_i)_{1 \le i \le n}$ be an orthonormal basis of $T_x S^n$, and $(E_1, \cdots, E_n, N)$ be a field of orthonormal frames in a neighborhood of x in $\mathbf{R}^{n+1}$ whose value at x is $(e_1, \cdots, e_n, N_x)$. For any point of S^n where this frame field is defined, we have

$$\tilde{\Delta} F = -\mathrm{tr} \tilde{D} df = -\sum_{i=1}^{n} \left(E_i \cdot df(E_i) - dF(\tilde{D}_{E_i} E_i) \right) - \tilde{D} dF(N, N).$$

Since $\tilde{D}_N N = 0$, we have $\tilde{D} dF(N, N) = -\frac{\partial^2 F}{\partial r^2}$. Therefore, we get

$$\tilde{\Delta} F = -\sum_{i=1}^{n} (E_i \cdot df(E_i) - df(D_{E_i} E_i)) - n dF(N) - \frac{\partial^2 F}{\partial r^2},$$

which is just the claimed formula. ∎

4.49 Corollary. *Let $\tilde{P}$ be any homogeneous harmonic polynomial of degree k on $\mathbf{R}^{n+1}$, and P be its restriction to S^n. Then*

$$\Delta P = k(k + n - 1) P.$$

We are going to see that we have got all the eigenfunctions of Δ for the sphere. Denote by $\tilde{P}_k$ the vector space of degree k homogeneous polynomials on $\mathbf{R}^{n+1}$, by $\tilde{H}_k$ the subspace of harmonic polynomials, by P_k and H_k the spaces obtained by restricting these polynomials to S^n. Since polynomials are uniformly dense in $C^0(S^n)$, our claim follows from the following lemma.

4.50 Lemma. *We have the decompositions*

$$\tilde{P}_k = \bigoplus_{i=0}^{[k/2]} r^{2i} \tilde{H}_{k-2i}$$

$$P_k = \bigoplus_{i=0}^{[k/2]} H_{k-2i}$$

Proof. Consider the natural action of $G = SO(n+1)$ on $C^\infty(\mathbf{R}^{n+1})$ and $C^\infty(S^n)$. These actions leave the spaces $\tilde{P}_k$ and P_k globally invariant. Since the Laplacian commutes with isometries, the spaces $\tilde{H}_k$ and H_k are also invariant. Indeed, in the more general situation of a Riemannian homogeneous space

$M = G/H$, we have an orthogonal representation of G in the prehilbertian space $(C^\infty(M), \langle,\rangle)$ and this representation leaves every eigenfunctions vector space invariant.

4.51 Definition. An H_0-invariant smooth function f on G/H is said to be *zonal* (H_0 denotes as usual the neutral component of H).

If V is a G-invariant subspace of $C^\infty(G/H)$, we shall denote by $Z(V)$ the subspace of zonal functions.

4.52 Lemma. *If V is a non-zero finite dimensional vector space, then $Z(V) \neq \{0\}$. If* $\dim Z(V) = 1$*, then V is G-irreducible.*

Proof. Let $[e]$ be the base-point of $M = G/H$. Since V is G-invariant and G transitive on M, as soon as $V \neq \{0\}$ there exists a function $f \in V$ such that $f([e]) \neq 0$. Take the linear form

$$\phi : f \to f([e]).$$

Its kernel is H-invariant, and so is $\operatorname{Ker}\phi^\perp$. Therefore, we have a one-dimensional orthogonal representation of H_0 in $\operatorname{Ker}\phi^\perp$. This representation is trivial, since H_0 is connected, which proves the first part.

Suppose that V is not irreducible, and take a decomposition $V = V' \oplus V''$ into orthogonal non trivial G-invariant subspaces. We have just seen that each of them contains a line of zonal functions, therefore $\dim Z(V) \geq 2$. ∎

4.53 Exercise. When did we use the finite dimension assumption ? Let us come back to $(S^n, \operatorname{can})$. Remark that $\tilde{P}_k$ and P_k (resp. $\tilde{H}_k$ and H_k) can be identified as representation spaces of $C^\infty(S^n)$. Denote by $\tilde{Q}_k$ the orthogonal of $r^2 \tilde{P}_{k-2}$ in $\tilde{P}_k$, and by Q_k the corresponding subspace of P_k.

4.54 Lemma. *The representation space Q_k is irreducible.*

Proof. Recall that S^n is the homogeneous space $SO(n+1)/SO(n)$ where $SO(n)$ is (for instance) the subgroup of $SO(n+1)$ which leaves the point $a = (1, 0, \cdots, 0)$ fixed. Since $SO(n+1)$ acts transitively on pairs of equidistant points in S^n, the zonal functions are just the functions which only depend on the distance to a.

In particular, $Z(P_k)$ is generated by the restrictions to S^n of the polynomials

$$x_0^k, x_0^{k-2}\left(r^2 - x_0^2\right), \cdots, x_0^{k-[k/2]}\left(r^2 - x_0^2\right)^{[k/2]},$$

and $\dim Z(P_k) = [k/2]$. Therefore $\dim Z(Q)_k = 1$ and our claim follows ∎. Now lemma 4.50 is a consequence of the following.

4.55 Lemma. *For any integer k, we have the orthogonal decomposition*

$$P_k = H_k \oplus P_{k-2}.$$

Proof. Proceed by induction. The lemma is plainly true for $k = 0$ and $k = 1$. Suppose it is true up to $k-2$. Since the spaces H_l are pairwise orthogonal (use 4.11 and the elementary properties of eigenfunctions), H_k is then orthogonal to P_{k-2}. On the other hand, $\tilde{H}_k$ is the kernel of the linear map

$$\tilde{\Delta} : \tilde{P}_k \rightarrow \tilde{P}_{k-2}.$$

Since $\dim \tilde{P}_k > \dim \tilde{P}_{k-2}$, it is non-trivial, and so is H_k. Then the preceding lemma shows that $H_k = Q_k$. ∎

4.56 Remark. * The study of the Laplace operator for (possibly non-compact) Riemannian symmetric spaces has been growing for may years into a very important domain of Mathematics, Harmonic Analysis, cf.[Hn] for a beginning. It has its own object and methods, but a geometric point of view can be useful, cf. [Ro] for instance. *

4.F The minimax principle

Now we come back to the general case. As we have already explained, all that we can hope is to be able to give *estimates* on the eigenvalues. Let (M, g) be a compact Riemannian manifold. First consider for simplicity the case where M is closed (i.e. without boundary). When writing down the spectrum $(\lambda_i)_{0 \leq i < +\infty}$ (either for the Dirichlet or the Neumann problem if $\partial M \neq \emptyset$), we make the convention that any eigenvalue is counted as many times as its multiplicity. Let $(\phi_i)_{0 \leq i < +\infty}$ be a corresponding sequence of eigenfunctions, supposed to be orthonormal Then, for any $f \in C^\infty(M)$, we have the L^2 decomposition

$$f = \sum_i a_i \phi_i \quad \text{where} \quad a_i = \langle f, \phi_i \rangle.$$

Writing the same decomposition for Δf, we see that

$$\Delta f = \sum_i \lambda_i a_i \phi_i.$$

Hence, for any smooth f,

$$\langle \Delta f, f \rangle = \langle df, df \rangle = \sum_i a_i^2 \lambda_i^2.$$

Set

$$E(f) = \langle df, df \rangle = \int_M |df|^2 v_g \qquad \text{(Dirichlet or energy integral)}.$$

The following crucial property follows immediately.

4.57 Proposition. *Then the k-th eigenvalue λ_k is characterized by*

$$\lambda_k = \inf \{ E(f), f \in C^\infty(M), \|f\| = 1, \langle f, \phi_0 \rangle = \cdots \langle f, \phi_{k-1} \rangle = 0 \}.$$

4.58 Remark. An analyst could say "we have bridled the horse by the tail", and he (or she) would be right. Indeed, the spectral decomposition of Δ is proved by Hilbert spaces methods (using spectral theory of non bounded self-adjoint operators), and the existence of eigenvalues is obtained by using property 4.57!

Once the appropriate Hilbert spaces have been introduced, the case of manifolds with boundary does not present any difference (see e.g. [Be 2] for a neat discussion).

4.59 Definition. The *admissible space* for the Neumann (resp. the Dirichlet) problem, denoted by $H^1(M)$ (resp. $H_0^1(M)$) is the Hilbert space obtained by completing $C^\infty(M)$ (resp. the subspace $C_0^\infty(M)$ of smooth functions with compact support contained in the interior of M) for the "norm" $\langle df, df \rangle$. When making statements which are valid for both Dirichlet and Neumann problems, it will be convenient to denote the admissible space by $H_*^1(M)$.

Now (cf. [Be2] for instance), the characterization of λ_k for the Neumann (resp. the Dirichlet) problem is the same as in 4.57, after replacing $C^\infty(M)$ by $H^1(M)$ (resp. $H_0^1(M)$).

4.60 Remark. From this point of view, in the Neumann or in the closed case, the first eigenvalue $\lambda_0 = 0$ is ruled out.

4.61 Definition. The *Rayleigh quotient* of an admissible function f is the ratio

$$R_g(f) = \frac{\int_M |df|_g^2\, v_g}{\int_M f^2 v_g} = \frac{E(f)}{\|f\|_g^2}.$$

As usually, g will be omitted if there is no ambiguity.

The following property is very useful. As the proof comes from 4.57 using linear algebra, we leave it to the reader.

4.62 Theorem (the minimax principle): *For any finite dimensional subspace of $H_*^1(M)$, set*

$$m_g(E) = \max\{R_g(f),\ f \in E\}$$
$$M_g(E) = \min\{R_g(f),\ f \perp E\}$$

Then,

$$\lambda_k = \min\{m_g(E),\ \dim E = k+1\}$$
$$= \max\{M_g(E),\ \dim E = k\}$$

The minimum (resp. the maximum) is achieved when E is the subspace generated by the first $k+1$ (resp. k) eigenfunctions.

4.63 Corollary. *Let g and g_0 be two metrics on a closed manifold M such that*

$$a^2 g_0 \le g \le b^2 g_0 \qquad (\text{with } 0 < a^2 < b^2).$$

Then for any k,

$$\frac{a^n}{b^{n+2}}\lambda_k(g_0) \leq \lambda_k(g) \leq \frac{b^n}{a^{n+2}}\lambda_k(g_0).$$

Hence, if a sequence g_n of metrics converges to g_0 for the C^0-topology, then $\lambda_k(g_n)$ converges to $\lambda_k(g_0)$ for any k.

Proof. We have

$$a^n v_{g_0} \leq v_g \leq b^n v_{g_0} \quad \text{and}$$

$$\frac{1}{b} \mid df \mid_{g_0} \leq \mid df \mid_g \leq \frac{1}{a} \mid df \mid_{g_0},$$

hence

$$\frac{a^n}{b^{n+2}} R_{g_0}(f) \leq R_g(f) \leq \frac{b^n}{a^{n+2}} R_{g_0}(f).$$

Take the vector space E generated by the first $k + 1$ eigenfunctions of Δ_{g_0}. Then the minimax principle says that

$$\lambda_k(g) \leq m_g(E) \leq \frac{b^n}{a^{n+2}} m_{g_0}(E) = \frac{b^n}{a^{n+2}}\lambda_k(g_0).$$

In the same way,

$$\frac{a^n}{b^{n+2}}\lambda_k(g_0) = \frac{a^n}{b^{n+2}} M_{g_0}(E) \leq M_g(E) \leq \lambda_k(g). \blacksquare$$

Remark. This property illustrates the strength of the minimax principle: indeed, if g_n converges to g_0 in the C^0-topology, it is not always true that Δ_{g_n} converges to Δ_{g_0}.

The minimax principle as stated in 4.62 is valid for any quadratic form Q on a Hilbert space, provided Q admits a Hilbertian basis of eigenvectors (as the example of the Dirichlet integral shows, in general Q is neither continuous nor everywhere defined). The following property is useful for proving comparison theorems.

4.64 Lemma. *Let Φ be a linear injective map from a Hilbert space $(H_1, \langle,\rangle_1)$ into $(H_2, \langle,\rangle_2)$, and let Q_1 and Q_2 be quadratic forms on H_1 and H_2 respectively. Suppose that*

$$\langle f, f \rangle_1 \leq C_1 \langle \Phi(f), \Phi(f) \rangle_2$$
$$\text{and} \quad Q_1(f) \geq C_2 Q_2(\Phi(f)).$$

Then

$$\lambda_k(Q_1) \geq \frac{C_2}{C_1}\lambda_k(Q_2).$$

Proof. Take for E the space generated by the first k eigenfunctions of Q_1, and apply the minimax principle to $\Phi(E)$. Since Φ is injective, we have

$$\lambda_k(Q_2) \leq m_{Q_2}(\Phi(E)) \leq \frac{C_1}{C_2} m_{Q_1}(E) = \frac{C_1}{C_2}\lambda_k(Q_1),$$

and our claim is proved. $\blacksquare$

This property has the following important application:

4.65 **Corollary.** *Let (M, g) be a closed Riemannian manifold. Consider pairwise disjoint domains $D_1, \cdots, D_k$, and domains $\Omega_1, \cdots, \Omega_j$ whose interiors cover M. Then*

$$\lambda_{kl-1}\big((M, g)\big) \leq \max_i \{\lambda_{l-1}^D(D_i)\}$$

$$\lambda_{jl}\big((M, g)\big) \geq \frac{1}{m}\min_l \{\lambda_l^N(\Omega_i)\}$$

where λ_l^D (resp. λ_l^N) denote the eigenvalues of the Dirichlet (resp. Neumann) problem, and m the maximum number of domains Ω_i whose interiors have non-empty intersection.

Proof. Take $H_1 = \oplus_{i=1}^k H_0^1(D_i)$, and for $f = (f_1, \cdots, f_k) \in H_1$, set

$$\langle f, f \rangle_1 = \sum_{i=1}^k \int_{D_i} f_i^2 v_g$$

$$Q_1(f) = \sum_{i=1}^k \int_{D_i} |df_i|^2 v_g.$$

Take $H_3 = \oplus_{i=1}^k H^1(\Omega_r)$, and equip it in the same way with a scalar product $\langle, \rangle_3$ and a quadratic form Q_3. The space H_2 will be the usual Hilbert space $L^2(M)$ equipped with $Q_2(f) = E(f)$.

Now, define $\Phi : H_1 \to H_2$ and $\Psi : H_2 \to H_3$ by

$$(f_1, \cdots, f_k) \to \sum_{i=1}^k \bar{f}_i$$

(where $\bar{f}_i$ is just f_i on D_i and zero outside), and

$$f \to (f_{|\Omega_1}, \cdots, f_{|\Omega_j}).$$

For any $f \in H_1$ and $h \in H_2$, we have

$$\langle \Phi(f), \Phi(f) \rangle_2 = \langle f, f \rangle_1, \quad Q_2(\Phi(f)) = Q_1(f),$$

$$\langle \Psi(h), \Psi(h) \rangle_3 \geq \langle h, h \rangle_2 \quad \text{and} \quad Q_3(\Psi(h)) \leq m Q_2(h).$$

Our claims then follow from the preceding lemma. ∎

4.66 **Exercise.** State and prove similar properties for a Riemannian manifold with boundary.

4.G Ricci curvature and eigenvalues estimates

4.G.1 Introduction

Weyl asymptotic formula for the spectrum can be derived from 4.65 (see for instance [Be 2]). But we are more interested in Riemannian aspects of spectral problems. If we keep in view Gromov compactness theorem (cf. 3.116) as a heuristic guide, it is natural to expect estimates which involve the diameter D and a lower bound k for the Ricci curvature (we mean that $\mathrm{Ric}_g \geq kg$); the case where k is negative is far more interesting, since there are may more possibilities for the topological type of the manifold, compare with 3.85). Let us begin with rough methods.

4.G.2 Bishop's inequality and coarse estimates

4.67 Definition. An ϵ-*filling* of a (closed) Riemannian manifold (M, g) is a maximal family of pairwise disjoint balls of radius ϵ. The maximum number of such balls is the *filling function* of the manifold, and will be denoted by $N(\epsilon, M, g)$ (or $N(\epsilon)$ if there is no ambiguity).

An ϵ-filling by balls B_i being given, the balls $\tilde{B}_i$ of same centers and radius 2ϵ cover M: otherwise, we could find $x \in M$ whose distance to all the centers would be greater than 2ϵ. Then the ball $B(x, \epsilon)$ would meet no B_i, a contradiction. Applying 4.65, and using the same notations, we get

4.68 Lemma. *Let* $(B_i)_{1 \leq i \leq N(\epsilon)}$ *be an* ϵ-*filling of* (M, g). *Then*

$$\lambda_{N(\epsilon)-1} \leq \max \lambda_0^D(B_i)$$

$$\lambda_{N(\epsilon)} \geq \frac{1}{m} \min \lambda_1^N(\tilde{B}_i),$$

where m *is the order of the covering by the* $\tilde{B}_i$.

In order to give significant applications of this lemma, we need the following information:

a) an estimate for the filling function $N(\epsilon, M, g)$;
b) an upper bound of m;
c) an upper bound of $\lambda_0^D(B_i)$;
d) a lower bound of $\lambda_1^N(\tilde{B}_i)$.

Such an information is nearly trivial to get for Euclidean domains. By the way, it gives a coarse version of the Weyl asymptotic formula: namely, $\lambda_k = O(k^{2/n})$. In the Riemannian case, a), b), c) are easily derived from Bishop's theorem. We send back to [Gr2] for d), which is much more delicate.

4.G.3 Some consequences of Bishop's theorem.

We are going to give geometric estimates for manifolds (M, g) whose diameter is smaller than D and Ricci curvature greater than $(n - 1)k$.

a) *An estimate of $N(\epsilon)$*. Since (M, g) is equal to any geodesic ball of radius D, we get, using the notations of 3.101 and 4.19

$$\frac{\mathrm{vol}(B(x, \epsilon))}{\mathrm{vol}(M)} \geq \frac{V^k(\epsilon)}{V^k(D)}.$$

Therefore

$$N(\epsilon, M, g) \leq \frac{\mathrm{vol}(M)}{\inf\{\mathrm{vol}(B_i)\}} \leq \frac{V_k(D)}{V_k(\epsilon)}.$$

We get an estimate of type $N(\epsilon) \leq C(k, D)\epsilon^{-n}$, with an explicit function C of k and D.

b) *An upper bound for m.* If $x \in \tilde{B}_i$, then $B_i \subset B(x, 3\epsilon)$. The order m of the covering by the balls $\tilde{B}_i$ is therefore smaller than

$$\sup_{x \in M,\, y \in B(x, 2\epsilon)} \left\{ \frac{\mathrm{vol}(B(x, 3\epsilon))}{\mathrm{vol}(B(y, \epsilon))} \right\},$$

and a fortiori smaller than

$$\sup_{y \in M} \left\{ \frac{\mathrm{vol}(B(y, 5\epsilon))}{\mathrm{vol}(B(y, \epsilon))} \right\},$$

or than $\frac{V^k(5\epsilon)}{V^k(\epsilon)}$.

c) *An upper bound for $\lambda_0^D(B_i)$.* Let x_i be the center of the ball B_i, and define

$$f : B_i \to \mathbf{R} \quad \text{by} \quad f(x) = 1 - \frac{1}{\epsilon} d(x_i, x).$$

Clearly $\mid df \mid = \frac{1}{\epsilon}$ almost everywhere (i.e. outside the cut-locus of x), and $f(x) \geq \frac{1}{2}$ for any $x \in B(x_i, \frac{\epsilon}{2})$. Furthermore, it is easy to check that $f \in H_0^1(B_i)$. Hence the minimax principle gives

$$\lambda_0^D(B_i) \leq \frac{\int_{B_i} \mid df \mid^2 v_g}{\int_{B_i} f^2 v_g} \leq \frac{4}{\epsilon^2} \frac{\mathrm{vol}(B(x_i, \epsilon))}{\mathrm{vol}(B(x_i, \epsilon/2))}.$$

Using Bishop's inequality again, we see that

$$\lambda_0^D(B_i) \leq \frac{4}{\epsilon^2} \frac{V^k(\epsilon)}{V^k(\epsilon/2)}.$$

More explicitly, we can write

$$\lambda_0^D(B_i) \leq \frac{4}{\epsilon^2} \left[\frac{\sinh(\epsilon\sqrt{\mid k \mid})}{\sinh(\epsilon\sqrt{\mid k \mid}/2)} \right]^n.$$

A consequence of these estimates is the

4.69 Theorem (Cheng, cf. [Ch]): *Let (M, g) be a compact Riemannian manifold with diameter d and volume V, whose Ricci curvature is greater than $(n-1)k$. Set $i_0 = \left(\frac{c(n)d^n}{V} \right)^{1/(n-1)}$, where $c(n)$ is the volume of the unit ball in $\mathbf{R}^n$. Then*
i) if $i \leq i_0$

$$\lambda_i(M, g) \leq 16 \frac{i^2}{d^2} A_1 \left(\frac{d\sqrt{|k|}}{i} \right) ;$$

ii) if $i \geq i_0$

$$\lambda_i(M, g) \leq 16 \left[\frac{(i+1)c(n)}{V} \right]^{2/n} A_2 \left[\left(\frac{|k|^{n/2} V}{(i+1)c(n)} \right)^{1/n} \right],$$

where

$$A_1(x) = \left[\frac{\sinh(x/2)}{\sinh(x/4)} \right]^n \quad and \quad A_2(x) = \left(\frac{\sinh x}{x} \right)^2 A_1(x).$$

Remark. To see why we must have different estimates for $i \leq i_0$ and $i \geq i_0$, the reader is invited to check the explicit example of $(M, g) = S^{n-1}(\alpha) \times S^1(D)$, with $\alpha < D$. (Use the fact that the spectrum of a Riemannian product is the sum of the spectrums of the factors). Anyhow, for i big, the estimating function has the same qualitative growth as the growth given by Weyl's theorem.

Proof. i) For any integer i, take $\epsilon = \frac{d}{2i}$. Then $N(\epsilon)$ is greater than $i+1$. Indeed, there exists a minimizing geodesic γ of length d. Divide γ into i equal segments of lengths $\frac{d}{i}$. Then the balls of radius ϵ whose centers are the ends of these segments are pairwise disjoint. Now, the claimed estimate follows from 4.68 together with the estimate c) above.

ii) Take

$$\epsilon = \frac{1}{2\sqrt{|k|}} \text{Argsh} \left[\left(\frac{V|k|^{n/2}}{c(n)i} \right)^{1/n} \right]$$

and show that $N(\epsilon) \geq i$. Indeed, take an ϵ-filling of M by balls B_i, and recall that the balls $\tilde{B}_i$ of same centers and radii 2ϵ cover M. Therefore,

$$N(\epsilon) \geq \frac{V}{\sup[\text{vol}(\tilde{B}_i)]} .$$

Bishop's theorem gives

$$N(\epsilon) \geq \frac{V}{V^k(2\epsilon)} \geq \frac{V|k|^{n/2}}{c(n)\sinh^n(2\epsilon\sqrt{|k|})} = i.$$

The estimate follows from a direct computation. ∎

4.G.4 Lower bounds for the first eigenvalue

Lower bounds are more difficult to get than upper bounds: indeed, if we want to use the minimax principle, we must find estimates which hold for any function (say with zero integral on M if we are interested on λ_1). The oldest Riemannian result is due to A. Lichnerowicz.

4.70 Theorem. *Let (M, g) be a compact Riemannian manifold, and suppose that* $\mathrm{Ric} \geq (n-1)kg$, *with* $k > 0$. *Then*

$$\lambda_1 \geq nk.$$

Proof. Using 4.36, for any smooth f, we know that

$$\Delta df = D^* D df + \mathrm{Ric}(df).$$

Taking the scalar product with df and integrating by parts, we get

$$\langle \Delta d, \Delta f \rangle = \langle D df, D df \rangle + \int_M \mathrm{Ric}(df, df) v_g$$

so that

$$\|\Delta f\|^2 - \|D df\|^2 \geq (n-1)k\|df\|^2.$$

Now, Schwarz's inequality, when applied to the Euclidean spaces $S_m^2 M$, gives

$$(\Delta f)^2 \leq |g|^2| \, Ddf \, |^2 = n \mid Ddf \mid^2,$$

hence $\|\Delta f\|^2 \geq kn\|df\|^2$. If $\Delta f = \lambda f$ with $\lambda \neq 0$, we see that

$$\|\Delta f\|^2 = \lambda \langle \Delta f, f \rangle = \lambda \|df\|^2 \geq (n-1)k\|df\|^2. \quad \blacksquare$$

Remark. This result just means that $\lambda_1(M, g) \geq \lambda_1(S^n(k))$. Furthermore, if the equality is achieved, (M, g) is isometric to $S^n(k)$ (Obata's theorem, cf. [B-G-M], III.D.1).

4.71 This argument is very similar to the argument of theorem 4.37, where it was proved that the first Betti number of a manifold with positive Ricci curvature vanishes. On the other hand, this last result can be obtained by geometric methods, using the Myers theorem. Now, the question is: is it possible to get lower bounds for the first eigenvalues by "geometric" methods ? The answer is yes, and the basic idea is to use the so-called "isoperimetric inequalities". A typical result in that direction is the following.

4.72 Theorem (Cheeger's inequality). *For a compact Riemannian manifold (M, g), set*

$$h = \inf \frac{\mathrm{vol}(S)}{\min\{\mathrm{vol}(M_1), \mathrm{vol}(M_2)\}},$$

where S runs over codimension one submanifolds of M which divide M into two pieces M_1 and M_2. Then

$$\lambda_1 \geq \frac{h^2}{4}.$$

The argument relies on the following lemma, called "coarea formula", whose statement is quite natural.

4.73 Lemma. *Let f be a smooth positive function on a compact Riemannian manifold (M, g). Then*

$$\int_M f v_g = \int_0^{\sup f} \mathrm{vol}_n f^{-1}([t, +\infty[) dt$$

$$\int_M |df| v_g = \int_0^{\sup f} \mathrm{vol}_{n-1}(f^{-1}(t)) \, dt$$

Sketch of proof. The first equality is straightforward for the so-called step-functions, which are constant on pairwise disjoint measurable sets covering M, and the general case follows by approximation in the L^1 norm.

For the second one, take a regular value t of f. Then $f^{-1}(t)$ is a $(n-1)$-dimensional submanifold. Denote by $v_{f^{-1}(t)}$ its volume form. Then at any point of $f^{-1}(t)$, we have

$$|df| v_g = v_{f^{-1}(t)} \otimes dt \,.$$

Since almost every value is regular by Sard's theorem, the result follows by integration, using Fubini's theorem. For a more general statement, see [B-Z], p.103. ∎

Proof of the theorem. Take an eigenfunction f for the eigenvalue λ_1. We can assume that 0 is a regular value for f (if not, apply the following argument to a sequence $f_n = f + \epsilon_n$ where $\epsilon_n \in \mathbf{R}$ goes to zero). Set

$$M_+ = \{f \geq 0\}, \quad M_0 = \{f = 0\}, \quad \text{and} \quad M_- = \{f \leq 0\},$$

and assume that $\mathrm{vol} M_+ \leq \mathrm{vol} M_-$ (if not, work with $-f$). Since f vanishes on the boundary M_0 of the smooth manifold M_+, integration by parts (4.3) yields

$$\int_{M_+} |df|^2 v_g = \int_{M_+} <\Delta f, f> v_g = \lambda_1 \int_{M_+} f^2 v_g,$$

hence, using Schwarz inequality,

$$\lambda_1 = \frac{\left(\int_{M_+} f^2 v_g\right)\left(\int_{M_+} |df|^2 v_g\right)}{\left(\int_{M_+} f^2 v_g\right)^2}$$

$$\geq \frac{\left(\int_{M_+} f |df| v_g\right)^2}{\left(\int_{M_+} f^2 v_g\right)^2} \geq \frac{1}{4}\left(\frac{\int_{M_+} |d(f^2)| v_g}{\int_{M_+} f^2 v_g}\right)^2.$$

The claimed inequality will now follow from applying the coarea formula 4.73. to $F := f^2$ on M_+. Actually, for any positive value of t,

$$\int_{M_+} |\, d(f^2)\, |\, v_g = \int_0^{\sup f} \mathrm{vol} f^{-1}(\sqrt{t})dt$$

$$\int_{M_+} f^2 v_g = \int_0^{\sup f} \mathrm{vol} f^{-1}([\sqrt{t}, \sup f])dt$$

Noticing that

$$\mathrm{vol} f^{-1}([\sqrt{t}, \sup f]) \leq \mathrm{vol} M_+ \leq \mathrm{vol} M_- \leq \mathrm{vol} f^{-1}[\inf f, \sqrt{t}],$$

the very definition of h yields, for any regular value $\sqrt{t}$ of f, hence for almost every t,

$$\mathrm{vol} f^{-1}(\sqrt{t}) \geq h\, \mathrm{vol} f^{-1}([\sqrt{t}, \sup f]),$$

so that $\int_{M_+} |\, d(f^2)\, |\, v_g \geq h \int_{M_+} f^2 v_g$, hence the result.∎

Up to now, we do not even know whether h is strictly positive. Lower bounds for h, under suitable geometric assumptions on the manifold (M, g), are examples of so-called "isoperimetric inequalities".

4.H Paul Lévy's isoperimetric inequality

4.H.1 The statement

The topic of isoperimetric inequalities (indeed the denomination of "geometric inequalities", which is the title of the wonderful book [B-Z] should be preferred) appeals to many domains of mathematics. Here we deal with the aspect which is the most relevant to our point of view. Namely, the generalization by M. Gromov of an inequality discovered by Paul Lévy for positively curved surfaces. Not surprisingly, in higher dimension, the relevant positivity assumption is about Ricci curvature.

4.74 **Definition.** An *isoperimetric profile* for a Riemannian manifold (M, g) is a function I from $\mathbf{R}^+$ to $\mathbf{R}^+$ such that, for any compact domain D of M,

$$\mathrm{vol}(\partial D) \geq I(\mathrm{vol}(D))\cdot$$

For example, the function $I(t) = \sqrt{4\pi t}$ is an isoperimetric profile (the best one in fact) for the Euclidean plane. The classical isoperimetric inequality just says that. See for instance [B-G] for a proof, and [Os] or [B-Z] to convince you that the case of the Euclidean plane is already very rich.

4.75 **Remark:** The classical isoperimetric inequality is still valid in any dimension. Namely, for any compact domain with boundary in $\mathbf{R}^n$,

$$\frac{\big(\mathrm{vol}(\partial D)\big)^n}{\big(\mathrm{vol}(D)\big)^{n-1}} \geq \frac{\big(\mathrm{vol}(S^{n-1}, \mathrm{can})\big)^n}{\big(\mathrm{vol}(B^n)\big)^{n-1}}$$

(here B^n denote the unit Euclidean ball). Moreover, the equality is achieved if and only if D is an Euclidean ball. In particular,

$$I(t) = \mathrm{vol}(S^{n-1}, \mathrm{can})^{\frac{1}{n}} \, (nt)^{\frac{n-1}{n}}$$

is an isoperimetric profile for the Euclidean space. Indeed, as soon as $n > 2$, this results is not so easy (cf. [B-Z], ch. 2). An indirect proof will be provided below.

Now, if we consider the standard sphere, it should be expected that an isoperimetric inequality still holds, the "best" domains being geodesic balls. Namely, take any β in $]0, 1]$, and consider a geodesic ball B_β of volume $\beta\mathrm{vol}(S^n, \mathrm{can})$. Set

$$Is(\beta) = \frac{\mathrm{vol}(\partial B_\beta)}{\mathrm{vol}(S^n, \mathrm{can})}.$$

4.76 Theorem *Let (M, g) be a compact Riemannian manifold whose Ricci curvature is bounded from below by $n - 1$. Let D be any domain with smooth boundary in M such that*

$$\frac{\mathrm{vol}(D)}{\mathrm{vol}(M)} = \beta = \frac{\mathrm{vol}(B_\beta)}{\mathrm{vol}(S^n, \mathrm{can})}$$

Then

$$\frac{\mathrm{vol}(\partial D)}{\mathrm{vol}(M)} \geq Is(\beta) := \frac{\mathrm{vol}(\partial B_\beta)}{\mathrm{vol}(S^n, \mathrm{can})}$$

Moreover, the equality is achieved if and only if (M, g) is isometric to the standard sphere and D to the geodesic ball B_β.

In other words, we have *both* an optimal isoperimetric profile for the standard sphere and an explicit (not optimal of course) isoperimetric profile for any compact manifold with positive Ricci curvature.

4.H.2 The proof

Assume first we can find a domain D with smooth boundary $\partial D = H$ which achieves the infimum of

$$\frac{\mathrm{vol}(\partial D)}{\mathrm{vol}(M)}$$

for domains with volume $\beta\mathrm{vol}(M)$. Denote by $h(\beta)$ this infimum. (This is just a normalized isoperimetric profile for (M, g)). Then, using the first variation formula for the volume (see 5.20 below) and a standard Lagrange multiplier argument, we see that H has *constant* mean curvature η. Clearly H divides M into two components, one of which is D. We can estimate the volume of D using Heintze-Karcher inequality. Assuming that the unit normal to H points outwards, by integrating inequality 4.21, we get

$$\mathrm{vol}(D) \leq \mathrm{vol}(H) \int_0^r (\cos t - \eta \sin t)^{n-1} dt \tag{4.3}$$

where r is the first zero of $\cos t - \eta \sin t$. (We just take into account the domain on which the function $\overline{h}$ of 4.21 is positive). Now, the following property is easy to check.

4.77 **Lemma.** *The geodesic ball $B(r)$ of radius r in the standard n-sphere has mean curvature $\cot r$, and*

$$\frac{\text{vol}(B(r))}{\text{vol}(\partial B(r))} = \int_0^r (\cos t - \eta \sin t)^{n-1} dt.$$

Proof. The first claim is straightforward, since the induced metric on $\partial B(r)$ is $\sin^2 r g_0$, where g_0 is the standard metric of the S^{n-1}. The second follows from the volume computations of chapter 3, taking into account that $\eta = \cot r$ and making the change of variable $t' = r - t$. ∎

If we denote by $\frac{1}{a(r)}$ this ratio, inequality (4.3) can be rewritten as

$$h(\beta) = \frac{\text{vol}(H)}{\text{vol}(M, g)} \geq \frac{a(r)\text{vol}(D)}{\text{vol}(M, g)}.$$

Now, replace D by the other component D^c in these estimates. Then η is replaced by $-\eta$ and r by $\pi - r$ (we know that by Myers' theorem 3.85 that the diameter of (M, g) is smaller than π). We also have

$$h(\beta) \geq \frac{a(\pi - r)\text{vol}(D^c)}{\text{vol}(M, g)}.$$

Therefore

$$h(\beta) \geq \max\{\beta a(r), (1 - \beta)a(\pi - r)\}$$
$$\geq \inf_{0 \leq t \leq \pi} \max\{\beta a(t), (1 - \beta)a(\pi - t)\}$$

On $[0, \pi]$, the function $a(t)$ is monotone decreasing, so that this infimum is realized for a t_0 such that

$$\beta a(t_0) = (1 - \beta)a(\pi - t_0)$$

For this t_0 we have

$$\beta \frac{\text{vol}(\partial B(t_0))}{\text{vol}(B(t_0))} = (1 - \beta)\frac{\text{vol}(\partial B(\pi - t_0))}{\text{vol}(B(\pi - t_0))} = \frac{\text{vol}(\partial B(t_0))}{\text{vol}(S^n, \text{can})},$$

so that

$$\beta = \frac{\text{vol}(B(t_0))}{\text{vol}(S^n, \text{can})}.$$

In other words, $B_\beta = B(t_0)$, proving the claimed inequality. If the equality is achieved, it must also be achieved in all the steps of the proof, and the last conclusion follows easily. ∎

4.78 Remark. We have been very allusive concerning a crucial step in the beginning of this proof, namely the existence of a *smooth* hypersurface realizing the minimum. This step requires a suitable generalization of manifolds: in general the minimizing object is a so called *integral current* (cf. [Mo]). Currents are for manifolds what distributions are for smooth functions. Then one must use a regularity theorem, or point out, following Gromov, that for any $m \in M$, the distance to the support of the minimizing current can only be realized at a *regular* point of this support.

4.79 Remark. As a by-product of the above inequality, we obtain, by comparing the Euclidean space to a sphere of large radius via the exponential map, the classical inequality. But with this proof, the equality case can not be taken into account. We also obtain the result we needed at the end of the preceding section.

4.80 Corollary. *If (M, g) is a compact Riemannian manifold whose Ricci curvature is bigger that $(n-1)g$, then*

$$h \geq 2Is(\frac{1}{2}).$$

Proof: One can check by a direct computation that the function $\frac{Is(\beta)}{\beta}$ is decreasing. ■

This method is just a point of departure. If the Ricci curvature is no more positive, but just bounded from below, the method of theorem 4.76 still applies. Indeed, if we have an upper bound on the diameter, we can still apply the Heintze-Karcher inequality. The comparison function will involve hyperbolic functions (compare with 4.21). The reader can switch to [Ga3] to see how the story goes on. See also [Be4].

5

Riemannian submanifolds

In this chapter, we study the relations between the Riemannian Geometry of a submanifold and that of the ambient space. It is well known that surfaces of the Euclidean space were the first examples of Riemannian manifolds to be studied. In fact, the first truly Riemannian geometry result is due to Gauss, and roughly says the following.

Assume $N \subset \mathbf{R}^3$ is a smooth surface. To measure how much N is curved in $\mathbf{R}^3$ at a point, a natural *extrinsic* quantity to study is the *Gaussian curvature* (defined as the infinitesimal area swept on $\mathbf{S}^2$ by a unit normal vector field): it measures how fast the normal vector varies. If N is a plane, a cylinder or a cone, the Gaussian curvature vanishes, while if N is a sphere of radius r, its Gaussian curvature is $1/r^2$.

It turns out, and this is the famous Theorema Egregium of Gauss, that the Gaussian curvature is actually an *intrinsic* quantity. Namely, it does not depend on the isometric immersion of (N, h) into $\mathbf{R}^3$.

We begin with Gauss theorem, which is used to give explicit computations of curvatures.

The next section is devoted to a global result of Hadamard concerning positively curved hypersurfaces of Euclidean space.

We end the chapter with a short introduction to the theory of minimal submanifolds. We study minimal isometric immersions of irreducible homogeneous spaces into spheres.

We refer to [dCa1] for an introduction to Riemannian geometry putting the emphasis on the extrinsic view-point.

5.A Curvature of submanifolds

5.A.1 Second fundamental form

Let $(\tilde{M}, \langle , \rangle)$ be a Riemannian manifold, and (M, g) be a Riemannian submanifold of $\tilde{M}$. We want to compute the curvature of M in terms of the curvature of $\tilde{M}$. We first consider the case where M is an hypersurface of $\tilde{M}$ (the reader can keep in mind the example of surfaces in $\mathbf{R}^3$). The general case is treated in exercise.

Let M be an n-dimensional hypersurface of $\tilde{M}$, D and $\tilde{D}$ be their respective covariant derivatives.

5.1 Definition. The *vector valued second fundamental form* of M is defined, for $m \in M$ and $u, v \in T_m M$, by:

$$II(u,v) = \left(\tilde{D}_U V - D_U V \right)_m ,$$

where U and V are vector fields defined in a neighborhood of m in $\tilde{M}$, tangent to M at any point of M, and with respective values u and v at m.

One checks easily (use criterion 1.114) that this quantity does not depend on the fields U and V, and that II is symmetric (the value on M of the bracket -in $\tilde{M}$- of U and V equals the bracket -in M- of the restrictions of U and V to M). From 2.56, we have

$$II(u,v) = (\tilde{D}_U V)^{\perp}_m,$$

where $X^{\perp}$ denotes the component of X normal to $T_m M$.

It is always possible to build locally on M a normal unit field ν (around any point of M, there exists a submersion $f : U \subset \tilde{M} \to \mathbf{R}$ such that $f^{-1}(0) = U \cap M$: take then $\nu = \frac{\mathrm{grad} f}{|\mathrm{grad} f|}$). This field is unique up to sign. Just recall that if both manifolds M and $\tilde{M}$ are orientable, such a field exists globally (that is on the whole M). Once such a normal is chosen, we set the following definitions.

5.2 Definitions. *The real valued second fundamental form l of M is defined for $u, v \in T_m M$ by*

$$II(u,v) = l(u,v)\nu_m.$$

The *shape operator* is the endomorphism S of $T_m M$ given by $S(u) = \tilde{D}_u \nu,$. Indeed, the map S takes its values in $T_m M$ since

$$\langle \tilde{D}_u \nu, \nu \rangle = \frac{1}{2} u.\langle \nu, \nu \rangle = 0.$$

On the other hand,

$$\langle S(u), v \rangle = \langle \tilde{D}_u \nu, v \rangle = -\langle \nu, \tilde{D}_u V \rangle = -l(u,v),$$

so that $S = -l^{\sharp}$ (this relation is known as *Weingarten equation*).

If we change the choice for the unit normal vector field, S and l are changed into their opposite.

Remarks. i) This name comes from classical surface theory in 3-space: the induced Riemannian metric on a surface was traditionally called the "first fundamental form."

ii) The second fundamental form of an hypersurface of the Euclidean space is just the infinitesimal variation of the induced metric in the space. Indeed, suppose that M is locally defined by an immersion f, choose a normal vector field ν, and consider $f_t = f + t\nu$. For t small enough, it is an immersion. If g_t is the induced metric, the preceding argument shows that

$$\frac{d}{dt}(g_t)_{|t=0} = -\frac{1}{2}l$$

This interpretation still works with any ambient space, cf. 5.4 a).

5.3 Definition. The *Gaussian curvature* $\bar{K}$ of the 2 dimensional subspace of T_mM generated by u and v is

$$\bar{K}(u,v) = \frac{l(u,u)l(v,v) - l(u,v)^2}{g(u,u)g(v,v) - g(u,v)^2}.$$

When (u,v) is an orthonormal system, then

$$\bar{K}(u,v) = l(u,u)l(v,v) - l(u,v)^2.$$

Note that the Gaussian curvature of the hypersurface does not depend on the choice of the unit normal vector field.

5.4 Exercises. a) Let M be a Riemannian hypersurface of $(\tilde{M}, \tilde{g})$, ν be a unit normal vector field on M defined in a neighborhood of $m \in M$, and c_m be the geodesic of $\tilde{M}$ with initial conditions $c_m(0) = m$ and $c_m'(0) = \nu_m$. Consider for $u \in T_mM$ a curve γ on M such that $\gamma(0) = m$ and $\gamma'(0) = u$. We define a variation of c by

$$H(s,t) = c_{\gamma(t)}(s) = \exp_{\gamma(t)} s\nu_{\gamma(t)}.$$

Denote by $\tilde{Y}$ the vector field $\frac{\partial H}{\partial t}$ associated to this variation. Show that $Y = \tilde{Y}(.,0)$ is the Jacobi field along c_m satisfying to $Y(0) = u$ and $Y'(0) = S(u)$. If $v \in T_mM$ and X is the field associated in the same way to v, prove that

$$l(u,v) = -\langle X, Y' \rangle(0).$$

This yields a method to compute l.

b) *Another approach for the Gaussian curvature.*
Let $f : U \subset \mathbf{R}^2 \to \mathbf{R}^3$ be an immersion, and $M = f(U)$ be equipped with the induced metric. We denote by $\frac{\partial}{\partial x^1}$ and $\frac{\partial}{\partial x^2}$ the coordinate vector fields on M, and by $N : M \to S^2$ the unit normal field on M defined by

$$N = \frac{\frac{\partial}{\partial x^1} \wedge \frac{\partial}{\partial x^2}}{\| \frac{\partial}{\partial x^1} \wedge \frac{\partial}{\partial x^2} \|},$$

where $\wedge$ is the vector product in $\mathbf{R}^3$. Define volume forms ω and σ on S^2 and M respectively by

$$\text{for} \quad u, v \in T_m M \qquad \sigma(u, v) = (u, v, \nu),$$

$$\text{and for} \quad \xi, \eta \in T_x S^2 \qquad \omega(\xi, \eta) = (\xi, \eta, x),$$

(mixed products). Show that $\nu^*(\omega) = \bar{K} \cdot \sigma$, where $\bar{K}$ is the Gaussian curvature of M. Hence $\bar{K}$ measures the quotient of the oriented areas respectively described by the normal ν to M in S^2, and a point on the manifold.

5.A.2 Curvature of hypersurfaces

Let $M^n \subset \tilde{M}^{n+1}$ be a Riemannian submanifold, R, K (resp. $\tilde{R}$ and $\tilde{K}$) be the curvature tensor and the sectional curvature of M (resp. $\tilde{M}$), and $\bar{K}$ be the Gaussian curvature of M.

5.5 Theorem (Gauss equation). *For $x, y, u, v \in T_m M$, we have*

$$R(x, y, u, v) = \tilde{R}(x, y, u, v) + l(x, u)l(y, v) - l(x, v)l(y, u)$$
$$K(x, y) = \tilde{K}(x, y) + \overline{K}(x, y)$$

In particular, the sectional curvature of a Riemannian hypersurface of the Euclidean space is equal to its Gaussian curvature.

Proof. Let X, Y, U, and V be local vector fields defined in a neighborhood of $m \in M$ in $\tilde{M}$, tangent to M at each point of M, with respective values x, y, u and v at m. On M, we have $D_X Y = (\tilde{D}_X Y)^\top$, hence

$$\tilde{D}_X Y = D_X Y + II(X, Y) = D_X Y - l(X, Y) \cdot \nu.$$

Therefore

$$\begin{aligned}
\tilde{D}_X \tilde{D}_Y U &= \tilde{D}_X [D_Y U - l(Y, U)\nu] \\
&= D_X D_Y U - l(X, D_Y U).\nu - \tilde{D}_X [l(Y, U).\nu] \\
&= D_X D_Y U - l(X, D_Y U).\nu - l(Y, U)\tilde{D}_X \nu - [X \cdot l(Y, U)]\nu,
\end{aligned}$$

Taking the scalar product with V, we get

$$\begin{aligned}
\langle \tilde{D}_X \tilde{D}_Y U, V \rangle &= \langle D_X D_Y U, V \rangle - l(Y, U)\langle \tilde{D}_X \nu, V \rangle \\
&= \langle D_X D_Y U, V \rangle - l(Y, U)l(X, V),
\end{aligned}$$

which yields the result. ∎

5.6 Remarks. a) The eigenvalues $(\lambda_i)_{(i=1,\cdots,n)}$ of the second fundamental form are called *principal curvatures*, and their inverses are the *principal curvature radii*. They are changed into their opposite when we change of normal vector field. When M is 2-dimensional, the Gaussian curvature equals the product of the principal curvatures: $\bar{K} = \lambda_1 \lambda_2$. In the case of a surface in $\mathbf{R}^3$, the principal curvatures are the extremal values of the geodesic curvatures of normal sections of M at m (curves obtained by intersection of M with a 2-plane normal to the surface at this point).

b) Gauss called this result "Theorema Egregium". Indeed, he introduced in the extrinsic theory of surfaces in $\mathbf{R}^3$ the quantity $\bar{K}$ (Gaussian curvature), and proved that it actually does not depend on the isometric embedding of this (Riemannian) surface in the Euclidean 3-space. If indeed (M, g) is an n-dimensional manifold isometrically embedded in $\mathbf{R}^{n+1}$, its Gaussian curvature does not depend on the embedding: assume that M_1 and M_2 are two hypersurfaces of $\mathbf{R}^{n+1}$, and that $\phi : M_1 \to M_2$ is an isometry, then

$$\bar{K}^1_m(u, v) = \bar{K}^2_m(T_m\phi.u, T_m\phi.v) \quad \text{since} \quad K^1(m) = K^2(\phi(m)).$$

c) When the (hyper)surface is given by an immersion f, its curvature tensor a priori depends on the third derivatives of f. Gauss, who did not dispose of connections, proved his result by making a tricky elimination of some third derivatives.

5.7 Example. Let us compute the principal curvature radii, and the connection of the sphere of radius r in $\mathbf{R}^{n+1}$, let S^n_r. If $x \in S^n_r$, then $\nu_x = \frac{x}{r}$, hence for $u \in TS^n_r$:

$$\tilde{D}_u\nu = d\nu(u) = \frac{u}{r},$$

and $S = \frac{1}{r}\text{Id}$. Therefore all the curvature radii of S^n_r are equal to r, and $\bar{K} = K = \frac{1}{r^2}$. On the other hand, if X and Y are vector fields tangent to S^n_r, locally extended to $\mathbf{R}^{n+1}$,

$$D_X Y = \tilde{D}_X Y + l(X, Y).\nu = \tilde{D}_X Y + \frac{1}{r}\langle X, Y \rangle.\nu.$$

5.8 Exercises. a) Show that an hypersurface of $\mathbf{R}^{n+1}$ ($n \geq 3$) cannot have at a point all its sectional curvatures negative. When $n = 2$, see 5.11.

b) **Gauss theorem in codimension k:**
Let (M, g) be an n-dimensional submanifold of $(\tilde{M}, \langle, \rangle)$, with $\dim\tilde{M} = n + k$. Let $\nu_1, \cdots, \nu_k$ be vector fields defined in a neighborhood of $m \in M$ in $\tilde{M}$, and such that for $p \in M$, $(\nu_1, \cdots, \nu_k)$ is an orthonormal basis of the orthogonal complement of T_pM in $T_p\tilde{M}$. Define, for $i = 1, \cdots, k$, linear maps S_i from T_mM to itself by $S_i(u) = (\tilde{D}_u\nu_i)^\perp$, and set $l_i(u, v) = \langle S_i(u), v \rangle$. Show that, if R and $\tilde{R}$ are the curvature tensors of M and $\tilde{M}$, then for any x, y, u and v in T_mM,

$$R(x, y, u, v) = \tilde{R}(x, y, u, v,) + \sum_{i=1}^{k} \big(l_i(x, u)l_i(y, v) - l_i(x, v)l_i(y, u)\big).$$

c) A Riemannian submanifold $M \subset \tilde{M}$ is *totally geodesic* if any geodesic of M is also a geodesic of $\tilde{M}$. Show that M is totally geodesic if and only if its second fundamental form is zero.

d) Let (M, g) be an n-dimensional complete Riemannian manifold, and assume that the sectional curvature of M is strictly positive. Show that any two totally geodesic compact submanifolds, such that the sum of their respective dimensions is greater than n intersect. Show that if we only assume that Ric_M is strictly positive, the result is still true for two totally geodesic compact hypersurfaces.

5.9 Theorem (Gauss-Codazzi equation)._Let $M \subset \tilde{M}$ be a Riemannian hypersurface, with $\dim \tilde{M} = n + 1$. Let x, y, u be vectors in $T_m M$, ν be a unit normal vector at m, and l be the second fundamental form of M associated to ν. Then_

$$\tilde{R}(x, y, u, \nu) = D_y l(x, u) - D_x l(y, u).$$

Proof. Let X, Y, U be tangent vector fields to M, with respective values x, y, u at M. Write

$$\tilde{D}^2_{X,Y} U = \tilde{D}_X \tilde{D}_Y U - \tilde{D}_{\tilde{D}_X Y} U$$
$$= \tilde{D}_X (D_Y U + l(Y, U)\nu) - \tilde{D}_{D_X Y + l(X,Y)\nu} U$$

It follows that the ν-component of $\tilde{D}^2_{X,Y} U$ is given by

$$l(X, D_Y U) + X \cdot l(Y, U) - l(D_X Y, U).$$

Therefore the ν-component of $\tilde{R}(X, Y)U = \tilde{D}^2_{X,Y} U - \tilde{D}^2_{Y,X} U$ is just

$$Y \cdot l(X, U) + l(Y, D_X U) - l(D_Y X, U) - X \cdot l(Y, U) - l(X, D_Y U) + l(D_X Y, U),$$

that is $D_Y l(X, U) - D_X l(Y, U)$ as claimed. ∎

Here is a typical application.

5.10 Exercise. a) An hypersurface is said to be *totally umbilical* if $l = fg$ for some function $f \in C^\infty(M)$ (that is if the shape operator is proportional to the identity map). If $\dim M \geq 2$, show that f is constant.

b) Show that a totally umbilical hypersurface of $\mathbf{R}^{n+1}$ is a piece of sphere or a piece of hyperplane.

5.A.3 Application to explicit computations of curvatures

Let (M, g) be a Riemannian hypersurface in $(\mathbf{R}^{n+1}, \text{can})$. Note that for any curve γ drawn on M, we have

$$(\gamma'')^\perp = (D_{\gamma'} \gamma')^\perp = II(\gamma', \gamma'), \quad \text{hence} \quad l(\gamma', \gamma') = \langle \gamma'', \nu \rangle,$$

where ν is a unit normal vector field on M. In particular if c is a geodesic of M, then its acceleration vector field is normal, and $II(c', c') = c''$.

If u_i is an eigenvector for S and if γ_i is a curve on M such that $\gamma_i'(0) = u_i$, the corresponding principal curvature is $\lambda_i = -\langle \gamma_i''(0), \nu \rangle$.

5.11 Curvature of the image of an immersion.

We denote by $u \times v$ the cross product of two vectors in $\mathbf{R}^3$. Recall that the components of this vector are

$$\left(\begin{vmatrix} u_2 & v_2 \\ u_3 & v_3 \end{vmatrix}, \begin{vmatrix} u_3 & v_3 \\ u_1 & v_1 \end{vmatrix}, \begin{vmatrix} u_1 & v_1 \\ u_2 & v_2 \end{vmatrix} \right)$$

and that its norm is equal to the area of the parallelogram generated by u and v.

Let $f : U \subset \mathbf{R}^2 \to \mathbf{R}^3$ be an immersion, and (x^1, x^2) be the corresponding local coordinates for $M = f(U)$. We have

$$\frac{\partial}{\partial x^i} = df(e_i) = \frac{\partial f}{\partial x^i} \quad \text{and} \quad g_{ij} = g\left(\frac{\partial}{\partial x^i}, \frac{\partial}{\partial x^j} \right) = \langle \frac{\partial f}{\partial x^i}, \frac{\partial f}{\partial x^j} \rangle.$$

For $N = \frac{\partial f}{\partial x^1} \wedge \frac{\partial f}{\partial x^2}$, $\nu = \frac{N}{\|N\|}$ is a unit normal vector field on M. Note that

$$\left\| \frac{\partial f}{\partial x^1} \wedge \frac{\partial f}{\partial x^2} \right\| = [\det(g_{ij})]^{1/2}.$$

Let $l_{ij} = l(\frac{\partial}{\partial x^i}, \frac{\partial}{\partial x^j})$. Then $l_{ij} = -\langle \nu, \tilde{D}_{\frac{\partial}{\partial x^i}} \frac{\partial}{\partial x^j} \rangle = -\langle \nu, \frac{\partial^2 f}{\partial x^i \partial x^j} \rangle$, hence

$$l_{ij} = \det \left(\frac{\partial^2 f}{\partial x^i \partial x^j}, \frac{\partial f}{\partial x^1}, \frac{\partial f}{\partial x^2} \right) \cdot [\det(g_{kl})]^{-1/2},$$

and we finally obtain the sectional curvature with 5.5.

5.12 Applications: a) For the Gaussian curvature and the principal curvatures of the *helicoid*, namely the image of $f : \mathbf{R}^2 \to \mathbf{R}^3$ defined by

$$f(s, t) = (t \cos s, t \sin s, bs)\cdot,$$

we find

$$K = \frac{-b^2}{(b^2 + t^2)^2}, \quad \lambda_1 = \frac{b}{b^2 + t^2}, \quad \lambda_2 = \frac{-b}{b^2 + t^2}.$$

b) For the *one-sheeted hyperboloid*, defined by the parametrization

$$f(s, t) = (a \cos s, b \sin s, 0) + t(-a \sin s, b \cos s, c)$$

(surface of equation $\frac{x^2}{a^2} + \frac{y^2}{b^2} - \frac{z^2}{c^2} = 1$), the curvature is given by:

$$K = \frac{-1}{a^2 + b^2 + c^2} \left[\frac{x^2}{a^4} + \frac{y^2}{b^4} + \frac{z^2}{c^4} \right]^{-2}.$$

c) Compute the curvature of a *revolution surface,* obtained as the image of the map $f : I \times \mathbf{R} \to \mathbf{R}^3$ defined by

$$f(s,\theta) = \big(c_1(s)\cos\theta, c_1(s)\sin\theta, c_2(s)\big).$$

Show that, if the curve $c = (c_1, c_2)$ is parametrized by arclength (so that the metric is given by $g = (ds)^2 + c_1^2(s)d\theta^2$), then $K = -\frac{c_1''}{c_1}$.

Deduce that the principal curvatures, and the curvature of the *embedded torus,* namely the image of the map

$$f(s,\theta) = [(a + b\cos s)\cos\theta, (a + b\cos s)\sin\theta, b\sin s],$$

are given by

$$\lambda_1 = \frac{1}{b} \qquad \text{(curvature of the meridian circle)},$$

$$\lambda_2 = \frac{\cos s}{a + b\cos s} = \frac{\text{horizontal component of the normal vector}}{\text{radius of the parallel circle}},$$

$$\text{and} \quad K = \frac{\cos s}{b(a + b\cos s)}.$$

For more details, and the computations, see ([Sp], t.3).

d) Curvature of the *two-sheeted hyperboloid* defined by the equation

$$x^2 + y^2 - z^2 = -1, \quad \text{and} \quad z > 0,$$

equipped with the induced metric. It is a revolution surface, which can be parametrized by

$$(s,\theta) \to (s\cos\theta, s\sin\theta, \sqrt{1+s^2}).$$

Its curvature at (x, y, z) is given by $K = (x^2 + y^2 + z^2)^{-2} > 0$. This manifold is complete, but non-compact, and we can notice that its sectional curvature goes to zero at infinity. On the other hand, its volume is given by

$$\text{vol}(H) = \int_0^\infty \int_0^{2\pi} [\det(g_{ij})]^{1/2}\, ds\, d\theta$$

$$= \int_0^\infty \int_0^{2\pi} \frac{s\sqrt{1+2s^2}}{\sqrt{1+s^2}}\, ds\, d\theta = \infty.$$

Compare with Myers' and Bishop's theorems 3.85 and 3.101.

5.13 Curvature of a submanifold defined by a submersion:
Let $f : U \subset \mathbf{R}^3 \to \mathbf{R}$ be a submersion (that is $df \neq 0$), and $M = f^{-1}(0)$ be equipped with the induced metric. A unit normal vector is given by

$$\nu = \frac{\operatorname{grad} f}{\| \operatorname{grad} f \|} = \frac{\left[\frac{\partial f}{\partial x^1}, \frac{\partial f}{\partial x^2}, \frac{\partial f}{\partial x^3} \right]}{\left[\left(\frac{\partial f}{\partial x^1} \right)^2 + \left(\frac{\partial f}{\partial x^2} \right)^2 + \left(\frac{\partial f}{\partial x^3} \right)^2 \right]}.$$

For $u, v \in T_m M$, we have

$$\tilde{D}_u \nu = \frac{1}{\| \operatorname{grad} f \|} \left(\tilde{D}_u \operatorname{grad} f \right) - u \cdot \left(\frac{1}{\| \operatorname{grad} f \|} \right) \operatorname{grad} f,$$

$$\text{hence} \quad l(u, v) = \langle \tilde{D}_u \nu, v \rangle = \frac{1}{\| \operatorname{grad} f \|} \langle \tilde{D}_u \operatorname{grad} f, v \rangle.$$

We get if (e_1, e_2, e_3) is an orthonormal basis of $\mathbf{R}^3$ with $u = \sum u_i e_i$ and $v = \sum v_i e_i$:

$$l(u, v) = \frac{1}{\| \operatorname{grad} f \|} \sum_{j=1}^{3} \left(u \cdot \frac{\partial f}{\partial x^j} \right) v_j = \frac{1}{\| \operatorname{grad} f \|} \sum_{i,j} \frac{\partial^2 f}{\partial x^i \partial x^j} u^i v^j.$$

This finally yields

$$l(u, v) = \frac{1}{\| \operatorname{grad} f \|} \operatorname{Hess} f(u, v).$$

5.14 Exercise. The purpose of this exercise is to show that the curvature does not determine the metric, even locally. Let M_1 and M_2 be the submanifolds of $\mathbf{R}^3$ respectively parametrized by

$$f(s, \theta) = (s \sin \theta, s \cos \theta, \operatorname{Log} s)$$

$$\text{and} \quad g(s, \theta) = (s \sin \theta, s \cos \theta, s),$$

both equipped with the induced metrics. Show that the map F defined from M_1 to M_2 by

$$F\big(f(s, \theta)\big) = g(s, \theta)$$

preserves the curvature, but is not an isometry.

5.B Curvature and convexity

Let M be an n-dimensional Riemannian submanifold of $\mathbf{R}^{n+1}$. For $m \in M$, we denote by H_m the hyperplane tangent to M at m.

5.15 Definition. A submanifold M is *convex at* m if there exists a neighborhood U of m in M, such that the whole U lies on one side of H_m. If $U \cap H_m$ is reduced to the point m, M is said to be *strictly convex at* m.

M is *convex* (resp. *strictly convex*) if the previous conditions are satisfied at any point of M.

The following local result states a relationship between the convexity of M at a point, and the curvature of M at this point.

5.16 Proposition. *If the sectional curvature of M at m is strictly positive, then M is strictly convex at m.*

Conversely, if M is convex at m, then the sectional curvature of M at m is nonnegative.

Proof. i) Using a translation, we can assume that m is the origin. Let (e_i) $(i = 1, \cdots, n)$ be an orthonormal basis of H_m where l is diagonal. For two vectors in H_m, let

$$u = \sum_{i=1}^{n} u_i e_i \quad \text{and} \quad v = \sum_{i=1}^{n} v_i e_i,$$

$$\text{then} \quad R(u, v, u, v) = \sum_{(i,j,k,l)} u_i u_k v_j v_l R(e_i, e_j, e_k, e_l).$$

Using 5.5, and denoting by λ_i the eigenvalues of l, we get

$$R(u, v, u, v,) = \sum_{i,j} (u_i^2 v_j^2 - u_i v_i u_j v_j) \lambda_i \lambda_j = \sum_{i<j} (u_i v_j - u_j v_i)^2 \lambda_i \lambda_j.$$

Hence $K_m > 0$ if and only if all the $\lambda_i \lambda_j$ are strictly positive or, equivalently, if and only if all the λ_i are either strictly positive, or strictly negative. Hence

$$K_m > 0 \quad \Longleftrightarrow \quad l_m \quad \text{is definite positive or definite negative}$$

$$\text{and} \quad K_m \geq 0 \quad \Longleftrightarrow \quad l_m \quad \text{is positive or negative.}$$

ii) We now turn to the relationship between the position of M with respect to H_m, and the sign of K_m. Let e_0 be a unit vector, normal to H_m: there exists a function $f : \mathbf{R}^n \to \mathbf{R}$ such that, in a neighborhood of m, M is the hypersurface of equation

$$x_0 = f(x_1, \cdots, x_n)$$

(coordinates with respect to the basis (e_i)). By construction, $f(m) = 0$, and m is a critical point for f. From 5.12, we know that for $u, v \in H_m$,

$$l(u, v) = -\operatorname{Hess} f(u, v),$$

hence if l is definite positive or definite negative at m, the same is true for $(\operatorname{Hess} f)_m$, and (use a Taylor expansion at order 2 at m) M lies locally on one side of H_m.

Conversely, if M is convex at m, we get that $(\operatorname{Hess} f)_m$ is positive or negative, and hence that all its eigenvalues are either positive, or negative: this forces $K_m \geq 0$ (of course the fact that M is strictly convex at m does not force

$K_m > 0$, since the first non zero terms in the Taylor expansion could be of order 4, for example). ∎

The following theorem, of global nature, is due to J. Hadamard.

5.17 Theorem. *Let (M, g) be a compact and connected Riemannian hypersurface of $\mathbf{R}^{n+1}$ ($n \geq 2$). Then*

(1) *The following are equivalent:*

i) *the sectional curvature of M is never zero;*

ii) *the sectional curvature of M is strictly positive;*

iii) *M is orientable, and if $\nu : M \to S^n$ is a unit normal vector field on M, then ν (Gauss map) is a diffeomorphism between M and S^n.*

(2) *The previous conditions imply that M is strictly convex.*

5.18 Remarks: i) Note, for $n = 1$, that a plane embedded curve with non-negative geodesic curvature is convex. However, there exist closed *immersed* curves with strictly positive geodesic curvature. Draw such a curve!

ii) We have seen in 5.8 that there is no hypersurface of $\mathbf{R}^{n+1}$ ($n \geq 3$) with strictly negative sectional curvature. On the other hand, there is no complete hypersurface in $\mathbf{R}^3$ with $K \leq -a^2 < 0$ (the proof in the case of an hypersurface with constant negative sectional curvature is due to Hilbert). The case of the one-sheeted hyperboloid is typical: the curvature is negative, but it goes to zero at infinity.

iii) There are complete hypersurfaces in $\mathbf{R}^3$, with strictly positive sectional curvature, and which are not diffeomorphic to S^2: take for example one sheet of the 2-sheeted hyperboloid (5.11 d)).

Proof of the theorem. a) Let us show that *the second fundamental form of a compact hypersurface of $\mathbf{R}^{n+1}$ is positive (or negative) definite at one point at least.*

Consider the function defined on M by $f(m) = \|m\|^2$. Since M is compact, this function achieves its maximal value at $m_0 \in M$: by construction, there exists on a neighborhood of m_0 in M a unit normal vector field ν such that $\nu(m_0) = \frac{m_0}{\|m_0\|}$. On the other hand, if X is a tangent vector field on M with $X(m_0) = x \in T_{m_0}M$, then

$$\text{Hess} f(x, x) = x \cdot (X \cdot f) = x \cdot \langle X(m), m \rangle.$$

But $x.\langle X(m), m \rangle = \langle \tilde{D}_x X, m_0 \rangle + \langle x, \tilde{D}_x m \rangle = -\| m_0 \| l(x, x) + \langle x, x \rangle$. Since f is maximal at m_0, then $\text{Hess}_{m_0} f \leq 0$ and

$$l(x, x) = \frac{1}{\| m_0 \|} [\langle x, x \rangle - \text{Hess} f(x, x)] > 0$$

for $x \neq 0$: with our choice for the normal vector, the second fundamental form of M is positive definite.

b) We now prove that *if the sectional curvature of M is never zero, then l is never degenerate, and the sectional curvature is in fact strictly positive.*

If indeed $u, v \in T_m M$ are two eigenvectors of l for the eigenvalues λ_i and λ_j respectively, then $K(u, v) = \lambda_i \lambda_j \neq 0$, hence l is non degenerate. The set of points where l is positive definite or negative definite (according to a choice of normal) is an open subset of M. The set of points where l is non-positive, or nonnegative, is a closed subset of M. Since l is non degenerate, these two sets do coincide. Since they contain m_0, they are not empty, and are equal to M. Hence l is everywhere positive (or negative) definite, and since

$$R(u, v, u, v) = l(u, u)l(v, v) - l(u, v)^2,$$

(from 5.5), Cauchy-Schwarz inequality yields that the sectional curvature of M is strictly positive.

c) We now prove that *if $K > 0$, then M is orientable and the Gauss map is a diffeomorphism.*

If the sectional curvature of M is strictly positive, then all the eigenvalues of l at a point have the same sign and are non zero, hence l is positive (or negative) definite.

Let X be a non zero tangent vector field on a neighborhood of m in M. For any unit normal vector field N around m,

$$\langle \tilde{D}_X X, N \rangle = \pm l(X, X) \neq 0.$$

Set $\nu = \frac{-(\tilde{D}_X X)^\perp}{\|(\tilde{D}_X X)^\perp\|}$, where $(\tilde{D}_X X)^\perp$ is the normal component of $\tilde{D}_X X$. The vector field ν is a normal unit field, and does not depend on the choice of X: if indeed Y is another non zero tangent field around m, then $\langle \tilde{D}_X X, \nu \rangle$ and $\langle \tilde{D}_Y Y, \nu \rangle$ have the same sign (as $l(X, X)$ and $l(Y, Y)$)...

We can therefore define a smooth unit normal vector field on the whole manifold and M is orientable.

The Gauss map satisfies for $u \in T_m M$:

$$T_m \nu(u) = d\nu(u) = \tilde{D}_u \nu = S(u).$$

If $K > 0$, all the eigenvalues of S, and hence of $T_m \nu$ are non zero, and ν is a local diffeomorphism. Since M is compact and the sphere S^n is connected, it is a covering map (1.84). Now since the sphere is simply connected, ν is indeed a diffeomorphism.

d) Now, $m \in M$, the map $u \to T_m \nu \cdot u = \tilde{D}_u \nu$ is a linear isomorphism from $T_m M$ on itself: the associated bilinear form l is then never degenerate, and hence everywhere positive (or negative) definite, and the sectional curvature of M is everywhere strictly positive.

e) *Let us prove* **2)**. Let $m \in M$ be the origin, and $(e_0, \cdots, e_n)$ be an orthonormal basis of $\mathbf{R}^{n+1}$ such that $e_0 \perp T_m M$. Let f be the function defined on M by $f(x_0, \cdots, x_n) = x_0$. Since M is compact, f achieves its minimum and its maximum at p and q respectively. Since the equations of the tangent spaces $T_p M$ and $T_q M$ are both $x_0 = $ constant, the vectors ν_m, ν_p and ν_q are equal up to sign (ν is the Gauss map which exists by hypothesis). Since we assumed the Gauss map was bijective, we have $\nu_p = -\nu_q$, and $m = p$ for example: hence the whole $M \setminus \{m\}$ lies strictly under $T_m m$ since f admits 0 as a maximum

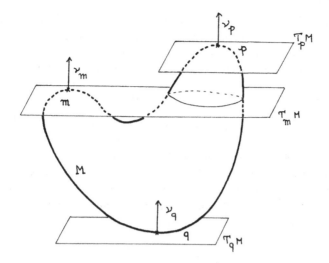

Fig. 5.1. Hadamard's theorem

and since this maximum can only be achieved at one point. Same conclusion if $m = q$ (see figure 5.1). ∎

Remark. The orientability has been proved as a free gift, since we know that any compact hypersurface in $\mathbf{R}^{n+1}$ is orientable (see [Gr]).

5.C Minimal surfaces

5.C.1 First results

We generalize in this section the problem of finding geodesics -that is locally minimizing curves- in a Riemannian manifold. We will just settle the first ideas and give some simple examples. For more examples and some results on minimal submanifolds in $\mathbf{R}^3$, see ([Sp], t.3). For further results, see [L]. Let $(\tilde{M}, \tilde{g})$ be an $(n + k)$-dimensional Riemannian manifold, and M be an n-dimensional compact submanifold of $\tilde{M}$, with boundary ∂M. The problem is to find among the submanifolds of $\tilde{M}$, with given boundary ∂M, those of minimal Riemannian volume (for the induced metric). We shall be content with the submanifolds having a given boundary ∂M, which are critical for the volume function.

5.19 Definition. A *variation with fixed boundary* of the submanifold M is a C^2 map $H : M \times] - \epsilon, \epsilon[\to \tilde{M}$, such that

$$\forall m \in M, \quad H(m, 0) = m \qquad \text{and} \qquad \forall (m, t) \in \partial M \times] - \epsilon, \epsilon[, \quad H(m, t) = m.$$

Note that for t small enough, the map $\phi_t : m \to H(m,t)$ is a local diffeomorphism from M onto its image M_t (indeed, for any $m \in M$, $T_m\phi_0 = Id$: use the compactness of M and a continuity argument). The image M_t of M is then an immersed submanifold of $\tilde{M}$. We define its volume by

$$\text{vol}(M_t) = \text{vol}(M, g_t), \quad \text{where} \quad g_t = \phi_t^* \tilde{g}$$

is the pull back of $\tilde{g}$ by ϕ_t (1.106). If M_t is a genuine submanifold of $\tilde{M}$ (that is embedded), the volume so defined coincides with the Riemannian volume of M_t. The following theorem is to be compared with 3.31.

5.20 Theorem. *Let $\tilde{M}$ and M be as above, and H be a variation of M, with fixed boundary. We chose at any point $m \in M$ an orthonormal basis $(N_1, \cdots, N_k)$ for the normal space to M at m, and denote by $(l_j)_{(j=1,\cdots,k)}$ the associated fundamental forms (that is $l_j(u,v) = -\tilde{g}(\tilde{D}_U V, N_j)$). If*

$$Y = \frac{\partial H}{\partial t} = \frac{\partial}{\partial t}$$

is the vector field on M defined by the infinitesimal variation associated to H, then:
1) if $\mathcal{V}_t$ is the volume of M_t and dv is the Riemannian volume element on M:

$$\frac{d\mathcal{V}_t}{dt}_{|t=0} = \int_M \sum_{j=1}^k tr(l_j)\tilde{g}(Y, N_j)dv \, ;$$

2) the submanifold M is critical for the volume function if and only if, for $j = 1, \cdots, k$, $tr(l_j) = 0$: in this case, the submanifold M is a minimal submanifold.

5.21 Remarks. 1) If we denote by II the vector valued second fundamental form of M (that is $II(u,v) = (\tilde{D}_u V)^\perp$, the trace of II at m is the vector $\sum_{i=1}^n II(e_i, e_i)$ where (e_i) is an orthonormal basis of $T_m M$, and the previous theorem can be restated as

$$\frac{d\mathcal{V}_t}{dt}_{|t=0} = -\int_M \tilde{g}(trII, Y)dv.$$

2) We define the *mean curvature vector* η of M by $\eta = -\frac{1}{n}trII$: M will be minimal if and only if its mean curvature is zero. For example, the equator S^n is minimal in S^{n+1}.
3) The previous theorem is still true in the case of an immersed submanifold $M \subset \tilde{M}$.

Proof of the theorem. Let $(x^1, \cdots, x^n)$ be a local coordinate system on M. The Riemannian volume element associated to g_t is

$$dv_t = \left[\det g_t\left(\frac{\partial}{\partial x^i}, \frac{\partial}{\partial x^j}\right)\right]^{1/2} dx^1 \cdots dx^n = \frac{\left[\det g_t\left(\frac{\partial}{\partial x^i}, \frac{\partial}{\partial x^j}\right)\right]^{1/2}}{\left[\det g\left(\frac{\partial}{\partial x^i}, \frac{\partial}{\partial x^j}\right)\right]^{1/2}} dv = J(m,t)dv.$$

By differentiating under the integral sign, we get

$$\frac{d\mathcal{V}_t}{dt} = \int_M \frac{\partial}{\partial t} J(m,t)_{|t=0} dv.$$

To compute the argument at m_0 we can assume that the coordinate vector fields are orthonormal at this point, and hence since

$$g_t(u,v) = \tilde{g}(T_m\phi_t u, T_m\phi_t v),$$

we get

$$\frac{\partial}{\partial t} J(m_0,t)_{|t=0} = \sum_{i=1}^{n} \frac{1}{2}\frac{\partial}{\partial t}\tilde{g}(\overline{\frac{\partial}{\partial x^i}}, \overline{\frac{\partial}{\partial x^j}}),$$

where $\overline{\frac{\partial}{\partial x^i}} = T\phi \cdot \frac{\partial}{\partial x^i}$.

Hence, since the brackets of the (coordinate) vector fields $\frac{\partial}{\partial t}$ and $\frac{\partial}{\partial x^i}$ are zero, and from the very definition of Y:

$$\frac{\partial}{\partial t} J(m_0,t)_{|t=0} = \sum_{i=1}^{n} \tilde{g}\left(\bar{D}_{\frac{\partial}{\partial x^i}}Y, \overline{\frac{\partial}{\partial x^i}}\right)_{|t=0}$$

$$= \sum_{i=1}^{n} \frac{\partial}{\partial x^i} \cdot \tilde{g}(\frac{\partial}{\partial x^i}, Y) - \sum_{i=1}^{n} \tilde{g}(\bar{D}_{\frac{\partial}{\partial x^i}}\frac{\partial}{\partial x^i}, Y).$$

Since

$$\tilde{D}_{\frac{\partial}{\partial x^i}}\frac{\partial}{\partial x^i} = D_{\frac{\partial}{\partial x^i}}\frac{\partial}{\partial x^i} - \sum_{j=1}^{k} l_j(\frac{\partial}{\partial x^i}, \frac{\partial}{\partial x^i})N_j,$$

we finally get

$$\frac{\partial}{\partial t} J(m_0,t)_{|t=0} = \sum_{j=1}^{k} \text{tr}(l_j)\tilde{g}(N_j, Y) - \text{div}(Y^\perp),$$

where the coordinate system chosen for the computation doesn't appear any more. Integrating, we get:

$$\frac{d\mathcal{V}_t}{dt}_{|t=0} = \int_M n\tilde{g}(H,Y)dv,$$

since $\int_M \text{div}(Y^\perp)dv = 0$ (use Stokes formula 4.9 and the fact that $Y \equiv 0$ on ∂M). $\blacksquare$

5.22 **Remarks.** i) To keep its balance under the action of the intern tensions ($\eta = 0$), a soap film bounded by a wire takes up the shape of a minimal surface.

ii) The only minimal revolution surfaces in $\mathbf{R}^3$ are domains in the plane or in catenoids. The only ruled minimal surfaces in $\mathbf{R}^3$ are domains in the plane or in helicoids. Note that the helicoid and the catenoid are locally isometric (2.12). Fore more details, and the proofs, see ([Sp], t.3).

The following theorem gives a simple description of the minimal submanifolds in the Euclidean spaces.

5.23 Theorem. *Let M be an n dimensional submanifold in $\mathbf{R}^3$. For $v \in \mathbf{R}^3$, we define a function f_v on M by $f_v(m) = \langle m, v \rangle$. We chose at each point of M an orthonormal basis $(N_1, \cdots, N_k)$ for the normal space to M, and let (l_j) be the associated second fundamental forms. If Δ is the Laplace operator on M equipped with the induced metric, then*

$$\Delta(f_v) = \sum_{j=1}^{k} tr(l_j)\langle N_j, v \rangle = n\langle \eta, v \rangle.$$

In particular, M is minimal if and only if the coordinate functions, restricted to M, are harmonic on M.

Proof. Let D and $\tilde{D}$ be the respective connections of M and the ambient space. For $x, y \in T_m M$, let X and Y be two vector fields defined in a neighborhood of m in $\mathbf{R}^{n+k}$, tangent to M at any point of M, and with respective values x and y at m. Since $df_v(x) = \langle x, v \rangle$, we have:

$$X.(Y.f_v) = X.\langle Y, v \rangle = \langle \tilde{D}_x Y, v \rangle + \langle y, \tilde{D}_x v \rangle$$

$$= \langle \tilde{D}_x Y, v \rangle = \langle D_x Y, v \rangle + \langle II(x, y), v \rangle.$$

Since $Ddf_v(x, y) = X.(Y.f_v) - df_v(D_x Y) = \langle II(x, y), v \rangle$ and $\Delta(f_v) = -tr Ddf_v$, this yields the first assertion.

Now, if M is minimal, its mean curvature η is zero and, for any $v \in \mathbf{R}^{n+k}$, f_v is harmonic on M, that is the coordinate functions are harmonic.

Conversely, if the coordinate functions are harmonic, then all the f_v are also harmonic on M, η is zero, and M is minimal. ■

5.24 Corollary. *There are no non trivial (i.e. not reduced to a point) compact minimal submanifolds without boundary in the Euclidean spaces.*

Proof. The coordinate functions would be harmonic functions on a compact manifold without boundary, and hence constant (4.12). In particular, if M is a compact submanifold of $\mathbf{R}^{n+k}$ without boundary, there exist variations of M which lessen the volume function, and others which increase it. ■

5.25 Exercise: minimal isometric immersions into spheres.

a) Let $F : M^n \to \mathbf{R}^m$ be an isometric immersion. Prove that $\Delta F = n\eta$, where η is the mean curvature of $F(M)$ in $\mathbf{R}^m$, and where the Laplacian must be considered component by component.

b) Let $F : M^n \to S^m \subset \mathbf{R}^{m+1}$ be an isometric immersion. Show that F, as an immersion in the sphere, is minimal if and only if there exists a smooth function ϕ on M such that $\Delta F = \phi F$, and that in this case we actually have $\Delta F = nF$.

c) Let $M = G/H$ be an irreducible isotropy homogeneous space, with G compact, and equipped with a left invariant metric h. If λ is a non zero eigenvalue of the Laplacian on M, let E_λ be the corresponding eigenspace:

$$E_\lambda = \{f \in C^\infty(M), \quad \Delta f = \lambda f\}.$$

E_λ is non zero, and of finite dimension N (IV). Let $(f_0, \cdots, f_N)$ be an orthonormal basis of E_λ, for the integral scalar product deduced from h. Denote by $F : M \rightarrow \mathbf{R}^{N+1}$ the map defined by $F = (f_0, \cdots, f_N)$.

i) Prove that F actually takes its values in a sphere $S^N(r)$.

ii) Prove that, if F is non trivial, it is an immersion and, up to a dilation, it is an isometry.

iii) Eventually using a dilation on M, conclude that F yields a minimal isometric immersion in the sphere S^N.

5.26 Remarks. In the case $M = S^n$ we get, for each eigenvalue λ_k of the Laplace operator on the sphere, a minimal isometric immersion $F_k : S^n \rightarrow S^{N_k}$, by using the eigenfunctions of the Laplace operator, which are the restrictions to the sphere of the homogeneous harmonic polynomial functions of degree k in $\mathbf{R}^{n+1}$.

In particular on S^2:

For $k = 1$, the eigenfunctions are, up to scalar factors, the restrictions to the sphere of the coordinate functions and we get the identity map of S^2.

For $k = 2$, the eigenspace is 5-dimensional, and we get a minimal isometric immersion

$$F_2 : S^2\left(\frac{1}{\sqrt{3}}\right) \rightarrow S^4.$$

Since the eigenfunctions used to build F_2 are homogeneous polynomial functions of degree 2. Quotienting by the antipodal map yields a minimal isometric immersion

$$F_2' : P^2\mathbf{R}\left(\frac{1}{\sqrt{3}}\right) \rightarrow S^4$$

which actually is the Veronese surface (2.46).

5.C.2 Surfaces with constant mean curvature

Let $M \subset \tilde{M}$ an hypersurface which bounds a compact domain D. An alternative variational problem is to find the M whose $(n-1)$-dimensional volume is extremal for deformations which preserves the volume of D. Let $\mathcal{H} : M\times]-\epsilon, \epsilon[$ be a variation of M, and $Y \in \Gamma(M, T\tilde{M})$ the corresponding infinitesimal variation. Then

$$\frac{d}{dt}\mathrm{vol}_n\left(\phi_t(M)\right)_{|t=0} = \int_M \tilde{g}(Y, \nu)v_g.$$

Therefore, M is extremal if and only if, for any vector field along M such that $\int_M \tilde{g}(Y, \nu) v_g = 0$, one has also $\int_M \tilde{g}(\mathrm{tr}(II), \nu) v_g = 0$. Take for Y a normal vector field $f\nu$. Then $\int_M f H v_g$ must vanish as soon as $\int_M f v_g$ vanishes, which forces H to be constant.

Of course, like for minimal surfaces, this property makes interest for non-compact surfaces also. Round spheres in $\mathbf{R}^n$ have of course constant mean curvature. Another trivial example is the circular cylinder. A more sophisticated example was found by C. Delaunay, an astronomer of the nineteenth century: consider a conic which rolls without sliding on a line. Then the trajectory of a focus generates, when rotating around the line, a surface of revolution with constant mean curvature.

The search for compact examples is a difficult problem.

Let us quote two beautiful results, due respectively to H. Hopf and A.D. Alexandrov (see [Ho] for a proof).

a) *Any* immersed sphere *in* $\mathbf{R}^3$ *with constant mean curvature is a round sphere.*

b) *Any* embedded surface *in* $\mathbf{R}^3$ *with constant mean curvature is a round sphere.*

Other compact examples were found a long time after. In the eighties, H. Wente (cf. [Wn]) found an immersed torus with constant mean curvature. This discovery was followed by a series of other examples, a fairly common phenomena in mathematics. See [M-S] for a lively expository paper.

A

Some extra problems

1. Let (M, g) be a complete Riemannian manifold whose sectional curvature K satisfies $K \geq k > 0$. Let γ be a closed geodesic in M. Show that, for any $m \in M$,

$$d(m, \gamma) \leq \frac{\pi}{2\sqrt{k}}.$$

Is it possible to extend this result to compact *minimal* submanifolds of M? *Hint:* Use the second variation formula, and mimic the proof of Myers' theorem.

2. Let γ be a closed geodesic of length l in a complete 2-dimensional Riemannian manifold (M, g). Suppose that (M, g) satisfies $K \geq 0$. Show that

$$\text{vol}(M, g) \leq 2l\,\text{diam}(M, g).$$

Is this inequality optimal? What can be said if $K \geq k$ (if k is any real number) ? What can be said if (M, g) has any dimension, and if γ is replaced by a compact minimal hypersurface?

3. i) Let f and g be two solutions of the differential equation $y'' + Ky = 0$, where K denotes a real function. Show that $f'g - g'f$ is a constant. Consequently, show that if $f(0) = 0, f'(0) = 1, g(a) = 0, g'(a) = -1$, then $f(a) = g(0)$.

ii) Let UM be the unit tangent bundle of a complete 2-dimensional manifold (M, g). For $v \in U_m M$ and t real positive, define $J(v, t)$ by

$$\exp_m^*(v_g) = J(v, t)dv \times dt,$$

where dt is the Lebesgue measure on $\mathbf{R}$, and dv the canonical measure of the unit sphere $U_m M$. Show that, for any v and t,

$$J(v, a) = J(w, a),$$

where $w = -\gamma_v'(a)$ (we have denoted by γ_v the geodesic such that $\gamma_v'(0) = v$).

iii) We want to extend this result to higher dimensions. First consider the differential system

$$X''(t) + R(t).X(t) = 0,$$

where $R(t)$ is a self-adjoint endomorphism of $\mathbf{R}^{n-1}$, and show that, for any pair X, Y of solutions of this system, the function $\langle X', Y \rangle - \langle X, Y' \rangle$ is constant. Then take two families (X_i) and (Y_i) (with $1 \leq i \leq n-1$ of solutions of this equation such that $X_i(0) = 0$ and $Y_i(a) = 0$, while $(X_i'(0))$ and $(Y_i'(a))$ are orthonormal basis of $\mathbf{R}^{n-1}$.

4. Equip the complex projective space $P^n\mathbf{C}$ with its canonical Riemannian metric g. Take $m \in P^n\mathbf{C}$, and for any $r \in]0, \frac{\pi}{2}]$, denote by M_r the set of points in $P^n\mathbf{C}$ whose distance to m is r.

i) Show that M_r is a submanifold of $P^n\mathbf{C}$, which is diffeomorphic to S^{2n-1} if $r < \frac{\pi}{2}$. Equip M_r with the induced Riemannian metric. To which (well known) Riemannian manifold is $M_{\frac{\pi}{2}}$ isometric?

In the sequel, we shall suppose that $r \in]0, \frac{\pi}{2}[$.

ii) Take m_1 and m_2 in $P^n\mathbf{C}$. Show that the corresponding Riemannian manifolds M_r^1 and M_r^2 are isometric.

iii) Show that there exists a unitary normal vector field N on M_r. We choose its orientation in such a way that N points outside the ball $B(m, r)$. Compute the eigenvalues and the eigenvectors of the second fundamental form of M_r. Give estimates for the sectional cuvature of M_r.

iv) Show that the integral curves of the vector field JN on M_r are closed geodesics, whose length is $\pi \sin 2r$. For a given p in M_r, let c be the geodesic such that $c(0) = p$ and $c'(0) = JN$. Give the value s_1 of the parameter of the first conjugate point of p along c. Show that c meets the cut-locus of p for a value $s_0 < s_1$ of the parameter.

v) In a compact simply connected even-dimensional Riemannian manifold with sectional curvature K belonging to $]0, k]$, the length of any closed geodesic is greater than $2\frac{\pi}{\sqrt{k}}$ (Klingenberg, cf. [C-E]). Use the preceding question to show that this theorem is not true for odd-dimensional manifolds.

vi) Show that M_r is a Riemannian homogenous space. (Hint: look at the isotropy group at m of the isometry group of $P^n\mathbf{C}$.)

vi) Recall that the natural actions of $SO(2n)$ and $U(n)$ on (S^{2n-1}, can) are transitive and isometric. Does there exist other $SO(2n)$-invariant metrics on S^{2n-1}? Other $U(n)$-invariant metrics?

iii) Take the map ϕ of M_r on $P^{n-1}\mathbf{C}$ such that, for any unit tangent vector v at m,

$$\phi\big(\exp_m(rv)\big) = \exp_m\big(\frac{\pi}{2}v\big).$$

Show that ϕ is a differentiable submersion. Equip $P^{n-1}\mathbf{C}$ with the metric $g_r = \sin^2 rg_0$, where g_0 denotes the canonical Riemannian metric of $P^{n-1}\mathbf{C}$. Show that ϕ is then a Riemannian submersion.

ix) Set $\Gamma_t = \exp itI \in U(n)$. Show that Γ acts freely and isometrically on M_r, and that the quotient Riemannian manifold is just $(P^{n-1}\mathbf{C}, g_r)$.

x) Describe the cut-locus of m in M_r.

B

Solutions of exercises

Chapter 1

1.11 Denote by R the equivalence relation which defines the Möbius band M. The differential structure will be given by two charts, whose domains are the open sets

$$U_1 = (]0,1[\times \mathbf{R})/R \quad \text{and} \quad U_2 = ([0,\frac{1}{2}[\cup]\frac{1}{2},1] \times \mathbf{R})/R$$

of M. Define maps ϕ_i (with $i = 1,2$) from U_i into $\mathbf{R}^2$ by $\phi_1(x,y) = (x,y)$ and $\phi_2(x,y) = (x,y)$ if $x > \frac{1}{2}$, $\phi_2(x,y) = (x+1,-y)$ if $x < \frac{1}{2}$. We get homeomorphisms from U_i onto open sets of $\mathbf{R}^2$. The coordinate change

$$f = \phi_1 \circ \phi_2^{-1} : (]\frac{1}{2},1[\cup]1,\frac{3}{2}[) \times \mathbf{R} \to (]0,\frac{1}{2}[\cup]\frac{1}{2},1[) \times \mathbf{R}$$

is given by $f(x,y) = (x,y)$ if $x < 1$ and $f(x,y) = (x-1,-y)$ otherwise. It is indeed a diffeomorphism, and the charts (U_i,ϕ_i) equip M with a differential structure.

1.13 a) For the atlas we have defined on M, the open set $U_1 \cap U_2$ is decomposed into two connected open sets V and W. On V (resp. W) the coordinate change has positive (resp. negative) determinant. If M were orientable, there should exist an atlas $(W_j, g_j)_{j \in J}$ with positive determinants for the coordinate changes f_{ij}. Define

$$U_1^+ = \{x \in U_1 \mid \forall j \in J \quad \text{Jacob}_{\phi_1(x)}(g_j \circ \phi_1^{-1}) > 0\}$$

(it amounts to the same to say that it occurs for some j) and

$$U_1^- = \{x \in U_1 \mid \forall j \in J \quad \text{Jacob}_{\phi_1(x)}(g_j \circ \phi_1^{-1}) < 0\}.$$

We have got a partition of the open connected set U_1 into two disjoint open subsets. Hence we must have for instance $U_1^+ = U_1$. Now, we do the same job for U_2. Either $U_2 = U_2^+$, or $U_2 = U_2^-$. The first equality forces $\mathrm{Jacob}(\phi_2 \circ \phi_1^{-1})$ to be positive on $U_1 \cap U_2$, a contradiction. The second equality forces this Jacobian to be negative on $U_1 \cap U_2$, a contradiction again.

1.13 b) An open subset of an orientable manifold is clearly orientable. The figure B.1, page 267 shows how to get $P^2\mathbf{R}$ from a Möbius band and a disk. In this construction, the complement of a closed disk in $P^2\mathbf{R}$ turns out to be a Möbius band. Therefore, $P^2\mathbf{R}$ cannot be orientable.

1.13 c) No. The submanifold $P^2\mathbf{R}$ of $P^3\mathbf{R}$ is not orientable. However, any compact codimension 1 submanifold of $\mathbf{R}^n$ is orientable. The example of the Möbius band shows that this compactness assumption is indeed necessary.

1.13 d) Take the product of two orientation atlases for M and N respectively. Then, writing the differential of any coordinate change as a block of matrices, we see that the determinant is still positive.

1.13 e) Equip $\mathbf{R}^n$ with an Euclidean structure and an orientation. Then the gradient of f gives a normal vector field to M. Then use definition ii) of submanifolds of $\mathbf{R}^n$. Restricting oneself to connected open sets, and composing the diffeomorphisms ϕ with symmetries if necessary, we can suppose that they are orientation preserving, and that the last component of $D\phi(\nabla f)$ is always positive. Restricting the ϕ to M gives the required atlas.

1.20 a) Take x in N, a chart (V, ϕ) in a neighborhood of x and a chart (U, ψ) in a neighborhood of $f(x)$ in M. Then the map $\psi \circ f \circ \phi^{-1}$ of $\phi^{-1}(V)$ into $\psi(V)$ is an immersion. Furthermore, this map gives an homeomorphism of $\phi^{-1}(V)$ onto its image $\psi(f(N)) \cap U$. Hence $\psi(f(N)) \cap U$ is a submanifold of $\psi(U)$. This is true for any $f(x)$ in M, therefore $f(N)$ is a submanifold of M.

1.20 b) From their definition, manifolds are locally compact spaces. Take x in N, and an open set U containing $y = f(x)$ in M with compact closure. Set $\overline{U} = L$. Now, $f^{-1}(U) = V$ is an open neighborhood of x, and since f is proper, $f^{-1}(L) = K$ is a compact part of M containing U. Then $f_{|K} : K \to L$ is an homeomorphism onto its image. Since this property holds locally for any x, and since f is injective, f is an homeomorphism onto its image. The properness assumption rules out the phenomena described on figure B.2, page 268

1.20 c) Suppose we have an immersion j of S^1 into $\mathbf{R}$, and take for instance the chart (U, ϕ) given by $\phi^{-1}(\theta) = \exp i\theta$, with $\theta \in\ ] - \pi, \pi[$. Then $j \circ \phi^{-1}$ is an immersion of $] - \pi, \pi[$ into $\mathbf{R}$. Consequently, it is strictly monotonous. Since j is defined on the whole circle, the limits when θ goes to $+\pi$ or $-\pi$ of $j \circ \phi^{-1}(\theta)$ must be equal, a contradiction.

1.20 d) Take charts for f, and apply the inverse function theorem. Now let f be an immersion of S^n into $\mathbf{R}^n$. On one hand, $f(S^n)$ is open in $\mathbf{R}^n$. On the other hand, it must be closed since S^n is compact. Since $\mathbf{R}^n$ is connected, we get a contradiction. For $n = 1$, we recover the result of the preceding question.

We want to get the
quotient
of the sphere $\mathbf{S}^2$
by the antipodal map.

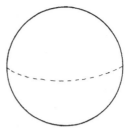

It is sufficient
to take the quotient
of this part
of the sphere.

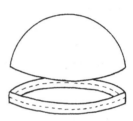

The upper part,
homeomorphic to a disk,
is left unchanged
under the quotient.

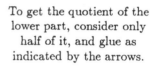

To get the quotient of the
lower part, consider only
half of it, and glue as
indicated by the arrows.

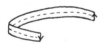

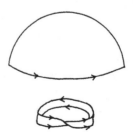

Hence, we get a
Möbius band,
to be glued with
the disk
along their boundaries.

Fig. B.1. The projective plane and the Möbius band

Fig. B.2. Non-proper immersions

1.31 Take an atlas (U_i, ϕ_i) of M. There corresponds to it the atlas $(\pi^{-1}(U_i), \Phi_i)$ of TM, whose coordinate changes are given by

$$\Phi_j \circ \Phi_i^{-1}(x, v) = \left(\phi_j \circ \phi_i^{-1}(x), D_x(\phi_j \circ \phi_i^{-1}) \cdot v\right).$$

The matrix of the differential of this coordinate change is given by

$$\begin{pmatrix} D_x(\phi_j \circ \phi_i^{-1}) & 0 \\ A & D_x(\phi_j \circ \phi_i^{-1}) \end{pmatrix}.$$

Therefore, whatever A may be,

$$\mathrm{Jacob}_{(x,v)}(\Phi_j \circ \Phi_i^{-1}) = \left[\mathrm{Jacob}_x(\phi_j \circ \phi_i^{-1})\right]^2 > 0.$$

1.33 Once TM has been equipped with the differential structure defined in theorem 1.30, the map ψ from $\pi^{-1}(U)$ into $U \times \mathbf{R}^n$ which is given by

$$\psi(\xi) = \left(\pi(\xi), \theta_{U,\phi,\pi(\xi)}^{-1}(\xi)\right)$$

is a diffeomorphism, and from its very definition, $\psi(T_xM) = \{x\} \times \mathbf{R}^n$. If (U', ϕ') is another chart, which defines in the same way a diffeomorphism ψ' of $\pi^{-1}(U')$ onto $U' \times \mathbf{R}^n$, then

$$\psi' \circ \psi^{-1}(x, v) = \left(x, D_x(\phi \circ \phi^{-1}) \cdot v\right)$$

is a diffeomorphism of $(U \cap U') \times \mathbf{R}^n$ into itself. The conditions prescribed in definition 1.32 are satisfied, and the triple (π, TM, M) is a vector bundle of rank n.

1.35 a) Let $\pi : E \to B$ a trivial vector bundle and $h : E \to B \times \mathbf{R}^n$ be a trivialization. Take a basis $(e_i)_{1 \leq i \leq n}$ of $\mathbf{R}^n$, and set $s_i(b) = h^{-1}(b, e_i)$. In that way, we get n linearly independent sections of B.
Conversely, let $\pi : E \to B$ be a vector bundle of rank n, and suppose this bundle has n linearly independent sections $(s_i)_{1 \leq i \leq n}$. We define a *bundle map* H of $B \times \mathbf{R}^n$ into E by

$$H(b; k_1, ..., k_n) = \sum_{i=1}^{n} k_i s_i(b).$$

This map is clearly bijective and smooth. To see that H is a diffeomorphism, just use local trivializations of E. The computations are quite analogous to the one in exercise **1.31.** Indeed, if

$$h : \pi^{-1}(U) \to U \times \mathbf{R}^n$$

is such a trivialization, the differential of $h \circ H$ at $(b, k) \in U \times \mathbf{R}^n$ is a matrix like

$$\begin{pmatrix} I_{T_b B} & 0 \\ * & I_{T_b B} \end{pmatrix}.$$

The details are left to the reader.

1.35 b) We have $S^1 = [0, 1]/R$, where R denotes the equivalence relation which identifies 0 and 1. Take the open sets $V_1 =]0, 1[/R$ and $V_2 = ([0, \frac{1}{2}[\cup]\frac{1}{2}, 1]/R$ which are diffeomorphic to $]0, 1[$ and $]\frac{1}{2}, \frac{3}{2}[$ respectively. Denoting by π the map from M into S^1 which is obtained by going to the quotient from the first projection of $[0, 1] \times \mathbf{R}$. Then, using the diffeomorphisms above, we get diffeomorphisms h_i $(i = 1, 2)$ of $\pi^{-1}(V_i)$ into $V_i \times \mathbf{R}$. The intersection $V_1 \cap V_2$ has two connected components W and W', and we have

$$(h_2 \circ h_1^{-1})(x, y) = (x, y) \quad \text{for} \quad x \in W$$

$$= (x, -y) \quad \text{for} \quad x \in W'.$$

As claimed, we have a vector bundle of rank 1 over S^1. This bundle is not trivial. Otherwise, there would exist a nowhere vanishing section s (1.35 a)). Over V_1 we have $h_1(s(x)) = (x, A(x))$. Since V_1 is connected and s does not vanish, we can suppose that A is positive everywhere. Over V_2, we have $h_2(s(x)) = (x, B(x))$, with B having constant sign. Now, we must have $A(x) = B(x)$ if $x \in W$ and $A(x) = -B(x)$ if $x \in W'$, a contradiction.

1.39 a) Take $y = \phi_2 \circ \phi_1^{-1}(x) = x/ \mid x \mid^2$. The required condition is just

$$X_2(y) = T_x(\phi_2 \circ \phi_1^{-1}) \cdot X_1(x).$$

More explicitely, it means that

$$X_2(y) = \mid y \mid^2 X_1\left(\frac{y}{\mid y \mid^2}\right) - 2\langle y, X_1\left(\frac{y}{\mid y \mid^2}\right)\rangle y$$

for $y \in \Omega_1 \cap \Omega_2$.

1.39 b) It works if and only if

$$Y(\theta + 2k\pi, \theta' + 2k'\pi) = Y(\theta, \theta')$$

for any $k, k' \in \mathbf{Z}$. In particular, we can take constant vector fields. Denote by E_1 and E_2 the vectors fields on T^2 which are defined by the fields $(1, 0)$ and

$(0, 1)$ on $\mathbf{R}^2$. Then the map

$$(m, x_1, x_2) \to x_1 E_1(m) + x_2 E_2(m)$$

is a diffeomorphism of $T^2 \times \mathbf{R}^2$ onto $T(T^2)$. It is also a vector bundle isomorphism.

1.43 a) Denote the coordinates of $m \in \mathbf{R}^{2n+2}$ by

$$(x^1, y^1, x^2, y^2, \cdots, x^{n+1}, y^{n+1}).$$

Then such a vector field is given by

$$X(m) = (-y^1, x^1, -y^2, x^2, \cdots, -y^{n+1}, x^{n+1}).$$

A better way to see that is to identify $\mathbf{R}^{2n+2}$ with $\mathbf{C}^{n+1}$ by setting $z_k = x_k + iy_k$. Then $X(m) = im$. This vector field is vertical for the submersion $S^{2n+1} \to P^n\mathbf{C}$, and will be used quite often in the sequel of this book.

1.43 b) Differentiating the relation ${}^t g g = I$ we see that

$$T_g O(n) = \{A \in \text{End}(\mathbf{R}^n), \, {}^t A g + {}^t g A = 0\}.$$

In particular, $T_I O(n)$ can be identified with the vector space of antisymmetric matrices. If $A \in T_I O(n)$ and $g \in O(n)$, then gA is a tangent vector at g, and the map $(g, A) \to gA$ is a diffeomorphism between $O(n) \times T_I O(n)$ and $TO(n)$. In fact, we shall see later that the tangent bundle to a Lie goup is always trivial.

1.48 Check directly that $\delta(f^3) = 3f(0)^2\delta(f)$. Since any continous function has a continuous cubic root, δ must vanish if $f(0) = 0$. Since a derivation is zero for constant functions, it follows that δ must be identically 0.

1.54 Test the brackets on decomposed functions $\phi(x) = f(r)g(\theta)$. One finds $[\tilde{X}, Y] = -X_1$ and $[X_1, Y] = 0$.

1.60 a) $\theta_t(x) = \frac{x}{(1-tx)}$.

1.62 One checks that $[X, Y] = Z$, $[Y, Z] = X$, and $[Z, X] = Y$. The flow of $aX + bY + cZ$ is given by $\theta_t(v) = \exp tM \cdot v$, where

$$v = \begin{pmatrix} x \\ y \\ z \end{pmatrix} \quad \text{and} \quad M = \begin{pmatrix} 0 & c & -b \\ -c & 0 & a \\ b & -a & 0 \end{pmatrix}.$$

1.65 b) It is just the vector field $x \to x\frac{\partial}{\partial x}$ on $\mathbf{R}^{+*}$.

1.74 The left invariant vector field associated with B is given by $B_g = gB$. The associated flow is $\theta_t(g) = g.\exp(tB)$. Since the differential of θ_t is just the right multiplication by $\exp(tB)$, we see that

$$(\theta_t * A)_I = T_{\theta_{-t}(I)}\theta_t(A_{\theta_{-t}(I)}).$$

The result follows from 1.68.

1.79 The elements of $\underline{O}(n)$ are just real antisymmetric matrices. We get a vector space of dimension $\frac{n(n-1)}{2}$. As for $\underline{U}(n)$, we get antihermitian matrices, namely matrices such that $A^t = -\overline{A}$. Since such a matrix has imaginary elements on the diagonal, we have $\dim \underline{U}(n) = n^2$. In the same way, $\underline{SU}(n)$ is composed with traceless antihermitian matrices, and $\dim \underline{SU}(n) = n^2 - 1$. Finally, the Lie algebra of $O(1, n)$ is composed with real matrices A such that

$$AJ + JA^t = 0$$

where J denotes the matrix $\mathrm{diag}(-1, 1, \cdots, 1)$. Hence $\dim O(1, n) = \frac{n(n+1)}{2}$.

1.84 a) Take $m \in M$, and as in 1.83 an open set containing U such that $p^{-1}(U) = \bigcup_{i \in I} U_i$. Clearly, any $q \in U$ has one and only one preimage in each U_i.

1.84 b) The map $p : z \to z^3$ is not a covering map. Indeed, $\mathrm{card}(p^{-1}(u)) = 3$ if $u \neq 0$, but $\mathrm{card}(p^{-1}(0)) = 1$.

1.84 c) The map p is surjective and is a local diffeomorphism. On the other hand, if $x \in \mathbf{R}$, then

$$p :]x - \pi, x + \pi[\to S^1 \setminus p(x + \pi)$$

is an homeomorphism, and

$$p^{-1}(S^1 \setminus p(x + \pi)) = \bigcup_{k \in \mathbf{Z}}]x + (2k - 1)\pi, x + (2k + 1)\pi[.$$

1.84 d) Using the inverse function theorem, we see that for any $x \in \tilde{M}$, there is an open set U_x containing x and an open set V_x containing $f(x)$ such that f is a diffeomorphism of U_x onto V_x (just take charts). In particular, $f(\tilde{M})$ is an open set. Since $\tilde{M}$ is compact, $f(\tilde{M})$ is also closed, and f is surjective. Now, for any $y \in M$, the preimage $f^{-1}(y)$ is discrete and non-empty from what we have seen. Using compactness of $\tilde{M}$ again, we see that it is a finite set $(x_1, \cdots, x_p)$. Take open sets U_{x_i} containing x_i as in the beginning, take a compact neighborhood V of y contained in $\bigcap_{i=1}^{p} f(U_{s_i})$, and set

$$U_i' = f^{-1}(V) \cap U_{x_i}.$$

Then f is a diffeomorphism between the interiors of U_i' and V. Now we must control $f^{-1}(V)$ (see the example below). The set $Z = f^{-1}(V) \setminus (\bigcup_i U_i')$ is compact, therefore $f(Z)$ is compact and $V' = V \setminus f(Z)$ is an open neighborhood of y. Setting $U_i = f^{-1}(V') \cap U_i'$, we have

$$f^{-1}(V') = \bigcup_{i=1}^{p} U_i$$

as required for having a covering map.

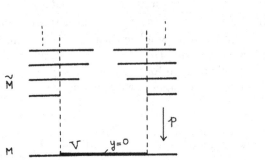
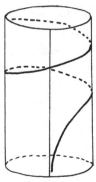

Fig. B.3. A local diffeomorphism which is not a covering map

Remarks: i) The following example shows that compactness is crucial for controlling the preimage $f^{-1}(V)$. Take

$$\tilde{M} = (\{0\} \times \mathbf{R}) \bigcup_n (\{n\} \times (\mathbf{R}\backslash] - 1/n, 1/n[)),$$

$M = \mathbf{R}$, the map f being the first projection.

ii) A counterexample with $\tilde{M}$ connected can be given as follows: take

$$\tilde{M} = \{\cos e^z, \sin e^z, z\} \subset \mathbf{R}^3,$$

$M = S^1 \subset \mathbf{R}^2$, the map f being the restriction to $\tilde{M}$ of the projection $(x, y, z) \to (x, y)$. The curve $\tilde{M}$ lies in a cylinder of $\mathbf{R}^3$, and the generatrix of this cylinder through $(1, 0)$ is asymptotic to $\tilde{M}$ when z goes to $-\infty$. Hence, for any small enough neighborhood of $(1, 0)$ in M, the preimage will have a connected component no point of which is projected onto $(1, 0)$.

1.89 a) The map $f : \mathbf{R}^n \to T^n$ given by

$$f(x_1, \cdots, x_n) = (\exp ix_1, \cdots, \exp ix_n)$$

goes to the quotient and gives a map $F : \mathbf{R}^n/G \to T^n$. This map is a bijective immersion, hence a diffeomorphism, since $\mathbf{R}^n/G$ is compact.

1.89 b) Take charts on S^n with domains

$$U_i^+ = \{x \in S^n, x_i > 0\}, \quad \text{and} \quad U_i^- = \{x \in S^n, x_i < 0\}.$$

Then π is a diffeomorphism of U_i^+ and U_i^- onto their common image, that we shall denote by V_i. The sets V_i can be taken as domains of charts for the differential structure of $P^n\mathbf{R}$. If $P^n\mathbf{R}$ were orientable for n even, using the V_i we could get an orientation atlas on S^n which should be invariant under the antipodal map. Proceeding as in 1.12, we see that it is impossible. If n is odd, the atlas of 1.10 is indeed an orientation atlas.

1.93 a) First note that a manifold is connected if and only if it is path-connected. Take $x \in E$, and let E_x be its connected component. We will show that $E_x = E$. The projection π is an open map, hence $B_x = \pi(E_x)$ is open in B. Take $z \in E_x$ and $b = \pi(z)$. Since the fiber $\pi^{-1}(b)$ is diffeomorphic to F, for any $z' \in \pi^{-1}(b)$ there is a continuous path from z to z', hence $z' \in E_x$, and $E_x = \pi^{-1}(B_x)$. Now, since E is partitioned by the E_x, this equality says that B is partitioned by the B_x. Since B is connected we are done.

1.93 b) Take the two charts ϕ_1 and ϕ_2 of $S^2 = P^1\mathbf{C}$ which were given in 1.19, and recall that the coordinate change is just $z \to 1/z$. Set

$$A_1 = \{(z_1, z_2) \in S^3, z_1 \neq 0\} \quad \text{and} \quad A_2 = \{(z_1, z_2) \in S^3, z_2 \neq 0\}.$$

The maps $f_i : A_i \to \mathbf{C} \times S^1$ $(i = 1, 2)$ given by

$$f_1(z_1, z_2) = (\frac{z_2}{z_1}, \frac{z_1}{|z_1|}) \quad \text{and} \quad f_2(z_1, z_2) = (\frac{z_1}{z_2}, \frac{z_2}{|z_2|})$$

are diffeomorphisms. By setting $h_i = (\phi_i, \mathrm{Id}) \circ f_i$, we get diffeomorphisms h_i from A_i into $U_i \times S^1$, and we have

$$h_2 \circ h_1(z, e^{i\theta}) = (\frac{1}{z}, \frac{z}{|z|} e^{i\theta}).$$

We have equipped S^3 with a structure of S^1 bundle over S^2. More globally, for $(z_1, z_2) \in S^3 \subset \mathbf{C}^2$ the projection π is given by

$$\pi(z_1, z_2) = (2\Re(z_2\bar{z}_1), 2\Im(z_2\bar{z}_1), |z_2|^2 - |z_1|^2).$$

The link between these two presentations is obtained by using stereographic projections of S^2.

Remark: This bundle is not trivial: otherwise, S^3 would be diffeomorphic to $S^1 \times S^2$. The former manifold is simply connected, while the latter is not.

1.115 Denote by $P_X S$ the second member of this equality. Then check that P_X is $C^\infty(M)$-linear with respect to the X_i (and defines therefore a tensor of $\Gamma(T_q^o)$), and furthermore that P_X is a derivation which coincide with L_X for functions and 1-forms.

1.118 a) Take $m \in S^2$. The sub-group of $SO(3)$ which leaves m fixed is isomorphic to $SO(2)$, and the action of this group on $T_m S^2$ or $T_m^* S^2$ (which is obtained by differentiation) is equivalent to the natural action of $SO(2)$ on $\mathbf{R}^2$ (see 2.40 for more details). Now, ω_m is $SO(2)$-invariant, and hence identically zero. It can be proved in the same way that on S^n any $SO(n+1)$-invariant k-form is zero if $0 < k < n$.

1.118 b) It amounts to the same to consider the fibration (that we shall still denote by p) of S^{2n+1} onto $P^n\mathbf{C}$, and the 2-form

$$\alpha = \sum_{k=0}^{n} dz^k \wedge d\bar{z}^k$$

restricted to S^{2n+1}. To get a 2-form ω on $P^n\mathbf{C}$ such that $p^*\omega = \alpha$, we shall proceede in the most simple minded way. Namely, take $m \in P^n\mathbf{C}$, tangent vectors u and v at m and chose $\tilde{m}$ such that $p(\tilde{m}) = m$, and tangent vectors $\tilde{u}$ and $\tilde{v}$ at $\tilde{m}$ such that

$$T_{\tilde{m}}p \cdot \tilde{u} = u \quad T_{\tilde{m}}p \cdot \tilde{v} = v$$

and set

$$\omega_m(u, v) = \alpha_{\tilde{m}}(\tilde{u}, \tilde{v}).$$

We must check that the result does not depend on our choices of $\tilde{m}, \tilde{u}, \tilde{v}$. This is true in view of the following properties of α.

i) α is invariant under the action of S^1 on S^{2n+1}. Namely, take $\lambda \in \mathbf{C}$ of modulus 1, and the map, that we shall still denote by λ, given by

$$(z_0, \cdots, z_n) \to (\lambda z_0, \cdots, \lambda z_n).$$

Since $\lambda^* dz^k = dz^k$ and $\lambda^* d\bar{z}^k = \bar{\lambda} d\bar{z}^k$, we have

$$\lambda^* \alpha = \alpha.$$

ii) α is vertical. Namely, if $T_{\tilde{m}}p \cdot w = 0$ then

$$\alpha_{\tilde{m}}(w, v) = 0 \quad \text{for any} \quad v.$$

But the vectors of the subspace $\mathrm{Ker}T_{\tilde{m}}p$ of $T_{\tilde{m}}S^{2n+1}$ can be written as $t.i\tilde{m}$ with t real. Now

$$\alpha_{\tilde{m}}(i\tilde{m}, v) = \langle \tilde{m}, v \rangle,$$

where $\langle , \rangle$ denotes the scalar product in $\mathbf{R}^{2n+2}$. The required property follows (see 2.29 for more information about the notations we used here).

There remains to proof that we get a smooth differential form. Once more (compare with 2.28 and 2.43) this comes from the fact that p has local sections. The algebraic properties of ω we claimed come from similar properties for α and from the same arguments as in the existence proof.

1.127 bis Let $p : S^n \to P^n\mathbf{R}$ be the canonical projection, and i be the antipodal map of S^n. Let ω be the volume form on $S^n \subset \mathbf{R}^{n+1}$ which is given by

$$\omega = i_x(dx^0 \wedge dx^1 \wedge \cdots \wedge dx^n)$$

(this form is said to be canonical because it is $SO(n+1)$-invariant): indeed, it will turn out to be the volume form associated with the standard Riemannian metric of the sphere).

If n is even, $i^*\omega = \omega$. Using similar arguments to those of 1.118 b) (but much more simple), we see that ω goes to the quotient and gives a volume form on $P^n\mathbf{R}$.

Now, take a volume form α on $P^n\mathbf{R}$. Then $p^*\alpha$ is a volume form on S^n, which is i-invariant since $p \circ i = p$. There exists on S^n a nowhere vanishing smooth function f such that $p_*\alpha = f\omega$. Then

$$i^*(f\omega) = f\omega = (f \circ i)i^*\omega = (-1)^{n+1}(f \circ i)\omega,$$

which is possible only if n is odd.

1.130 a) We find $\det(g)^{-n} \bigwedge_{1 \leq i,j \leq n} dx^{ij}$.

1.130 b) If ϕ denotes a symmetric bilinear form which is associated with a positive definite quadratic form on $\mathbf{R}^n$, and dk a Haar measure on K, the formula

$$\overline{\phi}(x, y) = \int_K \phi(kx, ky)dk$$

defines a positive definite K-invariant quadratic form. Take $g \in Gl_n\mathbf{R}$ such that $\overline{\phi}(x, y) = \langle gx, gy \rangle$. Then, for any $k \in K$,

$$\langle gkg^{-1}x, gkg^{-1}y \rangle = \overline{\phi}(kg^{-1}x, kg^{-1}y)$$
$$= \overline{\phi}(g^{-1}x, g^{-1}y) = \langle x, y \rangle,$$

which proves that $gKg^{-1} \subset O(n)$.

Chapter 2

2.3 By contradiction: assume there exists a Lorentzian metric g on S^2. Then g determines in each tangent space T_pS^2 two isotropic directions. Denote by $G^1(S^2)$ the bundle of non oriented 1-dimensional grassmannians of S^2.
We can choose a smooth section s of this bundle such that, for any $p \in S^2$, $s(p)$ is an isotropic direction for g: it is easy to build such sections s_1 and s_2 above the domains U and V of the stereographic charts for example. Chose $p \in U \cap V$, and prescribe $s_1(p) = s_2(p)$: the connectedness of $U \cap V$ ensures that s_1 and s_2 coincide on $U \cap V$ and hence yield a global section s.
Now, the bundle $G^1(S^2)$ has a two folded covering by the bundle of oriented 1-dimensional grassmannians, which can be identified with the unit tangent bundle US^2 of S^2. Since the sphere is simply connected, the section s has a lift $\tilde{s}$ which is a section of US^2, and we get a contradiction with 1.41: every vector field on the 2-sphere has a zero.

2.11 a) First note that f sends the set $\{x, \langle x - s, x - s \rangle \neq 0\}$ on itself and that f, restricted to this open set, is an involution ($f^2 = \text{Id}$). If $\langle x, x \rangle = -1$, since $\langle s, x \rangle = x_0$, we have

$$\langle x - s, x - s \rangle = -2(1 + x_0), \quad (f(x))_0 = 0, \quad \langle f(x), f(x) \rangle = |f(x)|^2 = \frac{x_0 - 1}{x_0 + 1}.$$

Also

$$f'(x) \cdot \xi = \frac{-2}{\langle x - s, x - s \rangle} \xi + \frac{4\langle x - s, \xi \rangle}{\langle x - s, x - s \rangle^2} (x - s)$$

$$= \frac{-2}{\langle x - s, x - s \rangle} \left[\xi - 2\frac{\langle x - s, \xi \rangle}{\langle x - s, x - s \rangle} (x - s) \right].$$

The map $\xi \to \xi - \frac{2\langle x - s, \xi \rangle}{\langle x - s, x - s \rangle}(x - s)$ is the Lorentzian symmetry with respect to the hyperplane $(x - s)^{\perp}$, and hence is an isometry. Then

$$\langle f'(x) \cdot \xi, f'(x) \cdot \xi \rangle = \frac{4}{\langle x - s, x - s \rangle^2}.$$

Now, if $x_0 = 0$, $\langle x - s, x - s \rangle = \sum_{i=1}^{n} x_i^2 - 1$ which yields the formula, since $f^{-1} = f$.

b) The same computations show that

$$f(x) = s + \frac{2}{\langle x - s, x - s \rangle} (x - s)$$

is an isometry between $S^n \backslash \{s\}$ and $\mathbf{R}^n$, equipped with the metric

$$\frac{\sum_{i=1}^{n} dx_i^2}{(1 + |x|^2)^2},$$

(which is not complete, see 2.102). This analogy should convince the reader that the "hyperboloid" model for the hyperbolic space is more natural than the Poincaré one.

c) is left to the reader. Note that in the disk model a point of H^n is distinguished, and a direction in the half-space model.

2.12 a) C and H are respectively given by

$$(r, \theta) \xrightarrow{\phi} (\cosh r. \cos \theta, \cosh r. \sin \theta, r) \quad \text{and} \quad (u, t) \xrightarrow{\psi} (u. \cos t, u. \sin t, t).$$

b) Confusing $g_{|C}$ and $\phi^*(g_{|C})$, we have

$$g_{|C} = \cosh^2 r(dr^2 + d\theta^2) \quad \text{and} \quad g_{|H} = du^2 + (1 + u^2)dt^2,$$

which proves a posteriori that ϕ and ψ are parametrizations. The map

$$(r, \theta) \to (u = \sinh r, t = \theta),$$

which is defined locally (determination of the angle), is clearly a local isometry. In fact $(u, t) \to (\text{Argsh} u, t)$ is the Riemannian universel covering of C, which is diffeomorphic to $S^1 \times \mathbf{R}$, by H which is diffeomorphic to $\mathbf{R}^2$.

2.13 a) The differential equation satisfied by these curves is

$$y'^2 + (y/y')^2 = 1 \quad \text{or} \quad y'^2 = y^2/(1-y^2).$$

With $y = 1/\cosh t$, we get $dx = \pm \tanh^2 t\, dt$, and the result we claimed.

b) The metric induced on the pseudo-sphere, which is singular for $t = 0$, is

$$g = \tanh^2 t\, dt^2 + \frac{d\theta^2}{\cosh^2 t}.$$

Just set $y = \cosh t$ and $x = \theta$.

2.16 The metric g is not the product metric, since the norm of the TS^{n-1}-component of a tangent vector depends on r: the product metric is $g_1 + dr^2$. The map $f : (S^{n-1} \times I, g) \to (\mathbf{R}^n \backslash \{0\}, \text{can})$ defined by $f(x, r) = rx$ is an isometry since first, it is a diffeomorphism and second

$$T_{(x,r)}f(X_i) = r.X_i \quad \text{and} \quad T_{(x,r)}f\left(\frac{\partial}{\partial r}\right) = x :$$

the norms and the right angles between these vectors are preserved.
One proves in the same way that the map $F : (S^{n-1} \times I, g_1 \times dr^2) \to (C, \text{can})$ defined y $F(x, r) = x + re_0$ is an isometry.
We have $g(Y, Y) = r^2 g_1(X, X)$ and $g_1 \times dr^2(Y, Y) = g_1(X, X)$.

2.25 a) The metric g_0 is given in local coordinates (θ, ϕ) by $g_0 = d\theta^2 + d\phi^2$. On the other hand, we have

$$\frac{\partial \Phi}{\partial \theta} = (-\cos \phi \sin \theta, -\sin \phi \sin \theta, \cos \theta)$$

$$\text{and} \quad \frac{\partial \Phi}{\partial \phi} = (-(2 + \cos \theta) \sin \phi, (2 + \cos \theta) \cos \phi, 0)$$

and hence $g = d\theta^2 + (2 + \cos \theta)^2 d\phi^2$: g is not the product metric.

b) The metric g is given in local coordinates by $g = d\theta^2 + d\phi^2$. The map Φ goes to the quotient and yields a diffeomorphism $\hat{\Phi}$ of T^2. Let $h = \hat{\Phi}^*(g)$. The lift of h to $\mathbf{R}^2$ is the Euclidean metric $\tilde{h} = dx^2 + dy^2$, hence (M, g) is isometric to the square torus (T^2, h).

c) To show that $\mathbf{R}^2/G$ is diffeomorphic to K, just notice that $\mathbf{R}^2/G$ is compact, and imitate the proof of 1.89. The Euclidean metric will only go to the quotient in the case G is a group of isometries of $\mathbf{R}^2$, that is if γ_1 and γ_2 are isometries. It is always the case for γ_1 which is a translation, but γ_2 which is up to a translation, a symmetry, is an isometry if and only if the vectors e_1 and e_2 are orthogonal.

d) An isometry of $\mathbf{R}^2$ goes to the quotient by a given lattice Γ if it sends any pair of Γ-equivalent points to another such pair. If A is such an isometry then $A\Gamma A^{-1} \subset \Gamma$, that is A belongs to the normalizer $N(\Gamma)$ of Γ in $\text{Isom}(\mathbf{R}^2)$. It

is clear that $A^{-1}\Gamma A \subset \Gamma$ if and only if the linear part α of A, considered as a rotation of center 0, globally preserves the lattice. In particular, α must send a vector of minimal norm on another vector of minimal norm.
Therefore:
- if Γ is hexagonal, $\mathrm{Isom}(T^2) = D_6 \times T^2$,
- if Γ is a square lattice, $\mathrm{Isom}(T^2) = D_4 \times T^2$,
- if Γ is a rectangle lattice, $\mathrm{Isom}(T^2) = D_2 \times T^2$,
- generically, $\mathrm{Isom}(T^2) = Z_2 \times T^2$.
Here, D_n denotes the diedral group of order $2n$, i.e. the isometry group of a regular polygon with n sides. e) Let p be the Riemannian covering map of $\mathbf{R} \times S^1$ by $\mathbf{R} \times \mathbf{R}$ defined by $(x, y) \to (x, e^{iy})$. We have

$$p(x, y + 2\pi) = (x, e^{iy}) \quad \text{and} \quad p(x + 1, y + \theta) = (x + 1, ze^{i\theta}),$$

which proves that the quotient of $\mathbf{R} \times S^1$ by the isometric action

$$T_n(x, z) = (x + n, e^{in\theta}z)$$

is isometric to the quotient of $\mathbf{R}^2$ by the lattice generated by $(1, \theta)$ and $(0, 2\pi)$. *Another method:* (more complicated, but instructive). Let ϕ be the diffeomorphism of $\mathbf{R} \times S^1$ defined by $\phi(x, z) = (x, e^{i\theta}z)$: it satisfies

$$\phi(x + n, z) = (x + n, e^{in\theta}e^{ix}z),$$

and hence goes to the quotient and gives a diffeomorphism $\bar{\phi}$ of $(\mathbf{R}/\mathbf{Z}) \times S^1$, which is the quotient of $\mathbf{R} \times S^1$ by the twisted action of $\mathbf{Z}$ we just defined. From the Riemannian point of view, the quotient metric is the pull-back by $\bar{\phi}$ of the product metric of $(\mathbf{R}/\mathbf{Z}) \times (\mathbf{R}/2\pi\mathbf{Z})$. We have

$$\bar{\phi}^*(dx^2 + d\sigma^2) = (1 + \theta^2)dx^2 + 2\theta dx d\sigma + d\sigma^2,$$

and we recover the previous results.
The Klein bottle can be obtained in a similar way, by quotient of $\mathbf{R} \times S^1$ by the isometry group generated by $(x, z) \to (x + 1, -z)$.

2.30 The space H_z is the orthogonal of an i invariant subspace of $\mathbf{R}^{2n+2}$. The multiplication by i goes to the quotient and gives an endomorphism of $T_x P^n \mathbf{C}$. Indeed, it commutes with the multiplication by $u \in S^1$, and $H_{uz} = uH_z$. It can be seen that we get a smooth field of endomorphisms, by using local sections of the fibration $p : S^{2n+1} \to P^n\mathbf{C}$.

2.32 It is immediate when noticing that (S^2, can) and $(P^1\mathbf{C}, \mathrm{can})$ are both homogeneous spaces, hence with constant curvature since these manifolds are 2-dimensional: use then 3.82.
But it is more instructive to consider $(P^1\mathbf{C}, \mathrm{can})$ as the homogeneous space $SU(2)/S^1$, which is isotropy irreducible, and to use the fact that the action of $SU(2)$ on $P^1\mathbf{C} \simeq S^2$ is get from the two fold cover $SU(2) \to SO(3)$ (see for

example [Br1], ch.8). The proportionality coefficient is determined by using 2.31.

2.37 The kernel of L is the subgroup G_0 of G such that $gG_0g^{-1} \subset H$ for any g. In this case, one checks directly that

$$G_0 = \mathrm{diag}(\lambda, \cdots, \lambda)$$

where λ goes through the roots of unity of order $n + 1$.

2.41 Identify $\underline{G/H}$ with the set of matrices of type

$$\begin{pmatrix} 0 & -v_1 & \cdots & -v_n \\ v_1 & 0 & \cdots & 0 \\ \vdots & \vdots & \ddots & \vdots \\ v_n & 0 & \cdots & 0 \end{pmatrix},$$

where $v_i \in \mathbf{C}$. The isotropy representation is the conjugation of these matrices by the $h \in H$. It is hence given by $(\lambda, A)v = \bar{\lambda}Av$, where $(\lambda, A) \in U(1) \times U(n)$, which proves irreducibility. This is still valid when replacing $U(n + 1)$ and $U(1) \times U(n)$ by $SU(n+1)$ and $S(U(1) \times U(n)) \simeq U(n)$ respectively.

2.44 Now $\underline{G/H}$ can be identified with the set of matrices of type

$$\begin{pmatrix} iy & -v_1 & \cdots & -v_n \\ v_1 & 0 & \cdots & 0 \\ \vdots & \vdots & \ddots & \vdots \\ v_n & 0 & \cdots & 0 \end{pmatrix},$$

where $y \in \mathbf{R}$ and $v_i \in \mathbf{C}$. The isotropy representation is hence given by $A \cdot (y, v) = (y, A \cdot v)$. It is decomposed into two irreducible factors, of respective dimensions 1 and $2n$.

It is even easier to understand this geometrically. Take $(1, 0, \cdots, 0)$ in $S^{2n+1} \subset \mathbf{C}^{n+1}$ as base point: the isotropy subgroup of $U(n+1) \subset SO(2n+2)$ preserves $(i, 0, \cdots, 0)$ which represents a tangent vector to the base point.

2.45 a) Let $u \in \mathbf{R}^3$. The orthogonal projection $u' = p_v(u)$ on the line $\mathbf{R}.v$ is characterized by the relations $u' = \lambda v$ ($\lambda \in \mathbf{R}$), and $\langle u - u', v \rangle = 0$. Since $\langle v, v \rangle = 1$, this yields the result.

b) Note first that the map $(x, y, z) \rightarrow (x^2, y^2, z^2, xy, yz, zx)$ is an immersion of $\mathbf{R}^3 \backslash \{0\}$ in $\mathbf{R}^6$: its Jacobian matrix is indeed

$$\begin{pmatrix} 2x & 0 & 0 & y & 0 & z \\ 0 & 2y & 0 & x & z & 0 \\ 0 & 0 & 2z & 0 & y & x \end{pmatrix},$$

which contains three non zero subdeterminants of order 3 (in particular $2x^3$, $2y^3$ and $2z^3$). By restriction to S^2, we get an immersion of the two sphere in $\mathbf{R}^6$. From a), two points $v = (x, y, z)$ and $v' = (x', y', z')$ have the same image under this immersion if only if the projections p_v and $p_{v'}$ are the same, that is if and only if $v = \pm v'$. This map therefore goes to the quotient and yields an injective immersion of $P^2\mathbf{R}$ in $\mathrm{EndSym}(\mathbf{R}) \simeq \mathbf{R}^6$, which is actually an embedding since $P^2\mathbf{R}$ is compact.

c) The group $SO(3)$ acts transitively on S^2 and this action goes to the quotient to give a transitive action on $P^2\mathbf{R}$. The isotropy subgroup of an element of $P^2\mathbf{R}$ (for example of $v = p(1, 0, 0)$, where p is the canonical projection) is the set of $A \in SO(3)$ such that $Av = \pm v$, and can be identified with the group of the matrices of type

$$\begin{pmatrix} \det B & 0 & 0 \\ 0 & & B \\ 0 & & \end{pmatrix},$$

where $B \in O(2)$. From the general properties of homogeneous spaces, $P^2\mathbf{R}$ is hence diffeomorphic to $SO(3)/O(2)$, the group $O(2)$ being embedded in $SO(3)$ as above.

The tangent spaces $T_v S^2$ and $T_{-v} S^2$ are both identified with the yOz plane in $\mathbf{R}^3$. Under these identifications, the differential map at v of B acting on S^2, is a linear map from $T_v S^2$ to itself when $\det B = 1$, and to $T_{-v} S^2$ when $\det B = -1$, which is given by $B \begin{pmatrix} y \\ z \end{pmatrix}$.

This proves that the isotropy representation of $SO(3)/O(2) = P^2\mathbf{R}$, which is obtained by differentiating at $p(v)$ the action of B on $P^2\mathbf{R}$, is also identified to the natural action of B on $\mathbf{R}^2$. In particular, $SO(3)/O(2)$ is an isotropy irreducible homogeneous space.

d) One checks easily that $\mathrm{tr}\,^t AA = \sum_{i,j} a_{ij}^2$. If $Q \in SO(3)$, $^t(Q^{-1}AQ) = Q^{-1\,t}AQ$, hence

$$\mathrm{tr}(^t c_Q(A).c_Q(A)) = \mathrm{tr}(Q^{-1\,t}AAQ) = \mathrm{tr}(^t AA).$$

We hence defined an isometric action of $SO(3)$ on $\mathrm{End}(\mathbf{R}^3)$ and $\mathrm{EndSym}(\mathbf{R}^3)$, for the metric given by $\mathrm{tr}(^t AA)$ (recall that for $Q \in O(3)$, then $c_Q(\mathrm{EndSym}) \subset \mathrm{EndSym}$). If $A = p_v$ is the orthogonal projection on v, $c_Q(A)$ is the orthogonal projection p_{Qv} on Qv. This proves that the c_Q preserve $V(P^2\mathbf{R})$. Since they are isometries of $\mathrm{EndSym}(\mathbf{R})$, they are also isometries of $V(P^2\mathbf{R})$ for the induced metric. Therefore the map $Q \to c_Q$ is just the transport by V on $V(P^2\mathbf{R})$ of the isometric action of $SO(3)$ on $P^2\mathbf{R}$.

Hence $V(P^2\mathbf{R})$, equipped with the induced metric, is still an homogeneous Riemannian space $SO(3)/O(2)$, but such a space has only one metric up to a scalar. To find out the coefficient of proportionality, just compare the length of a curve in $P^2\mathbf{R}$ with the length of its image in $V(P^2\mathbf{R})$. Let $c : t \to p(\cos t, \sin t, 0)$. Then

$$V(c(t)) = \begin{pmatrix} \cos^2 t & \cos t.\sin t & 0 \\ \cos t.\sin t & \sin^2 t & 0 \\ 0 & 0 & 0 \end{pmatrix}.$$

It is easy to check that $\text{length}(c) = \frac{1}{\sqrt{2}}\text{length}(V \circ c) = \pi$, which proves that V is a dilation whose ratio is $\left(\frac{1}{\sqrt{2}}\right)^2 = \frac{1}{2}$.

2.47 Let us treat the case of $U(n)$. We have

$$\text{tr}(XY) = \text{tr}(^tX^tY) = \text{tr}(^tY^tX) = \text{tr}(YX).$$

If $X = (x_{ij})_{(i,j=1,\cdots,n)}$, $-\text{tr}(X^2) = \sum_{i,j} \mid x_{ij} \mid^2$ (direct computation). Finally, the invariance under Ad is equivalent to the invariance of the trace under conjugation.

2.55 a) Let r and θ be the polar coordinates in $\mathbf{R}^2\backslash\{0\}$. Then $X = \frac{\partial}{\partial\theta}$ and $Y = \frac{\partial}{\partial r}$, hence $[X,Y] = 0$ and $D_X Y = D_Y X$. We compute directly $D_Y X$ in cartesian coordinates and find $D_Y X = \frac{X}{r} = \frac{X}{\|X\|}$.

b) For two connections D and D', the map $(X,Y) \to D_X Y - D'_X Y$ is also $C^\infty(M)$ linear with respect to Y, from the very definition of the connections.

c) Applying (2.52) to D and $\tilde{D}$, we get

$$2\tilde{g}(\tilde{D}_X Y, Z) - 2f^2 g(D_X Y, Z) = X \cdot (f^2 g(Y,Z)) + Y \cdot (f^2 g(Z,X))$$

$$-Z \cdot (f^2 g(X,Y)) - f^2 X \cdot g(Y,Z) - f^2 Y \cdot g(Z,X) + f^2 Z \cdot g(X,Y),$$

and finally

$$\tilde{D}_X Y = D_X Y + \left(\frac{1}{f}\right)[df(X)Y + df(Y)X - g(X,Y)\nabla f],$$

where ∇ is the *gradient* for the metric g.

2.57 a) We use 2.56 to avoid the computation of the Christoffel symbols. We find that

$$D_{\frac{\partial}{\partial\theta}}\frac{\partial}{\partial\theta} = 0, \quad D_{\frac{\partial}{\partial\phi}}\frac{\partial}{\partial\phi} = -\frac{1}{2}\sin(2\theta)\frac{\partial}{\partial\theta},$$

$$D_{\frac{\partial}{\partial\phi}}\frac{\partial}{\partial\theta} = D_{\frac{\partial}{\partial\theta}}\frac{\partial}{\partial\phi} = -\tan\theta\frac{\partial}{\partial\phi}.$$

b) The vector field $[\frac{\partial}{\partial\theta}, \frac{\partial}{\partial\phi}]$, which depends only on the differentiable structure, is zero in the two cases. We use the same method as in a) to find that for ϕ_1, $D_{\frac{\partial}{\partial\phi}}\frac{\partial}{\partial\theta} = 0$,

for ϕ_2, $D_{\frac{\partial}{\partial\phi}}\frac{\partial}{\partial\theta} = \frac{-\sin\theta}{2+\cos\theta}\frac{\partial}{\partial\phi}$.

2.63 a) Using the "Einstein convention" (that is we sum with respect to the so called "dumb" indices, which occur both in upper and lower position), we have:

$$(L_X g)_{ij} = g_{ik}\partial_j X^k + g_{jk}\partial_i X^k + g_{il}\Gamma^l_{kj}X^k + g_{jl}\Gamma^l_{ki}X^k.$$

b) It is an immediate consequence of the definition of the Lie derivative.

c) The flow of such a vector field is $\exp tA.x$, where A is the matrix given by $A_{ij} = -A_{ji} = 1$ $(i \neq j)$, the other coefficients being zero. Therefore this flow yields a group of isometries of $\mathbf{R}^{n+1}$ and of S^n.

2.65 a) Using the same convention as in 2.63 a), we have

$$(Ddf)_{ij} = \partial^2_{ij}f - \Gamma^k_{ij}\partial_k f.$$

b) Let $\tilde{g}$ be the Euclidean metric of $\mathbf{R}^{n+1}$: we have $\tilde{g} = dr^2 + r^2\bar{g}$. Since the Euclidean gradient $\tilde{\nabla}f$ generates a group of translations, $L_{\tilde{\nabla}f}\tilde{g} = 0$. Hence

$$L_{\tilde{\nabla}f}dr \otimes dr + dr \otimes L_{\tilde{\nabla}f}dr + 2rdr(\tilde{\nabla}f)\bar{g} + r^2 L_{\tilde{\nabla}f}\bar{g} = 0.$$

By restriction to S^n, this yields

$$L_{\nabla f}g = -2dr(\tilde{\nabla}f)g = -2fg,$$

since f is linear. But $L_{\nabla f}g = 2Ddf$, from 2.62.

We prove similarly for the hyperboloid model of H^n that the restriction to H^n of a linear form on $\mathbf{R}^{n+1}$ satisfies $Ddf = fg$.

Remark: We denoted by $\bar{g}$ the symmetric tensor of order 2, which is the dilation invariant extension to $\mathbf{R}^{n+1}$ of the metric g of S^n.

2.83 a) The development ϕ of the cone is an isometry between the cone with a meridian line taken away and an open angular sector in $\mathbf{R}^2$. The isometry ϕ exchanges the metrics, hence the connections and the geodesics. If c is a geodesic of the cone, then $\phi \circ c$ is a line segment in $\mathbf{R}^2$.

The loop of the lasso keeps its balance under the normal reaction of the cone: hence it yields on the cone a curve with normal acceleration vector field, that is a geodesic of the cone. This curve is symmetric with respect to the plane generated by the axis of the cone and the free strand of the lasso on which the traction operates. Cut the cone along the generatrix containing this strand, and develop: the loop yields a line segment of length L, normal to the axis

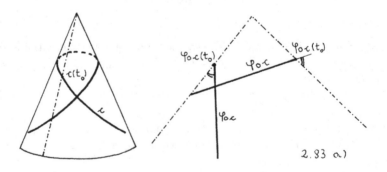

Fig. B.4. A geodesic on a revolution cone

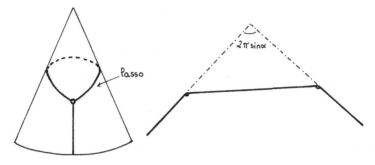

Fig. B.5. Catching a cone with a lasso

of the plane sector, two conditions which determine entirely its position. The lasso keeps its balance when $2\pi \sin \alpha < \pi$, that is when $\alpha < \pi/6$ (i.e. when the vertex angle is less than $\pi/3$).

b) Since the canonical projection p from $(\mathbf{R}^2, \text{can})$ to (K, can) is a Riemannian covering map, it lifts the parallel transport along the geodesics of the flat Klein bottle to the parallel transport along the corresponding geodesics in the plane. Assume that all the geodesics are parametrized by arclength.

The geodesic is periodic of period b, and the parallel transport along c between $c(t)$ and $c(t+b)$ is the identity of $T_{c(t)}K = T_{c(t+b)}$.

The geodesic γ_1 is periodic of period a, and the parallel transport along γ_1 between $\gamma_1(t)$ and $\gamma_1(t+a)$ preserves $\gamma_1'(t) = \gamma_1'(t+a)$ and reverses the orientation of $T_{\gamma_1(t)}K$.

The geodesic γ is periodic of period $2a$ and the parallel transport along γ between $\gamma(t)$ anf $\gamma(t+2a)$ is the identity of $T_{\gamma(t)}K$. See figure B.6.

c) If c is a geodesic of (M, g), then c' is parallel along c. Let $v \in T_{c(0)}M$ such that $\angle(c'(0), v) = \alpha$. Then the parallel transport $X(t)$ of v at $c(t)$ is also

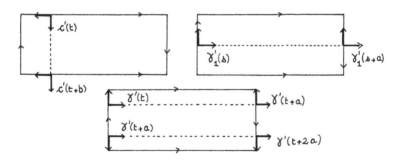

Fig. B.6. The geodesics of the Klein bottle

such that $\angle(X(t), c'(t)) = \alpha$. Since the manifold is 2 dimensional, this means that $X(t)$ can take only two values. If we assume also that the manifold is orientable, and oriented, the parallel transport preserves the orientation of the tangent plane. Let indeed (U, ϕ) be one of the charts of the orientation atlas of M, and $\frac{\partial}{\partial x^i}$ $(i = 1, 2)$ be the corresponding coordinate vector fields. If v and w are two linearly independant vectors in $T_{c(0)}M$, and if $X(t)$ and $Y(t)$ are their respective parallel transports along c, we can write

$$X(t) = X_1(t)\frac{\partial}{\partial x^1} + X_2(t)\frac{\partial}{\partial x^2} \quad \text{and} \quad Y(t) = Y_1(t)\frac{\partial}{\partial x^1} + Y_2(t)\frac{\partial}{\partial x^2}.$$

Since v and w are linearly independant, and X and Y are parallel, the determinant

$$\begin{vmatrix} X_1(t) & Y_1(t) \\ X_2(t) & Y_2(t) \end{vmatrix}$$

is never zero, and hence keeps a constant sign.

We finally proved that on a 2-dimensional orientable manifold, the parallel transport $X(t)$ of v along c is the unique vector in $T_{c(t)}M$ of same length as v such that $\angle(X(t), c'(t)) = \alpha$ and $(c'(t), X(t))$ has the same orientation as $(c'(0), X(0))$.

d) i) Let (e_1, e_2) be the canonical basis of $\mathbf{R}^2 = \mathbf{C}$. In the chart $\mathrm{Id} : P \to \mathbf{R} \times \mathbf{R}^+$, the coordinate vector fields are $\frac{\partial}{\partial x^i}$ $(i = 1, 2)$. The Christoffel symbols can be computed with formula 2.54. Since $g_{11} = g_{22} = \frac{1}{y^2}$ and $g_{12} = 0$, then $g^{11} = g^{22} = y^2$ and $g^{12} = 0$. This yields:

$$\Gamma_{11}^1 = 0, \quad \Gamma_{11}^2 = \frac{1}{2}g^{22}\left(-\frac{\partial g_{11}}{\partial y}\right) = \frac{1}{y},$$

$$\Gamma_{12}^1 = \frac{1}{2}g^{11}\frac{\partial g_{11}}{\partial y} = \frac{-1}{y}, \quad \Gamma_{12}^2 = 0,$$

$$\Gamma_{22}^1 = 0, \quad \text{and} \quad \Gamma_{22}^2 = \frac{1}{2}g^{22} = \frac{-1}{y},$$

hence the following covariant derivatives:

$$D_{e_1}e_1 = \frac{e_2}{y}, \; D_{e_1}e_2 = D_{e_2}e_1 = \frac{-e_1}{y}, \quad \text{and} \quad D_{e_2}e_2 = \frac{-e_2}{y}.$$

ii) Let c be the curve of P defined by $c(t) = (x_0, e^{at})$. Then $c'(t) = ae^{at}.e_2$, hence

$$D_{c'}c' = a^2 e^{at}.e_2 + ae^{at}\left[ae^{at}D_{e_2}e_2\right] = 0,$$

and c is a geodesic of P. Alternative argument: it is clear from the explicit formula for the metric that vertical lines are length-minizing.

Note that $Sl(2, \mathbf{R})$ acts transitively and isometrically on P by $z \to \frac{az+b}{cz+d}$, and hence that

$$P = Sl_2\mathbf{R}/\pm \mathrm{Id} \simeq SO_0(1, 2).$$

The isotropy group of the point i is isomorphic to $SO(2) \simeq S^1$ and, if $f \in SO(2)$ is represented by $\theta \in S^1$, $T_i f$ is the rotation of angle 2θ in $T_i P$. Hence the group $Sl(2, \mathbf{R})$ acts transitively on the unit bundle UP: all the geodesics of P are get as the images, under the action of $Sl_2\mathbf{R}$, of the particular geodesics described in i) (uniqueness theorem). The geodesics of P are hence of type

$$c(t) = \left[x_0 + r \tanh(at + b), \frac{r}{\cosh(at + b)} \right].$$

e) *Geometric proof:* Let us assume that the rotation axis is vertical. The meridians, when parametrized proportional to length, are geodesics: indeed, $c(t)$ remains in a given plane containing the rotation axis, hence $\langle c'', \frac{\partial}{\partial \theta} \rangle = 0$. On the other hand, since $| c'(t) |$ is constant, we have $\langle c''(t) = c'(t) \rangle = 0$, and the acceleration vector field is normal to the surface.

The parallels (that is $u =$constant), when parametrized proportionally to arclength, are geodesics if and only if $a'(u) = 0$. Since the norm of $c'(t)$ is constant, we have

$$\langle c''(t), c'(t) \rangle = 0$$

and hence $c''(t)$ is normal to $\frac{\partial}{\partial \theta}$. The curve c being horizontal, $c''(t)$ is also horizontal: it will be normal to $\frac{\partial}{\partial u}$ if and only if $\frac{\partial}{\partial u}$ is vertical, that is if $a'(u) = 0$.

In the general case, a normal geodesic $c(t) = (u(t), \theta(t))$ is such that $c''(t)$ belongs to the plane containing the axis and the point $c(t)$. The projection of c on an horizontal plane is a curve γ with central acceleration (that is $\gamma''(t)$ parallel to $O\gamma(t)$). Therefore

$$| \gamma(t) |^2 \cdot \frac{d\theta}{dt} = C \quad \text{(compare with Kepler area law)},$$

where C is a constant depending on the geodesic c. This yields, since $| \gamma(t) | = a(u)$,

$$a^2 (u(t)) \cdot \frac{d\theta}{dt} = C \quad \text{and} \quad \left(\frac{du}{dt} \right)^2 + a^2 (u(t)) \left(\frac{d\theta}{dt} \right) = 1.$$

These two conditions can be rewritten as

$$\frac{d\theta}{dt} = \frac{C}{a^2 (u(t))} \quad \text{and} \quad \frac{du}{dt} = \sqrt{1 - \frac{C^2}{a^2 (u(t))}}.$$

The solutions of this system being entirely determined as soon as $u(0)$ and $\theta(0)$ are given, we deduce from uniqueness and existence theorem 2.79 that the geodesics of M are exactly the solutions of this differential system.

e) *Analytic proof:* We do not need any more to assume that the revolution surface is embedded in $\mathbf{R}^3$. The same argument can be applied to any surface M parametrized by $F : I \times \mathbf{R} \to M$, with $F(u, \theta + 2\pi) = F(u, \theta)$ and $g = du^2 + a^2(u)d\theta^2$. We first compute the connection of (M, g). On one hand,

$$g\left(D_{\frac{\partial}{\partial u}}\frac{\partial}{\partial \theta},\frac{\partial}{\partial \theta}\right) = \frac{1}{2}\frac{\partial}{\partial u}\cdot g\left(\frac{\partial}{\partial \theta},\frac{\partial}{\partial \theta}\right) = a'(u)a(u),$$

$$g\left(D_{\frac{\partial}{\partial u}}\frac{\partial}{\partial \theta},\frac{\partial}{\partial u}\right) = g\left(D_{\frac{\partial}{\partial \theta}}\frac{\partial}{\partial u},\frac{\partial}{\partial u}\right) = \frac{1}{2}\frac{\partial}{\partial \theta}\cdot g\left(\frac{\partial}{\partial u},\frac{\partial}{\partial u}\right) = 0$$

$$\text{hence}\quad D_{\frac{\partial}{\partial u}}\frac{\partial}{\partial \theta} = \frac{a'(u)}{a(u)}\frac{\partial}{\partial \theta}.$$

On the other hand,

$$g\left(D_{\frac{\partial}{\partial u}}\frac{\partial}{\partial u},\frac{\partial}{\partial u}\right) = \frac{1}{2}\frac{\partial}{\partial u}\cdot g\left(\frac{\partial}{\partial u},\frac{\partial}{\partial u}\right) = 0,$$

$$g\left(D_{\frac{\partial}{\partial u}}\frac{\partial}{\partial u},\frac{\partial}{\partial \theta}\right) = \frac{\partial}{\partial u}\cdot g\left(\frac{\partial}{\partial u},\frac{\partial}{\partial \theta}\right) - g\left(D_{\frac{\partial}{\partial u}}\frac{\partial}{\partial \theta},\frac{\partial}{\partial u}\right) = 0,$$

$$\text{hence}\quad D_{\frac{\partial}{\partial u}}\frac{\partial}{\partial u} = 0 :$$

we recover the fact that the meridian lines are geodesics.
Finally,

$$g\left(D_{\frac{\partial}{\partial \theta}}\frac{\partial}{\partial \theta},\frac{\partial}{\partial \theta}\right) = \frac{1}{2}\frac{\partial}{\partial \theta}\cdot g\left(\frac{\partial}{\partial \theta},\frac{\partial}{\partial \theta}\right) = 0,$$

$$g\left(D_{\frac{\partial}{\partial \theta}}\frac{\partial}{\partial \theta},\frac{\partial}{\partial u}\right) = \frac{\partial}{\partial \theta}\cdot g\left(\frac{\partial}{\partial u},\frac{\partial}{\partial \theta}\right) - g\left(D_{\frac{\partial}{\partial \theta}}\frac{\partial}{\partial u},\frac{\partial}{\partial \theta}\right) = -a'(u)a(u),$$

$$\text{that is}\quad D_{\frac{\partial}{\partial \theta}}\frac{\partial}{\partial \theta} = -a'(u)a(u)\frac{\partial}{\partial u}.$$

A curve $c(t) = (u(t),\theta(t))$ is a geodesic if and only if $D_{c'}c' = 0$, that is

$$\frac{d^2 u}{dt^2} - a'(u)a(u)\left(\frac{d\theta}{dt}\right)^2 = 0,$$

$$\text{and}\quad \frac{d^2\theta}{dt^2} + 2\frac{a'(u)}{a(u)}\frac{d\theta}{dt}\frac{du}{dt} = 0.$$

(At this point, it is clear that the Hamiltonian method of 2.123 is much better!)
We recover the fact that the meridians lines, when parametrized proportional
to length ($\frac{d\theta}{dt} = 0$, $\frac{d^2 u}{dt^2} = 0$), are geodesics. The parallels, when parametrized
proportional to length, are geodesics if and only if $a'(u)a(u) = 0$, since $\frac{d^2\theta}{dt^2} = \frac{du}{dt} = 0$.
Integrating the second equation, we get

$$a^2(u(t))\frac{d\theta}{dt} = C.$$

Assuming that the geodesic is parametrized by length, we get a new system:

$$\left(\frac{du}{dt}\right)^2 + a^2(u(t))\left(\frac{d\theta}{dt}\right)^2 = 1,$$

$$\text{and} \quad a^2(u(t))\frac{d\theta}{dt} = C.$$

One proves easily that this system is equivalent to the first one. Eventually reversing the parametrization, we can assume that $\frac{d\theta}{dt} > 0$, so that C is positive. Since

$$a^2(u(t)) \cdot \left(\frac{d\theta}{dt}\right)^2 \le 1,$$

we always have $a(u(t)) \ge C$, with equality at time s if and only if $\frac{du}{dt}(s) = 0$. Generically, the geodesic oscillates between two consecutive parallels satisfying to $a(u) = C$ (fig.i), to which it is tangent. If one of these parallels is extremal (that is $a'(u) = 0$), the geodesic is asymptotic to this parallel (fig.ii), which is itself a geodesic.

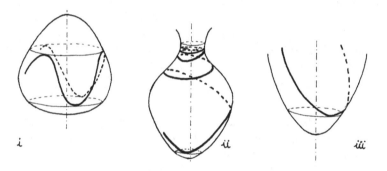

Fig. B.7. Geodesics on a revolution surface

2.90 a) Let $\langle,\rangle$ be a bi-invariant metric on G: it yields by restriction to T_eG an AdG-invariant scalar product. Let indeed $u, v \in T_eG$, and $g \in G$. Then

$$\langle gug^{-1}, gvg^{-1}\rangle_e = \langle D_{g^{-1}}L_g \cdot ug^{-1}, D_{g^{-1}}L_g \cdot vg^{-1}\rangle_e$$

$$= \langle ug^{-1}, vg^{-1}\rangle_{g^{-1}} = \langle u, v\rangle_e \quad,$$

since L_g and $R_{g^{-1}}$ are isometries.
Conversely, any scalar product $\langle,\rangle$ on T_eG comes from the left invariant metric on G defined for $g \in G$ by

$$\langle v, v\rangle_g = \langle X(e), X(e)\rangle_e \quad,$$

where X is the unique left invariant vector field with $X(g) = v$. This metric is clearly left invariant. If $\langle, \rangle_e$ is $\mathrm{Ad}G$-invariant, then for $v \in T_eG$:

$$\langle R_{g^{-1}}(v), R_{g^{-1}}(v) \rangle_e = \langle R_{g^{-1}}L_g.X(e), R_{g^{-1}}L_g.X(e) \rangle_e$$

$$\langle X(e), X(e) \rangle = \langle v, v \rangle_g \quad :$$

hence the metric $\langle, \rangle$ is also right invariant.

b) The map $g \to g.g^{-1}$ is constant, and its differential map is identically zero. For $g \in G$ and $v \in T_gG$, we have then

$$D_g i.v = -g^{-1}.v.g^{-1},$$

where we *denoted*, for $g, h \in G$ and $v \in T_hG$:

$$D_h L_g.v = g.v \quad \text{and} \quad D_h R_g.v = v.g.$$

Hence i is an isometry of $(G, \langle, \rangle)$, and $D_e i = -\mathrm{Id}$. The image under i of the geodesic c is then the unique geodesic γ with $\gamma(0) = e$ and $\gamma'(0) = -u$. Since by construction $\gamma(t) = c(t)^{-1}$ (for t in the interval of definition of c), we deduce that $c(-t) = c(t)^{-1}$.

Denote by $I =] - a, a[$ the maximal interval of definition of c. From 2.88, we know that there exists $\epsilon > 0$ such Exp_e is a diffeomorphism from $B(0, \epsilon) \subset T_eG$ on its image. Chose $0 < t_0 < \epsilon$: the left translation by $c(t_0)$ being an isometry, the curve Γ defined by $\Gamma(s) = c(t_0)c(s)$ is also a geodesic defined on $] - a, a[$, with

$$\Gamma(-t_0) = c(t_0)c(-t_0) = e \quad \text{and} \quad \Gamma(0) = c(t_0).$$

Now the uniqueness theorem claims that, for $0 \leq s \leq t_0$, $\Gamma(s) = c(t_0 + s)$: if $a < \infty$, the geodesic Γ extends c outside I, which is not maximal: finally c is defined on the whole $\mathbf{R}$, and more

$$\forall t_0 / \mid t_0 \mid < \epsilon, \quad \forall s \in \mathbf{R}, \quad c(t_0 + s) = c(t_0)c(s).$$

Now, for $t \in \mathbf{R}$, chose $n \in \mathbf{N}$ with $\left|\frac{t}{n}\right| < \epsilon$, and write:

$$c(t + s) = c(\frac{t}{n} + \cdots + \frac{t}{n} + s) = c(\frac{t}{n})c(\frac{n-1}{n}t + s) = \cdots = c(t)c(s) \quad :$$

the map $c : (\mathbf{R}, +) \to (G, .)$ is indeed a group homomorphism.

c) Let c be a geodesic with $c(0) = e$ and $c'(0) = u$, and X be the left invariant vector field such that $X(e) = u$. Differentiating the relation $c(t+s) = c(t)c(s)$, we get successively

$$\frac{d}{ds}(c(t + s))_{|s=0} = c(t).\frac{d}{ds}(c(s))_{|t=0} \quad ,$$

that is $\quad c'(t) = c(t).c'(0) = X(c(t)).$

Hence c is the integral curve of X with $c(0) = e$. The geodesics of G are the images under left translations of the geodesics through e. Hence all the geodesics of G are defined on the whole $\mathbf{R}$ and are exactly the integral curves of the left invariant vector fields on G.

d) If X is a left invariant vector field, then $D_X X = 0$, since its integral curves are geodesics. If Y is another left invariant vector field, we can write:

$$0 = D_{(X+Y)}(X + Y) = D_X X + D_Y Y + D_X Y + D_Y X = D_X Y + D_Y X,$$

hence, since $[X, Y] = D_X Y - D_Y X$,

$$D_X Y = \frac{1}{2}[X, Y].$$

2.90 bis a) It will be more convenient to denote the elements of H as $m = (x, y, z)$, the multiplication being

$$(x, y, z) \times (x', y', z') = (x + x', y + y', z + z' + xy').$$

If ϕ is the left translation by (a, b, c), the image under ϕ of the vector field B is the vector field

$$m \to (\phi_* B)_m = T_{\phi^{-1}(m)}\phi \cdot B_{\phi^{-1}(m)}.$$

The value of B at $\phi^{-1}(m) = (x - a, y - b, z - c + ab - ay)$ is $(0, 1, x - a)$. Since, at any point of $H \simeq \mathbf{R}^3$,

$$T\phi(\xi, \eta, \mu) = (\xi, \eta, \mu + a\eta),$$

this yields that $(\phi_* B)_m = B_m$. For the fields A and C, the computations are even simpler. It is immediate to prove that $[A, B] = C$, $[A, C] = [B, C] = 0$. Of course, the Lie algebra of vector fields on $\mathbf{R}^3$ generated by A, B and C is isomorphic to the Lie algebra of the group H, the images of A, B and C by this isomorphism being the matrices $E_{1,2}$, $E_{2,1}$ and $E_{1,3}$.

b) By construction, the metric g is left invariant. If ψ is the *right* translation by (a, b, c), a computation similar to the previous one shows that $\psi_* B = B - aC$: ψ is not automatically an isometry.

We can also note that, with respect to the standard coordinates (x, y, z) in $\mathbf{R}^3$,

$$g = dx^2 + dy^2 + (dz - x\,dy)^2.$$

If g was bi-invariant, the map $i : m \to m^{-1}$ would be an isometry. But

$$i(x, y, z) = (-x, -y, xy - z), \quad \text{hence} \quad i^*g = dx^2 + dy^2 + (dz - y\,dx)^2.$$

c) Let X, Y and Z be each any one of the three fields A, B and C. Their pairwise scalar products are constant, hence

$$2g(D_XY, Z) = g([X, Y], Z) + g([Z, X], Y) - g([Y, Z], X),$$

therefore $D_AB = -D_BA = \frac{1}{2}C$, $D_BC = -D_CB = \frac{1}{2}A$, $D_CA = -D_AC = \frac{-1}{2}B$, and $D_AA = D_BB = D_CC = 0$. Let $c(t) = ((x(t), y(t), z(t))$ be a curve on H. Then

$$c' = (x', y', z') = x'A + y'B + (z' - xy')C,$$

hence $\quad D_{c'}c' = x''A + x'D_{c'}A + y''B + y'D_{c'}B + z''C + z'D_{c'}C$

$$= x''A - \left(\frac{x'}{2}\right)(y'C + (z' - xy')B) + y''B + \left(\frac{y'}{2}\right)(x'C + (z' - xy')A)$$

$$+ (z' - xy')'C + \frac{1}{2}(z' - xy')(-x'B + y'A).$$

The equation of geodesics is given by the differential system

$$x'' + y'(z' - xy') = y'' - x'(z' - xy') = (z' - xy')' = 0.$$

Recall that, if we assume that the geodesics are parametrized by length,

$$g(c', c') = x'^2 + y'^2 + (z' - xy')^2 = 1.$$

It is sufficient to know the geodesics begining at the unit. It will be convenient to write the initial tangent vector v as

$$(\cos\theta \cos\phi, \sin\theta \cos\phi, \sin\phi).$$

The system can then be written as:

$$x'' + (\sin\phi)y' = y'' - (\sin\phi)x' = 0, \quad z' - xy' = \sin\phi,$$

with initial conditions $x(0) = y(0) = z(0) = 0$, $x'(0) = \cos\theta \cos\phi$ and $y'(0) = \sin\theta \cos\phi$. For $\sin\phi \neq 0$, we find

$$x(t) = \cot\phi\left[\sin(t\sin\phi + \theta) - \sin\theta\right], \quad y(t) = \cot\phi\left[\cos\theta - \cos(t\sin\phi + \theta)\right],$$

$$z = \frac{1}{2}\left(\sin\phi + \frac{1}{\sin\phi}\right)t - \frac{1}{4}\cot^2\phi\left[\sin 2(t\sin\phi + \theta) - \sin 2\theta\right]$$

$$+ \sin\theta \cot^2\phi\left[\cos(t\sin\phi + \theta) - \cos\theta\right].$$

The projections of these geodesics on the xOy plane are circles of radius $\cot\phi$ through the origin.

For $\cot\phi = 0$ (that is $v = (0, 0, \pm 1)$), we find that $x = y = 0$ and $z = \pm t$.

For $\sin\phi = 0$, we find that

$$x = t\cos\theta, \quad y = t\sin\theta, \quad z = \frac{1}{2}t^2 \cos\theta \sin\theta.$$

These geodesics, if $\cos\theta\sin\theta \neq 0$, are vertical parabolas which project on the horizontal plane as straight lines through the origin.

On the other hand, we can see by a direct computation that the one parameter group with infinitesimal generator $\alpha E_{1,2} + \beta E_{2,1} + \gamma E_{1,3}$ is

$$t \rightarrow (t\alpha, t\beta, t\gamma + \frac{t^2}{2}\alpha\beta).$$

We only get a geodesic if either $\gamma = 0$ (that is $\sin\phi = 0$), or $\alpha = \beta = 0$ (that is if $\cos\phi = 0$).

Conclusion: In the case of a Lie group equipped with a left invariant metric, which is not bi-invariant, the Riemannian exponential map and the exponential map in the Lie group do not always coincide.

d) Since the subgroup Z is the center of H, it is clear that H/Z is isomorphic to $\mathbf{R}^2$, and that the unique metric on $\mathbf{R}^2$ such that $p : H \rightarrow H/Z$ is a Riemannian submersion is translation invariant, hence flat. It is given by $dx^2 + dy^2$ at the origin, hence everywhere. For $m \in H$, the tangent space to the fiber at m is $\mathbf{R}C_m$; the fibers are geodesics, since $D_C C = 0$. The equation of the horizontal space, that is of the plane normal to C is $dz - x dy = 0$. The horizontal geodesics are then those for which $\sin\phi = 0$. Note that, in the equation of the geodesics, the condition $(z' - xy')' = 0$ says that a geodesic, which is horizontal at one point, is horizontal everywhere. Note also that the only geodesics from the unit which are also one parameter subgroups, are the horizontal and the vertical ones.

2.108 a) Since $T_m f$ is an isometry between $T_m M$ and $T_{f(m)} N$, it is a fortiori a vector spaces isomorphism, hence f is a local diffeomorphism and a local isometry. Proposition 2.106 claims that the map $f : M \rightarrow N$ is a covering map. But N is simply connected and M is connected, so f is a diffeomorphism on its image, which is equal to N since open and closed in N which is connected. This yields the conclusion.

b) Let (M, g) be a Riemannian homogeneous space, and $m \in M$. From 2.85, we know that $\exp_m$ is defined on $B(0_m, r)$, for r small enough. By hypothesis, the isometry group of (M, g) acts transitively on M, and hence for any $p \in M$, the map $\exp_p$ is also defined on the ball $B(0_p, r)$.

Let now c be a geodesic from m, and $I =]a, b[$ be its maximal interval of definition. If for example $b < \infty$, set $p = c(b - \frac{r}{2})$. The geodesic γ with initial conditions $\gamma(0) = p$ and $\gamma'(0) = c'(b - \frac{r}{2})$ extends c and is at least defined on the interval $] - r, r[$: c is defined on $]a, b + \frac{r}{2}[$, a contradiction.

c) Let G be a compact Lie group. From 2.90, we know that there exists a bi-invariant metric g on G, for which the maps $\exp$ (in the group sens) and Exp_e (of the Riemannian manifold) coincide. The metric space (G, d) is complete, since d defines on the compact group G its usual topology. Corollary 2.105 claims that the map $\mathrm{Exp}_e = \exp$ is surjective.

d) Assume there exists a bi-invariant Riemannian metric g on $SL_2\mathbf{R}$. Then, from 2.90 c), the map $\mathrm{Exp}_{\mathrm{Id}}$ is defined on $T_{\mathrm{Id}} SL_2\mathbf{R} = sl_2\mathbf{R}$, and coincide with the matrix exponential map. Hopf-Rinow's theorem claims that the map $\mathrm{Exp}_{\mathrm{Id}}$ is surjective. We get a contradiction by noticing that a matrix of $SL_2\mathbf{R}$ with trace less than -2 is not the exponential of a matrix of $sl_2\mathbf{R}$ ($sl_2\mathbf{R}$ is the set of traceless matrices of order two: just consider their eigenvalues...)

e) Two distinct geodesics from the same point $p(x)$ do not meet before time π, where they come back to $p(x)$ (2.82 c). Note also that if c is a normal geodesic, the curves $t \to c(t)$ and $t \to c(-t)$ ($t \in [0, \frac{\pi}{2}]$) are the only normal geodesics with length less or equal to $\frac{\pi}{2}$ from $c(0)$ to $c(\frac{\pi}{2})$. Since the manifold is compact, Hopf-Rinow's theorem claims that c is minimal between $c(0)$ and $c(\frac{\pi}{2})$, but not on any larger interval.

2.118 Since M is compact, the supremum of $d(x, y)$ is achieved. If p and q are two points such that $d(p, q) = \mathrm{diam}(M, g) = D$, Hopf Rinow theorem tells us that there is at least one geodesic segment from p to q. Denote by v the unit tangent vector at q of this geodesic. For any positive ϵ, there is a minimizing geodesic from p to $\exp_q v$, whose length is between $D - \epsilon$ and D. By compacity, there is a sequence (ϵ_k), $\epsilon_k \to 0$, such that the corresponding sequence geodesics converges to some geodesic segment from p to q. If this segment was the first one, we could build for big enough k a path from q to p with length strictly smaller than D, a contradiction.

2.127 The restriction to a geodesic c of this function is just $t \mapsto g_{c(t)}\big(X_{c(t)}, c'(t)\big)$, whose derivative is $g_{c(t)}\big(D_{c'(t)}X_{c(t)}, c'(t)\big) = DX^\flat\big(c'(t), c'(t)\big)$. It vanishes for every geodesic if and only if the bilinear form $DX^\flat$ is antisymmetric, that is if X is a Killing field.

2.132 For an unparemetrized geodesic $t \mapsto c(t)$, $D_{c'}c'$ is colinear to c'. This property is conformally invariant *for light-like curves*.

2.136 We use the technique of 2.25 e), and define a $\mathbf{Z}$-action on $(T^2 \times \mathbf{R}, g + dt^2)$ by

$$n \cdot (m, z) = (\overline{A}^n \cdot m, z + n).$$

Here g is the flat metric we have just defined. This action is clearly free, proper and isometric. The quotient is a flat compact Lorentzian manifold T_A^3 whose fondamental group is the semi-direct product $\mathbf{Z} \times_A \mathbf{Z}^2$. Making A vary, we obtain infinitely many non isomorphic groups.

2.140. It suffices to prove that, for any initial data (m_0, v_0) is contained in a compact submanifold of TM which is invariant under the geodesic flow.

First, it is easy to find a basis $(w_1, w_2, \cdots w_n)$ of $T_{m_0}M$ and numbers $(a_1, \cdots, a_n)$ such that

$$\{v, v \in T_{m_0}M, g_{m_0}(v, v) = g_{m_0}(v_0, v_0) \quad \text{and} g_{m_0}(v, w_k) = a_k \, (1 \le k \le n)\}$$

is a compact submanifold of $T_{m_0}M$.

From the lemma, we know that there are Killing fields $(X_1, \cdots, X_n)$ such that $(X_k)_{m_0} = w_k$ for any k. Now, the set

$$\{v_m, v_m \in TM, g_m(v_m, v_m) = g_{m_0}(v_0, v_0) \quad \text{and} \quad g(v_m, (X_k)_m) = a_k \, (1 \le k \le n)\}$$

is a compact submanifold of TM (the basis and the fiber are both compact submanifolds). It is left invariant by the geodesic flow according to 2.127.

2.143 The Hamiltonian is $H = -\frac{1}{2}f(x)p_x^2 + 2p_x p_y$, and we have the system

$$\frac{dp_x}{dt} = -f'(x)p_x^2 \quad \frac{dp_y}{dt} = 0$$

Coming back to the tangent space, we get $p_x = y'$ and $p_y = x' + f(x)y'$, Take a light-like geodesic such that $2x' + f(x)y' = 0$. Then x' is a non zero constant c, so that for this geodesic

$$2c + f(ct + x_0)y' = 0 \quad (\text{here } x_0 = x(0)).$$

It cannot be everywhere defined if f vanishes somewhere.

Chapter 3

3.10 Sum $df \cdot x\big(g(y,t)g(z,u) - g(z,t)g(y,u)\big)$ and the two other terms which are obtained by cyclic permutation of (x, y, z). From the second Bianchi identity, we get zero. Making x equal to ∇f, we get

$$\mid df \mid^2 R^o(y, z, t, u) = 0.$$

3.22 We have

$$\sum_{j,k \neq i} K(j, k) = Scal - 2Ric(e_i, e_i).$$

The answer to i) follows. Now, taking the trace of both members of the inequality $Ric \leq \frac{Scal}{2}g$, we get $Scal \leq n\frac{Scal}{2}$. In dimension 3, it is clear from what we have seen that condition (F), the positiveness of sectional curvature and the inequality $Ric \leq \frac{Scal}{2}g$ give three equivalent properties.

3.32 Take a sequence p_k of points of V such that $d(m, p_k)$ converges to $d(m, V)$. Using Hopf-Rinow's theorem, we can extract a convergent subsequence. For the limit point p of such a subsequence, $d(m, p) = d(m, V)$. Of course, p need not be unique: just think of the distance from the North pole to the equator in S^n.
Using Hopf-Rinow's theorem again, we get a geodesic c from m to p of length $d(m, p)$. We can suppose that c is parametrized by $[0, 1]$. To prove that c is orthogonal to V at p, we shall give, for any y in $T_p V$, a variation $H(s, t) = c_t(s)$ such that

$$c_t(0) = m, \quad c_t(1) \in V, \quad \text{and} \quad Y(1) = \overline{\frac{\partial}{\partial t}}(1) = y.$$

We proceed as follows. If $q \in c$ is close enough to p, there is a ball B with center 0 in $T_q M$ such that $\exp_{q|B}$ is a diffeomorphism. Moreover, we can take q

such that $\exp_q(B) \cap V$ contains a neighborhood of p in V. Take a curve $\gamma(t)$ in $V \cap \exp_q(B)$ such that $\gamma(0) = p$ and $\gamma'(0) = y$. We get a variation H_1 of c by following c from m to q, and by joining q to $\gamma(t)$, thanks to the minimizing geodesic given by $\exp_q$. This variation is only piecewise C^1. Using $\exp_q$ again, we can smooth it in a neighborhood of q. Such a variation is length decreasing whenever $\langle y, c'(1) \rangle < 0$. See figure B.8

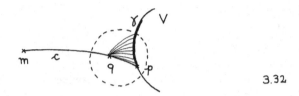

3.32

Fig. B.8. The shortest path from a point to a submanifold

3.34 The first variation will be given by a sum of k terms analogous to the one in 3.31 i). Apply the argument of 3.31 ii) to show that a critical point is piecewise a geodesic. But if there were angular points, there would exist a variation which lessens length and energy, see figure B.9

Fig. B.9. A curve with angular points is not length-minimizing

3.51 Remark that

$$\exp_m^* g = dr^2 + \sin^2 r d\theta^2 \quad \text{for} \quad S^2,$$

$$\exp_m^* g = dr^2 + \sinh^2 r d\theta^2 \quad \text{for} \quad H^2.$$

The computation of the curvature of (S^n, can) and (H^n, can) follows, since they have "many" totally geodesic 2-dimensional submanifolds.

3.58 We have to show that $J(D_X Y) = D_X(JY)$ for any vector fields X and Y. Using the notations of 2.33, recall that the multiplication by i leaves the horizontal space H_z stable. Denoting by D^o the connection of $\mathbf{R}^{2n+2}$, we have

$$\tilde{D}_{\tilde{X}} \tilde{Y} = (D_{\tilde{X}}^o \tilde{Y})^\top,$$

$$\tilde{D}_{\tilde{X}} i\tilde{Y} = (D_{\tilde{X}}^{o} i\tilde{Y})^{\top} = (iD_{\tilde{X}}^{o}\tilde{Y})^{\top}$$
$$= iD_{\tilde{X}}^{o}\tilde{Y} - \langle iD_{\tilde{X}}^{o}\tilde{Y}, z \rangle \cdot z,$$

therefore

$$i(\tilde{D}_{\tilde{X}}\tilde{Y}) - \tilde{D}_{\tilde{X}} i\tilde{Y} = (\text{sth.})z + (\text{sth.})iz$$

is vertical, using 3.56.

3.60 a) Take a complex 2-plane of $\mathbf{C}^{n+1}$. Then $P \cap S^{2n+1}$ is a totally geodesic 3-sphere in S^{2n+1}, and is globally invariant under the action of S^1. The quotient under this action will be a projective line $P^1\mathbf{C}$ in $P^n\mathbf{C}$. We get a totally geodesic submanifold. Indeed, any vector which is horizontal for the "small" fibration $S^3 \to P^1\mathbf{C}$ is also horizontal for the "big" fibration $S^{2n+1} \to P^n\mathbf{C}$. Using 2.110, we see it is a manifold with constant curvature 4. By the way, we get a new proof that $K(v, Jv) = 4$.

Now, take a real 3-plane (i.e. $E = \overline{E}$) in $\mathbf{C}^{n+1}$. Then $E \cap iE = 0$, therefore the tangent plane to $S^2 = E \cap S^{2n+1}$ is horizontal. The trace of the action of S^1 on this 2-sphere is just the antipodal map. Therefore, when going to the quotient, we get a totally geodesic $P^2\mathbf{R}$ with curvature $+1$. Now, if $u \in T_m S^{2n+1}$ is horizontal, and if v is horizontal and orthogonal to iu, the 3-plane

$$\mathbf{R} \cdot m + \mathbf{R} \cdot u + \mathbf{R} \cdot v$$

is real. Therefore $K(u, v) = 1$ if $g(Ju, v) = 0$. **b)** There are three distinct eigenvalues. The eigenvalue $n + 1$ has multiplicity 1 and the corresponding eigenspace is generated by $J^{\#} = \omega^{\#\#}$. The eigenvalue 1 has multiplicity $n^2 - 1$, and the eigenspace is generated by the 2-vectors $u \wedge v + Ju \wedge Jv$ and $u \wedge Ju - v \wedge Jv$. The eigenvalue 0 has multiplicity $n(n-1)$, and the eigenspace is generated by the 2-vectors $u \wedge v - Ju \wedge Jv$. Although the sectional curvature is strictly positive, the curvature operator has a zero eigenvalue. There is no contradiction: when computing the sectional curvature, only *decomposed 2-vectors* are involved.

3.74 Such a geodesic is a local minimum but not a global one.

3.83 a) Any $A \in SO(n + 1)$ has the eigenvalue $+1$. **b)** The two manifolds will be isometric if and only if the two actions of $\mathbf{Z}/p\mathbf{Z}$ are conjugate by an isometry of S^3. This means there exists an integer r such that $t_{k,p}$ and $t_{k',p}^r$ are conjugate as endomorphisms of $\mathbf{R}^4$. The claimed result follows by comparing the eigenvalues.

c) A Lie subgroup of a Lie group equipped with a bi-invariant metric is a totally geodesic submanifold. This is clear from 2.90. Now, if $P \subseteq M$ is totally geodesic, it can be checked directly (without using the Gauss equation that we shall see in chapter 5), that $R_P = R_{M|TP}$. It follows that T is flat. The Riemannian exponential map $\exp_e$ is the same as the exponential map for $SU(3)$ as a Lie group (cf. 2.90 again), and is both a Riemannian covering map and a group homomorphism from $\mathbf{R}^2 = \underline{T}$ to T. Remembering that

$$T = \left\{ \mathrm{diag}\big(\exp(ia), \exp(ib), \exp(ic)\big),\ a, b, c \in \mathbf{R},\ a + b + c = 0 \right\},$$

we see that $\exp_e^{-1}(e)$ is just the lattice generated by the vectors

$$v_1 = 2\pi \,\mathrm{diag}(i, -i, 0) \quad \text{and} \quad v_2 = 2\pi \,\mathrm{diag}(i, 0, -i)$$

of $\underline{T}$. Now, this lattice is hexagonal, since

$$g_e(v_1, v_1) = g_e(v_2, v_2) = 8\pi^2 \quad \text{and} \quad g_e(v_1, v_2) = 4\pi^2.$$

3.86 Take an orthonormal basis $(X_i)_{1 \leq i \leq n}$ of $\underline{G}$. Using 3.17, we see that

$$Ric(X, X) = \sum_{i=1}^{n} |\,[X, X_i]\,|^2 .$$

Therefore, if $\underline{G}$ has trivial center, $Ric(X, X) > 0$. It follows then from the compactness of the unit sphere that there is a constant $a > 0$ such that $Ric(X, X) \geq a g(X, X)$, and G is compact by Myers' theorem. Since its universal covering $\tilde{G}$ also satisfies $Ric \geq ag$, it is also compact (we need not know that the universal covering of a Lie group is a Lie group).

It can be checked directly that $\underline{SO}(n, 1)$ and $\underline{SL_n}\mathbf{R}$ have trivial center. Hence the corresponding groups carry no bi-invariant metric, since they are not compact.

3.89 a) We can suppose that the universal covering of M is given by $\exp_p :$ $T_pM \rightarrow M$. If c_1 and c_2 are two homotopic geodesics from p to q, the lift in T_pM of the loop $c_1 \cup c_2$, that we shall denote by $\tilde{c}_1 \cup \tilde{c}_2$, is again a loop. Now, $\tilde{c}_1$ and $\tilde{c}_2$ are rays from 0 in T_mM. That forces $\tilde{c}_1 = \tilde{c}_2$, therefore $c_1 = c_2$.

b) *In a space with strictly negative curvature, the energy, viewed as a function on the given free homotopy class, has a strict local minimum for any closed geodesic in that class. Intuitively, any smooth fonction whose critical points in a connected manifold are local minima has one critical point at most. This intuition is legitimated by a theorem in Morse theory (cf.[C-E] p.85), which we apply to the homotopy class.*

3.104 i) Just notice that the number of sequences of i positive integers such that

$$\sum_{i=1}^{k} n_i \leq s$$

is equal to $\binom{s}{k}$. To see that, make correspond to such a sequence the sequence

$$n_1,\ n_1 + n_2,\ \cdots,\ n_1 + n_2 + \cdots n_k.$$

ii) Any element of Γ has a unique reduced expression as a word involving the a_i and their inverses. Let $\lambda(s)$ be the number of elements of Γ whose reduced expression has length s. Then

$$\phi_S(s) = \phi_S(s-1) + \lambda(s).$$

Now, a reduced word of length s is written as $b_1 b_2 \cdots b_{s-1} b_s$, where the b_i are taken among the generators and their inverses, with the constraint that b_s is different from b_{s-1}. Therefore $\lambda(s) = (2k-1)\lambda(s-1)$, and the sequence $\phi_S(s)$ is forced to satisfy the recurrence relation

$$\phi_S(s) = 2k\phi_S(s-1) - (2k-1)\phi_S(s-2).$$

Since $\phi_S(0) = 1$ and $\phi_S(1) = 2k+1$, our claim follows.

3.129 A direct computation shows that

$$D_X^1 Y = D_X Y + df(X)Y + df(Y)X - g(X,Y)\nabla f$$

and

$$\exp(-2f)R(g_1) = R(g) + g \cdot \left(-Ddf + df \circ df - \frac{1}{2} \mid df \mid^2 g\right).$$

It follows that $W(g_1) = \exp(2f)W(g)$, since the terms like $g \cdot h$ do not affect the Weyl component. If $n = 2$, the curvature is determined by the scalar curvature, and

$$\exp(2f)Scal(g_1) = Scal(g) + 2\Delta f,$$

where $\Delta = -\text{tr}Dd$.

3.131 The product $S^p \times H^q$ is diffeomorphic to $\mathbf{R}^{p+q} \backslash \mathbf{R}^{q-1}$ (or to a connected component of this space if $p = 0$). On this space, the Riemannian metric

$$\frac{\sum_{i=1}^{p+q}(dx^i)^2}{r^2}$$

where $r^2 = \sum_{i=1}^{p+1}(x^i)^2$ is just, denoting by $d\sigma_p$ the canonical metric of S^p,

$$\left[dr^2 + r^2 d\sigma_p + \sum_{i=p+2}^{p+q}(dx^i)^2\right]/r^2,$$

that is

$$d\sigma_p + \frac{1}{r^2}\left[dr^2 + \sum_{i=p+2}^{p+q}(dx^i)^2\right]$$

The second term in this last expression is the half-space Poincaré metric.

3.133 a) Denote by g_+ and g_- the canonical metrics of S^p and H^q respectively. Then

$$2R(g_+ + g_-) = g_+ \cdot g_+ - g_- \cdot g_- = (g_+ + g_-) \cdot (g_+ - g_-).$$

Hence $R(g)$ is divisible by g and $W(g) = 0$! If p or q is equal to 1, this argument is still valid, since $dt^2 \cdot dt^2 = 0$! **b)** If s is the second fundamental form, the Gauss equation (see 5.5) gives $R^M = \frac{1}{2}s \cdot s$. Then a direct computation gives

$$c(s \cdot s) = 2\big(\mathrm{tr}(s)s - s^2\big).$$

Using 3.132, we see that M is conformally flat if and only if

$$s \cdot s = \frac{1}{n-2}g \cdot \left[2\mathrm{tr}(s)s - 2s^2 - \frac{1}{n-1}\big((\mathrm{tr}(s))^2 - \mid s \mid^2\big)g\right].$$

Diagonalize s with respect to an orthonormal basis. Then

$$s = \sum_{i=1}^{n} \lambda_i e_i \circ e_i,$$

and an elementary algebraic computation yields

$$g \cdot s = 2\sum_{i<j}(\lambda_i + \lambda_j)(e_i \wedge e_j) \circ (e_i \wedge e_j)$$

$$s \cdot s = 2\sum_{i<j}\lambda_i\lambda_j(e_i \wedge e_j) \circ (e_i \wedge e_j)$$

From now on, the details are left to the reader.

Chapter 4

4.17 If $f = \phi(r)$, where r is the distance function, then

$$\Delta f = \delta\big(\phi'(r)\big)dr = -\phi''(r) + \phi'(r)\Delta r.$$

Now, let T be the distribution on S^2 which is defined by the locally integrable function

$$y \to \log\big(1 - \cos d(x,y)\big) = \log(1 - \cos r).$$

Then, for any $f \in C^\infty(S^2)$,

$$\Delta T(f) = T(\Delta f) = \lim_{\epsilon \to 0} \int_{r \geq \epsilon} \log(1 - \cos r)\Delta f \Omega_2.$$

Now, for $r \neq 0$, we have

$$\Delta\big(\log(1 - \cos r)\big) = -\frac{1}{2}\sin^2\frac{r}{2} - \cot r = 1.$$

Therefore

$$\int_{r\geq\epsilon} (\log(1-\cos r))\, \Delta f - f)\omega_2 = \int_{r=\epsilon} (\log(1-\cos r))\frac{\partial f}{\partial r} - f\cot\frac{r}{2}\omega_1.$$

When ϵ goes to 0, the first term of this integral converges to 0, and the second one converges to $-4\pi f(x)$.

4.40 We know that

$$S\alpha(x,y) = D_x\alpha(y) + D_y\alpha(x).$$

Then, if h is a symmetric two-form, we have

$$< S\alpha, h >= 2 < D\alpha, h >= 2 < \alpha, \delta h > .$$

Taking the divergence and using Ricci identity, we get

$$\delta(S\alpha)(Y) = -\sum_{i=1}^{n}(D_{E_i}D_{E_i}\alpha(Y) + D_{E_i}D_Y\alpha(E_i))$$

$$= D^*D\alpha(Y) - \sum_{i=1}^{n}(D_Y D_{E_i}\alpha(E_i) + R(E_i,Y)E_i))$$

In other words,

$$\delta S = D^*D + d\delta - Ric.$$

If $\alpha^\#$ is a Killing vector field, $\delta\alpha = 0$, therefore

$$0 = \langle\delta S\alpha,\alpha\rangle = \langle D^*D\alpha,\alpha\rangle - \langle Ric(\alpha),\alpha\rangle.$$

Using the same argument as in theorem 4.37, we see that if Ric is strictly negative, α must be 0. Then $Isom(M,g)$, which is both discrete and compact, is finite.

If Ric is non-positive, the same argument as in 4.37 iii) shows that Killing vector fields are *parallel*. Then if (M,g) is Riemannian homogeneous, it must be a flat torus (using the same argument again), since the dimension of the space of Killing vector fields is at least n in that case.

4.47 The eigenvalues are the same as the eigenvalue of the rectangular torus which is a Riemannian covering of order 2 of the bottle. Only are the multiplicities different.

4.53 We implicitly used the continuity of the linear form $f \to f([e])$ for a L^2-norm!

Chapter 5

5.4 a) The vector field $Y = \tilde{Y}(.,0)$ is a Jacobi field along c since the curves $H(.,t)$ are geodesics, and we have:

$$Y(0) = \frac{\partial H}{\partial t}(0,0) = \frac{d}{dt}\left[\exp_{\gamma(t)}(0)\right]_{|t=0} = \frac{d}{dt}\gamma(t)_{|t=0},$$

that is $Y(0) = u$. On the other hand,

$$Y'(0) = \bar{D}_{\frac{\partial}{\partial s}}\frac{\partial}{\partial t} = \bar{D}_{\frac{\partial}{\partial t}}\frac{\partial}{\partial s} = \tilde{D}_u \nu = S(u).$$

In particular, the field Y does not depend on the choice of γ. We finally get:

$$l(v,u) = \langle v, S(u)\rangle = \langle X, Y'\rangle(0).$$

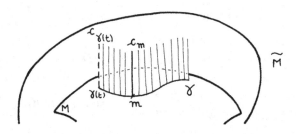

5.4 b) It is sufficient to compare at a given point the values of the 2-forms $N^*(\omega)$ and σ on the pair $\frac{\partial}{\partial x^1}, \frac{\partial}{\partial x^2}$. We can assume that at this point, the pair is orthonormal. Hence

$$N^*(\omega)\left(\frac{\partial}{\partial x^1}, \frac{\partial}{\partial x^2}\right) = \omega\left(dN(\frac{\partial}{\partial x^1}), dN(\frac{\partial}{\partial x^2})\right)$$

$$= \omega\left(\tilde{D}_{\frac{\partial}{\partial x^1}}N, \tilde{D}_{\frac{\partial}{\partial x^2}}N\right) = \left(\tilde{D}_{\frac{\partial}{\partial x^1}}N, \tilde{D}_{\frac{\partial}{\partial x^2}}N, N\right).$$

To compute this mixed product, we can work in the orthonormal basis defined by the two coordinate vectors and N, and compute the corresponding determinant. We get

$$N^*(\omega)\left(\frac{\partial}{\partial x^1}, \frac{\partial}{\partial x^2}\right) = \begin{bmatrix} l_{11} & l_{12} & 0 \\ l_{12} & l_{22} & 0 \\ 0 & 0 & 1 \end{bmatrix}$$

$$= \bar{K} = \bar{K}\sigma\left(\frac{\partial}{\partial x^1}, \frac{\partial}{\partial x^2}\right)$$

since $\left(\frac{\partial}{\partial x^1}, \frac{\partial}{\partial x^2}\right)$ is orthonormal at the point where we are computing.

5.8 a) Let us assume that all the sectional curvatures of the hypersurface M are negative at $m \in M \subset \mathbf{R}^{n+1}$, and let $(e_1, \cdots, e_n)$ be an orthonormal basis of $T_m M$ where l is diagonal, with corresponding eigenvalues (λ_i). From Gauss theorem we know that:

$$K(e_i, e_j) = \tilde{K}(e_i, e_j) + \bar{K}(e_i, e_j) = \lambda_i \lambda_j,$$

since the ambiant space is flat. Our hypothesis forces for $i \neq j$: $\lambda_i \lambda_j < 0$, which is impossible when $n \geq 3$.

5.8 b) The proof is analogous to the proof of 5.5. Just write:

$$\tilde{D}_X Y = D_X Y + \sum_{i=1}^{k} \langle \tilde{D}_X Y, \nu_i \rangle \nu_i = D_X Y - \sum_{i=1}^{k} l_i(X, Y) \nu_i.$$

5.8 c) Let c be a geodesic of M, that is $D_{c'} c' = 0$. This curve will also be a geodesic of $\tilde{M}$ if and only if $\tilde{D}_{c'} c' = 0$, that is if $\tilde{D}_{c'} c'$ is everywhere tangent to M. But since c is drawed on M, we have $\langle c', \nu \rangle = 0$ for any normal vector field ν on M, and hence

$$c' \cdot \langle c', \nu \rangle = \langle \tilde{D}_{c'} c', \nu \rangle + \langle c', \tilde{D}_{c'} \nu \rangle = 0.$$

All the geodesics of M will also be geodesics of $\tilde{M}$ if, for any normal vector field ν and any $u \in T_m M$:

$$\langle c', \tilde{D}_{c'} \nu \rangle_m = \langle u, \tilde{D}_u \nu \rangle = 0,$$

where c is the geodesic of M satisfying to $c'(0) = u$. This condition will be satisfied if and only if the second fundamental form of M is zero (5.8 b)).

5.8 d) Let P^p and Q^q be two compact totally geodesic submanifolds of M^n, and assume that P and Q do not intersect. By compactness, there exist $x \in P$ and $y \in Q$ with $d(x, y) = d(P, Q) \neq 0$. Let $c : [0, d] \rightarrow M$ be a minimal geodesic from x to y: the first variation formula just says that c is normal at x to P, and at y to Q. By assumption, for any variation $H(s, t) = c_t(s)$ of c with $H(t, O) \in P$ and $H(t, d) \in Q$,

$$\frac{d^2}{dt^2} (L(c_t))_{|t=0} \geq 0$$

(c is minimal from P to Q).
Assuming that the variation is normal (that is $y \perp c'$), and that $c(0, t)$ and $c(d, t)$ are geodesics (of P, resp. Q, or equivalently of the ambient space), the second variation formula 3.34 yields, with $Y = \frac{\bar{\partial}}{\partial t}$:

$$\frac{d^2}{dt^2} (L(c_t))_{|t=0} = \int_0^d (|Y'|^2 - R(Y, c', Y, c')) \, ds.$$

a) If $p+q > n$, the parallel transport along c of T_yQ intersects T_xP: chose $v \neq 0$ in the intersection, let Y be the parallel vector field along c with $Y(0) = v$, and H be the variation defined by

$$H(s,t) = \exp_{c(s)} tY(s).$$

Then $\frac{d^2}{dt^2}(L(c_t))_{|t=0} = \int_0^d -R(Y,c',Y,c')ds < 0$ when $K > 0$: contradiction.

b) If $p = q = n-1$, the parallel transport of T_yQ along c coincides with T_xP (c is normal to both P and Q). Hence, if $(e_i)_{(i=1,\cdots,n-1)}$ is an orthonormal basis of T_xP, Y_i are the parallel vector fields along c with $Y_i(0) = e_i$, and $H_i(s,t) = c_t^i(s)$ are the associated variations of c (constructed as above), then

$$\sum_{i=1}^n \frac{d^2}{dt^2}(L(c_t^i))_{|t=0} = \int_0^d \left(\sum_{i=1}^n -R(Y_i,c',Y_i,c') \right) ds,$$

which is negative as soon as the Ricci curvature is positive, a contradiction.

5.8 e) Let X, Y and U be vector fields tangent to M at any point of M, and with respective values x, y and u at m. Let N be a unit normal vector field with $N(m) = \nu$. The computations in 5.5 yield:

$$\tilde{D}^2_{X,Y} U = \tilde{D}_X \tilde{D}_Y U - \tilde{D}_{\tilde{D}_X Y} U$$

$$= D^2_{X,Y} U - l(X, D_Y U)N - \tilde{D}_X [l(Y,U)N] + l(D_X Y, U)N + l(X,Y)\tilde{D}_N U,$$

hence, by noticing that $\langle \tilde{D}_X N, N \rangle = 0$ and symmetrization:

$$\langle \tilde{R}(X,Y)U, N \rangle = \langle R(X,Y)U, N \rangle + l([X,Y],U) + l(Y, D_X U)$$

$$- l(X, D_Y U) - X.l(Y,U) + Y.l(X,U).$$

This yields the result, since $R(x,y)u$ is tangent to M, and hence normal to ν.

5.13 The manifold M_1 is obtained by revolving around the z-axis the curve $c(s) = (s, \text{Log} s)$. Its curvature at $f(s,\theta)$ is hence given by $K_1 = \frac{-1}{(1+s^2)^2}$. The manifold M_2 is an helicoid, and its curvature at $g(s,\theta)$ is $K_2 = \frac{-1}{(1+s^2)^2}$ (see 5.11). Hence F preserves the curvature.

On the other hand, $TF\left(\frac{\partial f}{\partial \theta}\right) = \left(\frac{\partial g}{\partial \theta}\right)$, with

$$\frac{\partial f}{\partial \theta} = (\sin\theta, \cos\theta, \frac{1}{s}) \quad \text{and} \quad \frac{\partial g}{\partial \theta} = (\sin\theta, \cos\theta, 1),$$

and F cannot be an isometry.

5.24 a) Let $(e_i)_{(i=1,\cdots,n)}$ be an orthonormal basis of $T_m M$, and $f_i = F_*(e_i)$. If Δ, D and $\tilde{D}$ are the respective connections of M, $F(M)$ and $\mathbf{R}^{n+k}$, we have at $m \in M$:

$$\Delta F = -\sum_{i=1}^{n}[e_i \cdot (e_i \cdot F) - (\nabla_{e_i} e_i \cdot F)] = -\sum_{i=1}^{n}\left[\tilde{D}_{f_i} f_i - D_{f_i} f_i\right] = n\eta.$$

In particular, $F(M)$ is minimal if and only if F is harmonic.

b) Let $\tilde{D}$ and D_0 be the respective connections of $\mathbf{R}^{m+1}$ and S^m ($D_0 = (\tilde{D})^{\top S^m}$). If η and η_0 are the respective mean curvatures of $F(M)$ in $\mathbf{R}^{m+1}$ and S^m, then

$$\eta_0 = (\eta)^{\top S^m} = \frac{1}{n}(\Delta F)^{\top S^m}.$$

Hence F will be minimal in S^m if and only if there exists a function $f : M \to \mathbf{R}$ with $\Delta F = f.F$, and then

$$\langle \Delta F, F \rangle = f \mid F \mid^2 = f.$$

For $v \in T_m M$, $\langle F_*(v), F \rangle = 0$: differentiating and using the notations of a), we get $\langle f_i, F \rangle \equiv 0$, hence

$$0 \equiv f_i \cdot \langle f_i, F \rangle = \langle \tilde{D}_{f_i} f_i, F \rangle + \langle f_i, f_i \rangle,$$

where $\langle , \rangle$ is the scalar product in $\mathbf{R}^{m+1}$. Since $\langle \tilde{D}_{f_i} f_i, F \rangle = d\tilde{D}_{f_i}(f_i)^{\perp}$, we finally get:

$$0 = \langle \mathrm{Tr}(II), F \rangle + n = -\langle \Delta F, F \rangle + n.$$

c) i) Let L_g be the left action of $g \in G$ on G/H. Since the L_g are isometries of $(G/H, h)$, the $N+1$ functions $f_i \circ L_g$ are also an orthonormal basis for E_λ. Hence there exists for any $g \in G$ a matrix $A = (a_{ij}(g)) \in O(N+1)$ such that

$$f_i \circ L_g = \sum_{j=0}^{N} a_{ij}(g) f_j.$$

Now for any $x, y \in G/H$ there exists $g \in G$ with $y = L_g(x)$ and hence:

$$\sum_{j=0}^{N} f_i^2(y) = \sum_{i=0}^{N}(f_i \circ L_g(x))^2 = \sum_{i=0}^{N} f_i^2(x),$$

since $A \in O(N+1)$. Eventually scaling, we can assume that $\sum f_i^2 \equiv 1$.

ii) Denote by h' the field of symmetric bilinear forms on M defined by

$$h'(u, v) = \sum_{i=0}^{N} \langle df_i(u), df_i(v) \rangle.$$

Since G/H is isotropy irreducible and h' is left invariant, h and h' are equal up to a scalar factor. In particular, if $h' \neq 0$, then the forms df_i are everywhere linearly independant, and F is an immersion.

iii) The property that F is colinear to ΔF is preserved under the homothetie we eventually use to make F an isometry. Just apply the criterion of b).

Bibliography

[Ad]   S. ADAMS, Dynamics on Lorentz Manifolds, World Scientific 2001

[Ar]   V.I. ARNOLD, Mathematical methods of classical mechanics, Graduate texts in Mathematics 60, Springer, New-York 1978

[A1]   T. AUBIN, Non-linear Analysis on Manifolds, Monge-Ampère Equations, Springer, Berlin 1982

[A2]   T. AUBIN, Equations différentielles non linéaires et probleme de Yamabe concernant la courbure scalaire, *J. Math. Pures Appl.* **55** (1976) 269–296

[A-L]   M. AUDIN, J. LAFONTAINE (eds.), Holomorphic curves in Symplectic Geometry, Progress in Math. 117, Birkhäuser

[Bb]   I. BABENKO, Souplesse intersystolique forte des variétés fermées et des polyèdres, *Ann. Institut Fourier* **52** (2002), 1259–1284

[BGS]   W. BALLMANN, M. GROMOV, V. SCHROEDER, Manifolds of non-positive curvature, Progress in Math., Birkhäuser, Basel 1985

[Bv]   C. BAVARD, Inégalité isosystolique pour la bouteille de Klein, *Math. Ann.* **274** (1986), 439–441

[B-P]   R. BENEDETTI, C. PETRONIO, Lectures on Hyperbolic Geometry, Universitext, Springer, Berlin-Heidelberg 1991

[Be1]   P. BERARD, Spectres et groupes cristallographiques I: domaines euclidiens, *Inv. Math.* **58** (1980), 179–199

[Be2]   P. BERARD, Spectral geometry: direct and inverse problems, Lecture Notes in Math. **1207,** Springer, Berlin-Heidelberg 1986

[Be3]   P. BERARD, Variétés riemanniennes isospectrales non isométriques, In: Séminaire Bourbaki Mars 89, exposé 705, *Astérisque* **177-178** (1989), 127–154

[Be4]   P. BERARD, From vanishing theorems to estimating theorems: the Bochner's technique revisited, *Bull. A.M.S.*, **19** (1988), 371–406

[BBG1]   P. BERARD, G. BESSON et S. GALLOT, Sur une inégalité isopérimétrique qui généralise celle de Paul Lévy, *Inv. Math.* **80** (1985), 295–308

[Be-G]   P. BERARD et S. GALLOT, Inégalités isopérimétriques pour l'équation de la chaleur et application à l'estimation de quelques invariants, In: Séminaire Goulaouic-Meyer-Schwartz 1983-84, Exposé XV, Ecole Polytechnique, Palaiseau 1984.

[B-M]   P. BERARD et D. MEYER, Inégalités isopérimétriques et applications, *Ann. Sc. Ec. Norm. Sup.* **15** (1982), 513–542

[Br1]   M. BERGER, La Géométrie Métrique des Variétés Riemanniennes, in Colloque Elie Cartan, *Astérisque* (supplementary issue), S.M.F., Paris 1985

[Br2]   M. BERGER, Géométrie, Cedic-Nathan, Paris 1979, or Geometry I & II (English translation), Universitext, Springer, Berlin-Heidelberg 1987

[Br3]   M. BERGER, Riemannian Geometry during the second half of twentieth century, University lecture Series 27, Amer. Math. Soc. 2000

[Br4]   M. BERGER, Systoles et applications selon Gromov, Séminaire Bourbaki, *Astérisque* **216** 1993, 279–310

[Br5]   M. BERGER, A panoramic view of Riemannian geometry, Springer, Berlin-Heidelberg 2003

[Bg]   Me. BERGER, On the conformal equivalence of compact 2-dimensional manifolds, *J. of Math. and Mech.* **19** (1969), 13–18

[B-G]   M. BERGER et R. GOSTIAUX, Géométrie différentielle, P.U.F., Paris 1986, English translation: Differential Geometry: Manifolds, Curves and Surfaces, GTM 115, Springer

[BGM]   M. BERGER, P. GAUDUCHON, E. MAZET, Le spectre d'une variété riemannienne, Lecture Notes in Math. **194,** Springer, Berlin-Heidelberg 1971

[B1]   A. BESSE, Manifolds all of whose geodesics are closed, Springer, Berlin-Heidelberg 1978

[B2]   A. BESSE, Einstein manifolds, Springer, Berlin-Heidelberg 1986

[B3]   A. BESSE (ed.), Actes de la table ronde de Géométrie Différentielle en l'honneur de Marcel Berger, Collection SMF "Séminaires et Congrés" 1, Société Mathématique de France 1996

[Bes]   G. BESSON, Volumes and Entropies, in [BLPP]

[BLPP]   G. BESSON, J. LOHKAMP, P. PANSU, P. PETERSEN, Riemannian Geometry, Fields Institute Monograph 4, Amer. Math. Soc. 1996

[B-C]   R.L. BISHOP and R.J. CRITTENDEN, Geometry of Manifolds, Academic Press, New York 1964

[Bo]   W. BOOTHBY, An Introduction to Differentiable Manifolds and Riemannian Geometry, Academic Press, New York 1975

[Bt]   R. BOTT, Some aspects of invariant theory in differential geometry, In: Differential operators on manifolds (C.I.M.E.), Ed. Cremonese, Roma 1975

[B-T]   R. BOTT and L. TU, Differential forms in algebraic topology, Graduate Texts in Maths., Springer, New-York 1982

[Bi]   N. BOURBAKI, Eléments de Mathématique, Masson, Paris, English edition: Elements of Mathematics, Springer

[B-H]   M. BRIDSON, A. HAEFLIGER, Metric spaces of non-positive curvature, Grundlehren der mathematischen wissenschaften 319, Springer 1999

[BBI]   D. BURAGO, Y. BURAGO, S. IVANOV, A Course in metric geometry, Graduate studies in mathematics (33), Amer. Math. Soc., Providence, R.I. 2001

[B-Z]   Y.D. BURAGO and V.A. ZALGALLER, Geometric inequalities, Grundlehren Math. 285, Springer, Berlin-Heidelberg 1988

[Bu1]   P. BUSER, A geometric proof of Bieberbach theorem on crystallographic groups, L'Enseignement Math. **31** (1985), 137

[Bu2]   P. BUSER, Geometry and Spectra of compact Riemann Surfaces, Progress in Math. 106, Birkhäuser 1990

[B-K]   P. BUSER and H. KARCHER, Gromov's almost flat manifolds, Astérisque **81**, S.M.F., Paris 1981

[Ca]   Y. CARRIERE, Autour de la conjecture de L. Markus sur les variétés affines, Invent. Math. **95** (1989), 615–628

[C-R]   Y. CARRIERE, L. ROZOY, Complétude des métriques lorentziennes de T^2 et difféormorphismes du cercle. Bol. Soc. Brasil. Mat. (N.S.) **25** (1994), 223–235

[Ca]   E. CARTAN, La géométrie des espaces de Riemann, Memorial Sc. Math., **IX**.

[Cha]   I. CHAVEL, Eigenvalues in Riemannian geometry, Academic Press, Orlando 1984

[Cn]   J. CHAZARAIN, Spectre des opérateurs elliptiques et flots hamiltoniens, In: Séminaire Bourbaki 1974-75, Lecture Notes in Math. **514,** Springer, Berlin-Heidelberg 1976

[C]   J. CHEEGER, Finiteness theorems for Riemannian manifolds, Amer. J. of Math. **92** (1970), 61—74

[C-E]   J. CHEEGER and D. EBIN, Comparison theorems in Riemannian Geometry, North-Holland, Amsterdam 1975

[Ch]   S.Y. CHENG, Eigenvalue comparison theorems and geometric applications, Math. Z. **143** (1975), 289–297

[CV]   Y. COLIN de VERDIERE, Quasi-modes sur les variétés Riemanniennes, Inv. Math. **43** (1977), 15–52

[Cr]   C.B. CROKE, Some isoperimetric inequalities and eigenvalues estimates, Ann. Sc. Ec. Norm. Sup. **13** (1981), 419–436

[D'A]   G. D'AMBRA, Isometry groups of Lorentz manifolds, Invent. Math. **92** (1988) 555–565

[D'A-G]   G. D'AMBRA, M. GROMOV, Lectures on transformation groups: geometry and dynamics, Surveys in Differential Geometry **1** (1991), 19–112

[D'A-Z] J.E. d'ATRI and W. ZILLER, Naturally reductive metrics and Einstein metrics on compact Lie groups, *Memoirs Amer. Math. Soc.* **18** (no. 215)(1979)

[Ci] E. CALABI, Extremal isosystolic metrics for compact surfaces, in [B3], 167–204

[D-M] E. DAMEK and F. RICCI, Harmonic analysis on solvable extensions of H-type groups, *J. Geom. Anal.* **2** (1992), 213–248.

[dCa1] M. DO CARMO, Differentiable curves and surfaces, Prentice Hall, New-Jersey, 1976

[dCa2] M. DO CARMO, Riemannian Geometry, Birkhaüser, Basel 1992

[DDM] M. DO CARMO, M. DAJCZER and F. MERCURI, Compact conformally flat hypersurfaces, *Trans. Amer. Math. Soc.* **288** (1985), 189–203

[Dom] P. DOMBROWSKI, 150 years after Gauss' "Disquitiones generales circa superficies curvas", *Astérisque* **62** (1979)

[D-G] J.J. DUISTERMAAT, V. GUILLEMIN, The spectrum of positive elliptic operators and periodic bicharacteristics, *Inv. Math.* **29** (1975), 39–79

[FLP] A. FATHI, F. LAUDENBACH, V. POENARU, Travaux de Thurston sur les surfaces, *Astérisque* **66–67** (1979)

[Fr] C. FRANCES, Géométrie et dynamique lorentzienne conformes, thesis, Ecole Normale Supérieure de Lyon, 2002

[F] J. FRANKS, Rotation numbers and instability sets, *Bull. of Amer. Math. Soc.* **40** (2003), 263–280

[F-U] D. FREED and K. UHLENBECK, Instantons and Four-Manifolds, M.S.R.I. Publications, Springer, New-York 1984

[F-H] W. FULTON, J. HARRIS, Representation theory, a first course, Graduate Texts In Mathematics 129, Springer, New-York 1991

[Ga1] S. GALLOT, Minorations sur le λ_1 des variétés riemanniennes, In Séminaire Bourbaki 1980-81, *Lecture Notes in Math.* **901,** Springer, Berlin-Heidelberg 1981

[Ga2] S. GALLOT, Sobolev inequalities and some geometric applications, In: Spectra of Riemannian manifolds, Kaigai Publications, Tokyo 1983

[Ga3] S. GALLOT, Inégalités isopérimétriques, courbure de Ricci et invariants géométriques I,II, *C.R.A.S. Paris* **296** (1983) 333–336, 365–368

[Ga4] S. GALLOT, Inégalités isopérimétriques et analytiques sur les variétés riemanniennes, *Astérisque* **163-164** (1988) 33–91

[Ga5] S. GALLOT, Isoperimetric inequalities based on integral norms of Ricci curvature, *Astérisque* **157-158** (Colloque Paul Lévy), 191–216

[Ga6] S. GALLOT, Volumes, courbure de Ricci et convergence des variétés [d'après T.H. Colding et Cheeger-Colding], Séminaire Bourbaki 1997-1998, *Astérisque* **252** (1998), 7–32

[G-M]   S. GALLOT et D. MEYER, Opérateur de courbure et Laplacien des formes différentielles d'une variété riemannienne, *J. Math. Pures et Appliquées* **54** (1975), 259–284

[G-Y]   L.Z. GAO and S. T. YAU, The existence of negatively Ricci curved metrics on three manifolds, *Inv. Math.* **85** (1986) 637–652

[G-H]   E. GHYS, P. de la HARPE (eds.), Sur les groupes hyperboliques d'après Gromov, Progress in Math. 83, Birkhaüser 1990

[G-T]   D. GILBARG, N. TRUDINGER, Elliptic Partial Differential Equations of Second Order, Classics in Mathematics, 2nd edition, Springer Berlin-Heidelberg 2002

[Gi]   P. GILKEY, Invariance theory, the heat equation and the Atiyah-Singer Index theorem, Publish or Perish

[G]   R. GODEMENT, Introduction à la théorie des groupes de Lie, Springer, Berlin-Heidelberg 2004

[G-M]   W. GOLDMAN, G. MARGULIS, Flat Lorentz 3-Manifolds and Cocompact Fuchsian Groups, Proceedings of the workshop "Crystallographic Groups and their Generalizations II," (Kortrijk, Belgium), *Contemporary Mathematics* **262**(2000), 135–146

[Gr]   A. GRAMAIN, Topologie des Surfaces, P.U.F., Paris 1970

[Gy]   A. GRAY, The volume of small geodesic balls, *Mich. J. of Math.* **20** (1973), 329–344

[Gr-Va] A. GRAY and L. VANHECKE, Riemannian Geometry as determined by the volume of small geodesics balls, *Acta Math.* **142** (1979), 157–198

[Gg]   M. GREENBERG, Lectures on Algebraic Topology, Benjamin, New York 1967

[G-H]   P. A. GRIFFITHS and J. HARRIS, Principles of Algebraic Geometry, J.Wiley, New-York 1978

[Gr1]   M. GROMOV, Metric structures for Riemannian and Non-Riemannian spaces, edited by J. Lafontaine and P. Pansu, Birkhäuser 1998

[Gr2]   M. GROMOV, Paul Lévy's isoperimetric inequality, Appendix C of [Gr1].

[Gr3]   M. GROMOV, Volume and bounded cohomology, *Publ. I.H.E.S.* **56** (1982) 5–99

[Gr4]   M. GROMOV, Filling Riemannian manifolds, *J. Diff. Geom.* **18** (1983), 1–147

[Gr5]   M. GROMOV, Systoles and intersystolic inequalities, in [B3]

[Gr6]   M. GROMOV, Asymptotic invariants of infinite groups, vol. 2 of "Geometric Group Theory, Sussex 1991", G.A. Niblo and M.A. Roller eds., Cambridge University Press 1993

[G-T]   M. GROMOV and W. THURSTON, Pinching constants for hyperbolic manifolds, *Inv. Math.* **89** (1987), 1–12

[G-R]   M. GROMOV and I. ROKHLIN, Embeddings and immersions in Riemannian geometry, *Russian Math. Surveys* **25** (1970), 1–57

[Gu]   V. GUILLEMIN, Lectures on spectral theory of elliptic operators, *Duke Math. J.* **44** (1977), 485–517

[Ha]   R.S. HAMILTON, Three-manifolds with positive Ricci curvature, *J. Diff. Geom.* **17** (1982), 165–222

[H-K]   E. HEINTZE and H. KARCHER, A general comparison theorem with applications to volume estimates for submanifolds, *Ann. Sci. Ec. Norm. Sup.* **11** (1978), 451–470

[Hn]   S. HELGASON, Differential Geometry, Lie groups and symmetric spaces, Academic Press, New York 1978

[H]   M. HIRSCH, Differential topology, Springer, New-York 1976

[Hf]   E. HOPF, Statistik der Lösungen geodatischer Probleme vom unstabilen Typus, II, *Math. Annalen* **117** (1940), 590–608

[Ho]   H. HOPF, Differential Geometry in the Large, Springer, Lecture notes in Math. 1000, 1969

[H-T]   D. HULIN and M. TROYANOV, Prescribing curvature on open surfaces, *Math. Ann.* **293** (1992), 277–315

[Hu]   D. HULIN, Pinching and Betti numbers, *Ann. Global Anal. Geom.* **3** (1985), 85–93

[J1]   J. JOST, Riemannian geometry and geometric analysis, Universitext, Springer, Berlin-Heidelberg 1995

[J2]   J. JOST, Compact Riemann surfaces, Universitext, Springer, Berlin-Heidelberg 1995

[Ka]   M. KAC, Can one hear the shape of a drum, *Amer. Math. Monthly* **73** (1986)

[K]   M. KAROUBI, *K*-theory, Springer, Berlin-Heidelberg 1978

[K-H]   A. KATOK and B. HASSELBLATT, Introduction to the modern theory of Dynamical systems, Cambridge University Press

[KW1]   J. KAZDAN and F. WARNER, Curvature functions for compact 2-manifolds, *Ann. of Maths.* **99** (1974), 14–47

[KW2]   J. KAZDAN and F. WARNER, Curvature functions for open 2-manifolds, *Ann. of Maths.* **99** (1974), 203–219

[Kl]   W. KLINGENBERG, Riemannian Geometry, DeGruyter Studies in mathematics, DeGruyter, Berlin 1982

[K-N]   S. KOBAYASHI and K. NOMIZU, Foundations of Differential Geometry, Wiley Interscience, New York-London 1963-1969

[Kr]   N. KUIPER, On conformally flat spaces in the large, *Ann. of Math.*, **50** (1949) 916–924

[K-P]   R. KULKARNI and U. PINKALL (eds.), Conformal Geometry, Vieweg, Aspects of Maths, 1988

[La1]   J. LAFONTAINE, Remarques sur les variétés conformément plates, *Math. Ann.* **259** (1982), 313–319

[La2]   J. LAFONTAINE, The theorem of Lelong-Ferrand and Obata, in [K-P]

[La3]   J. LAFONTAINE, Introduction aux Variétés Différentielles, collection Grenoble Sciences, Grenoble 1996

[L1]   B. LAWSON, Lectures on minimal submanifolds, Publish or Perish, Berkeley 1980

[L2]   B. LAWSON, Gauge fields in four dimensions, Regional Conferences Series A.M.S. n° 58

[L-M]   B. LAWSON, M.L. MICHELSOHN, Spin Geometry, Princeton Math. Series 38, Princeton 1989

[Le]   I. LEE, Introduction to Smooth manifolds, Grad. Text. Math 218, Springer, New-York 2002

[L-P]   J.M. LEE and T.H. PARKER, The Yamabe problem, Bull. Amer. Math. Soc. 17 (1987), 37–92

[Li]   P. LI, On the Sobolev constant and the p-spectrum of a compact Riemannian manifold, Ann. Sc. Ec. Norm. Sup. 13 (1981), 451–457

[Lo]   J. LOHKAMP, Global and local curvatures, in [BLPP]

[Ma]   J. MARTINET, Perfect lattices in Euclidean spaces, Grundlehren der Mathematischen Wissenschaften 327, Springer, Berlin-Heidelberg 2003

[My]   W. MASSEY, Algebraic topology: An introduction, GTM 127, Springer 1991

[M-S]   M. MELKO and I. STERLING, Integrable systems, harmonic maps and the classical theory of surfaces, in A.P. FORDY and J.C. WOOD (eds), Harmonic maps and integrable systems, Aspects of Mathematics 23, Vieweg, Wiesbaden 1994

[Mi1]   J. MILNOR, Whitehead torsion, Bull. Amer. Math. Soc. 72 (1966), 351–426

[Mi2]   J. MILNOR, Morse theory, Princeton University Press, Princeton 1963

[Mi3]   J. MILNOR, A note on curvature and the fundamental group, J. Diff. Geom. 2 (1968), 1–7

[Mi4]   J. MILNOR, Hyperbolic geometry: The first 150 years, Bull. Amer. Math. Soc. 6 (1982), 9–24

[MTW]   C. MISNER, K. THORNE, J. WHEELER, Gravitation, Freeman, San Francisco 1973

[M-Z]   D. MONTGOMERY and L. ZIPPIN, Topological transformations groups, New-York London, Interscience Publishers, 1955

[Mo]   F. MORGAN, Geometric measure theory, a beginner's guide, Academic Press 1988

[My]   S.B. MYERS, Connections between differential geometry and topology, Duke Math. J. 1 (1935), 376–391

[N-S]   M.A. NAIMARK and A. STERN, Theory of groups representations (translated from Russian), Grundlehren der Math. Wiss. 282, Springer, Berlin-Heidelberg 1982

[N]   R. NARASIMHAN, Analysis on real and complex manifolds, North-Holland, Amsterdam 1968

[Ot]   J. P. OTAL, Le spectre marqué des longueurs des surfaces courbure négative, Preprint Univ. Orsay 1989

[Pa]   R. PALAIS, Natural operations on differential forms, *Trans. Amer. Math. Soc.* **93** (1959), 125–141

[Pe]   P. PETERSEN, Riemannian Geometry, GTM 171, Springer, New-York 1998

[Pr]   A. PREISSMANN, Quelques propriétés globales des espaces de Riemann, *Comment. Math. Helv.* **15** (1943) 175–216

[Ra]   J.G. RATCLIFFE, Foundations of Hyperbolic Manifolds, GTM 149, Springer, New-York 1994

[Ro]   F. ROUVIÈRE, Espaces de Damek-Ricci, géométrie et analyse, in "Analyse sur les groupes de Lie et théorie des représentations, Séminaires et Congrès **7,** Société Mathématique de France

[Sc]   D. SCHÜTH, Continuous families of isospectral metrics on simply connected manifolds, *Ann. of Math.* **149** (1999), 287–308

[Sc]   P. SCOTT, The Geometries of 3-manifolds, *Bull. London Math. Soc.* **15** (1983), 401-487

[SB]   SEMINAIRE ARTHUR BESSE, Geométrie riemannienne en dimension 4, Cedic-Nathan, Paris 1980, available by Société Mathématique de France

[Se]   J.P. SERRE, Cours d'arithmétique, P.U.F. Paris 1971, English translation: A Course in Arithmetic, GTM 5, Springer

[Spa]   E. SPANIER, Algebraic Topology, Mc Graw Hill, New York 1966

[Sa]   B. SPINOZA, Ethica, ordine geometrico demonstrata, Amsterdam 1677

[Sp]   M. SPIVAK, Differential Geometry (5 volumes), Publish or Perish, Berkeley 1979

[St]   N. STEENROOD, The topology of fiber bundles, Princeton University Press, Princeton 1951

[Sr]   P. STREDDER, Natural differential operators on Riemannian manifolds, *J. Diff. Geom.* **10** (1975), 647–660

[Sz]   Z. SZABÓ, The Lichnerowicz conjecture on harmonic manifolds, *J. Diff. Geom.* **31** (1990),1– in28

[Ta]   M. TAYLOR, Partial differential Equations: Basic theory, Texts in applied mathematics (23), Springer, New-York 1999

[Th]   W. THURSTON, Three dimensional Geometry and Topology, Princeton University Press 1997

[T]   M. TROYANOV, Prescribing curvature on compact surfaces with conical singularities, to appear in *Trans. Amer. Math. Soc.* **324** (1991) 793–821

[VCN]   Variétés à courbure négative, Publications Mathématiques de l'Université Paris 7

[Wa]   F. WARNER, Foundations of differential manifolds and Lie groups, Springer, New-York 1983

[We]   A. WEINSTEIN, The cut-locus and conjugate locus of a Riemannian manifold, *Ann. Math.* **87** (1968), 29–41

[We′]   T. WEINSTEIN, An introduction to Lorentz Surfaces, de Gruyter
        Exp. in Math. 22, Berlin 1996

[Wn]   H. WENTE, Counterexample to a conjecture of H. Hopf, *Pacific J.
        of Math.* **121** (1986),193–243

[W1]   H. WEYL, The classical groups, their invariants and representations,
        Princeton University Press, Princeton 1939

[W2]   H. WEYL, Zur Infinitesimalgeometrie: Einordnung der projektiven
        und der konformen Auffassung (1921). In Gesammelte Abhandlun-
        gen, vol.II, 195–207

[Wo1]   J.A. WOLF, Spaces of constant curvature, Publish or Perish, Boston
        1974

[Wo2]   J.A. WOLF, Growth of finitely generated solvable groups and cur-
        vature of Riemannian manifolds, *J. Diff. Geom.* **2** (1968), 421–446

Index

List of figures

Printing and Binding: Strauss GmbH, Mörlenbach